Innovation Dynamics and Policy in the Energy Sector

Innovation Dynamics and Policy in the Energy Sector

Building Global Energy Markets, Institutions, Public Policy, Technology and Culture on the Texan Innovation Example

MILTON L. HOLLOWAY

Resource Economics, Inc., Austin, TX, United States

ACADEMIC PRESS

An imprint of Elsevier

ELSEVIER

Academic Press is an imprint of Elsevier
125 London Wall, London EC2Y 5AS, United Kingdom
525 B Street, Suite 1650, San Diego, CA 92101, United States
50 Hampshire Street, 5th Floor, Cambridge, MA 02139, United States
The Boulevard, Langford Lane, Kidlington, Oxford OX5 1GB, United Kingdom

Notices
Knowledge and best practice in this field are constantly changing. As new research and
experience broaden our understanding, changes in research methods, professional practices, or
medical treatment may become necessary.

Practitioners and researchers must always rely on their own experience and knowledge in
evaluating and using any information, methods, compounds, or experiments described herein. In
using such information or methods they should be mindful of their own safety and the safety of
others, including parties for whom they have a professional responsibility.

To the fullest extent of the law, neither the Publisher nor the authors, contributors, or editors,
assume any liability for any injury and/or damage to persons or property as a matter of products
liability, negligence or otherwise, or from any use or operation of any methods, products,
instructions, or ideas contained in the material herein.

British Library Cataloguing-in-Publication Data
A catalogue record for this book is available from the British Library

Library of Congress Cataloging-in-Publication Data
A catalog record for this book is available from the Library of Congress

ISBN: 978-0-12-823813-4

For Information on all Academic Press publications
visit our website at https://www.elsevier.com/books-and-journals

Publisher: Joe Hayton
Acquisitions Editor: Graham Nisbet
Editorial Project Manager: Naomi Robertson
Production Project Manager: Prasanna Kalyanaraman
Cover Designer: Mark Rogers

Typeset by MPS Limited, Chennai, India

Working together
to grow libraries in
developing countries

www.elsevier.com • www.bookaid.org

Contents

List of figures

List of tables

About the author

My 45-year participation in this uniquely Texas story has mostly been to oversee Texas energy policy-planning processes and to direct work programs to study, test, and demonstrate new technologies. And, my responsibilities have included promotion of Texas energy policy positions and responses to inquiries about new energy ideas from the legislative and executive branches of Texas and the federal government. Outside of my years of directing work in public policy development and advocacy, and management of related R&D programs, I have also consulted with both public and private sector entities on the economic impacts of policy and R&D efforts. I participated in, and led Texas energy policy planning organizations from 1973 to 1983 following the Arab Oil Embargo of 1973, and a private nonprofit electric industry R&D organization from 2005 to 2015 following the deregulation of the Texas electric sector. The years in between these two periods of major policy change were filled with teaching economics and consulting on energy, water resources, and environmental issues flowing from changes in these natural resource development periods of change. The consulting efforts often included expert witness assignments in law cases that are typical during and following periods of major market and public policy changes. I have taught managerial economics at three Texas universities. My career has been an adventure in reconciling the two worlds—applied economics and public policy of natural resources.

Preface

This book is about innovation in the Texas energy sector and the lessons learned from key Lone Star State energy events that provide important value to the international community. This story beginning with Spindletop in 1901 is one of a changing set of practices, policies, institutions, markets and technologies. The big events of energy innovation in the 20th century started with a simple process that grew to the complex, integrated system we see today. The dynamics of this multifaceted process developed since Spindletop has rolled out changes in institutions, markets, and public policy, and most recently collaborative organizations, an outcome that one could scarcely have imagined a 100 years ago.

The move from a simple to a complex innovation process has not been a linear phenomenon since 1901. It has accelerated greatly following the Arab Oil Embargo of 1973. The Embargo created an "energy crises" in the United States that dominated the Texas and US political debate for a decade as the country struggled to gain a new direction as the event laid obvious the economic and national security vulnerabilities unique to oil. Although the Embargo was about oil, the entire energy sector and the world economy were fundamentally impacted.

The book reviews several key events in Texas over a 117-year period that are the hallmarks of the unique energy sector contributions of the Lone Star State. These key events remain as turning points in markets, institutions, policy, and technology that have brought us the current explosion of energy sector innovation. Some of these events were driven heavily by market conditions, some by technological change, some by external events, some by research organizations, and others by the insights of unique individuals and state political leaders. The events all highlight the role of a number of key individuals that have represented the somewhat unique cultural qualities of risk-taking Texans.

The modern process of innovation is the combined result of public policy, markets, technological advancement via R&D organizations and private companies, institutional change, and more recently, collaborative organizations that drive the innovation process by generating and entertaining new ideas. Among all of these factors the singular most important aspect going forward is the centrality of a process for the generation of ideas. The unfolding of a process that developed after the Embargo was slow to mature because the political will to create institutional structures that can survive the ebb and flow of the markets is a real challenge. Political will is driven by short-term support of the voters and politicians that reflect the current market conditions. The will to change energy systems seems urgent after an abrupt market change or a national security challenge, but the initiatives often dissolve after the market and political urgency passes. The State and National willingness to make long term commitments is rare.

The events that gave rise to some of the most important energy innovations of the last century have taken place in the State of Texas because of a culture of rugged individualism, independence, and free market commitments of governments matched with giant-size deposits of oil and natural gas, abundant renewable wind and solar, and other interesting natural resources. But the Texas experience is also filled with practical considerations that have resulted in selective governmental regulation of markets, restrictions on property rights that are fundamental to the operation of private competitive markets, and dealing with the inevitable dark side of new innovation. In addition to reliance on the Texas experience, the State leadership has shown a willingness to build on the experience of others in the United States and abroad and to import technologies developed elsewhere.

The challenge going forward is to develop and maintain a process that generates and funds lots of ideas. The recommendation in the last chapter of the book is to set up and fund an institutional mechanism that employs a "crowd-sourcing" process to flood the landscape of energy sector prospects with ideas—an open and transparent process—and to fund early stage R&D of the best of the lot with a combination of public and private capital. Such an institution has to have stability that reaches beyond the current governor, legislature and presidential terms of office. The other part of the recommendation is to divide and fund the winning ideas into risky and not so risky prospects and to look to different strategies to fund them. The recommendation is the subject of Chapter 13, The Road Ahead: Ideas Are Key.

But as we try to accelerate the race down the innovation path we must for ever recognize that a new innovation almost always brings an unexpected dark side that must be dealt with. As a culture in this country and abroad we are constantly struggling to maintain tradition on the one hand, always defending its contribution to stability, while trying mightily to change our practices and technologies on the other hand which will, often unwittingly, change our future traditions. The irony of this circular process is that today's tradition is itself a product of past innovation. But innovate we must, as time marches on, because as Francis Bacon once said, "time is the greatest innovator." And without a thoughtful process we most likely will not like what "time" alone has to offer.

Following the completion of the final draft of this book in early 2021, a major cold-weather event spread across the mid-United States and throughout all of Texas in mid-February, creating one of the most severe winter conditions in history. It was so impactful that it challenges the fundamental theme of the book. So an addendum has been added to explore the likely changes in innovation that can now be expected to unfold. The addendum reviews the challenges now confronting Texas leadership and related policy directives now being debated. Finally, the addendum explores the prospects for a new round of innovation in the Lone Star state energy sector.

Addendum

Innovation Challenges for the Texas Energy Market
Following the Extreme Weather Event of February 2021

Introduction

Following the completion of the final draft of this book in early 2021 a major cold weather event spread across the mid U.S. and throughout all of Texas creating one of the most severe winter conditions in history. The event of mid February is summarized in some detail below. This weather event I will call the "Big Freeze." It was so impactful that it challenges the fundamental theme of the book. That theme, simply stated, is that Texas has been and remains today an exemplary economic, political state and culture for energy sector innovation.

As I write this addendum, the post modem and political responses regarding failures, responsibilities, and new directions following the Big Freeze are still unfolding in Texas and in Washington D.C. My accounting of events and the leadership responses therefore, will be incomplete. Assessments and predictions of new innovation directions are primarily my own. I rely in part on the work of Mark Taylor for my assessments of the key drivers of innovation (Taylor, 2016). Assessments of future innovation outcomes from the Big Freeze event are my own.

The following sections first summarize the polar vortex that spread into Texas in mid February, followed by a section on the economic and demographic impacts of the event. A third section contains an overview of the Texas electric grid and its development over the last 25 years. Finally, I review the current ideas for reforming the Texas electric grid and my predications of the innovation outcomes of the Big Freeze event.

The Polar Vortex and Arctic cold air outbreaks of February 2021

A severe cold weather event spread across the central U.S. during mid February 2021 that thrust Texas and states north into a cold chill at near record low temperatures extending several days without relief. The event lasted seven days beginning February 15 and finally ending on February 22. This crippling event was likely the result of an unusually expansive polar vortex that pushed a large cold air mass downward from the North Pole extending all the way into northern Mexico. The geographical extent of the extreme weather event is shown graphically below (Fig. 1).

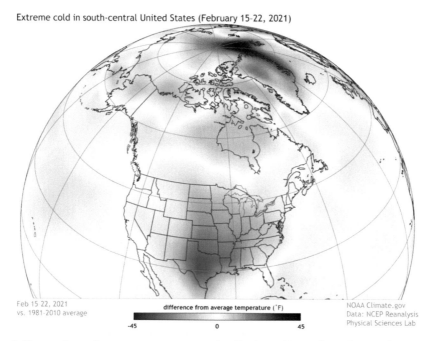

Figure 1 Near-surface air temperatures across the Northern Hemisphere from February 15—22, 2021, compared to the 1981—2010 average. *WeatherNation@ https://www.weathernationtv.com/news/the-polar-vortex-and-februarys-extreme-cold/.*

According to the National Oceanic and Atmospheric Administration, a polar vortex is a large area of low pressure and cold air surrounding both the North and South Poles. The vortex weakens in summer and strengths in winter. As is typical on other occasions, this 2021 event expanded from the northern hemisphere sending cold air southward with the jet stream. Such events are not unique to the U.S. but also occur in portions of Europe and Asia. Such extreme winter events are not necessarily the result of a polar vortex, but in this case the other winter conditions joined with the jet stream and the polar vortex produced an unusual southern push into the mid U.S. NOAA declared that February had the coldest temperatures in the lower 48 states since 1989 (*NOAA*, March 5, 2021).

The February weather event is testing the fabric of the Texas notorious energy-only competitive electric market. As one might expect, the severity of winter storm impacts has raised the attention of the State's political leaders and energy experts and is generating numerous legislative and executive proposals for reform. A central question now being revisited is whether an energy-only wholesale electric market design (including managed emergency protocols and ancillary services) can be expected to provide the market incentives adequate to the tasks posed by once-in-a-century extreme weather events.

There have been other similar severe freezing winter events in 1977, 1982, 1985, 1989, 2011, and 2014, but most notably the 2011 look–alike winter event. But the 2021 event was unusual because of the combination of below freezing temperatures, and at places, below or near zero degree F, joined with a large geographical and time period extent. It hung over the region unabated for seven days. The cold air mass spread like no other in recorded history across almost all of Texas and the surrounding states and remained there long enough to freeze gas pipeline and gas well equipment, electric generator facilities and water systems throughout the states of Texas, Louisiana, Oklahoma, and eastern New Mexico. NOAA Climate reported:

> In late February, as the Southern Plains and Gulf Coast suffered through an unusually strong blast of wintry weather, weather talk turned to the polar vortex and the possibility that the extreme cold was yet another example of weather-gone-wild due to global warming. In this article, we're talking to NOAA experts about the devastating extreme cold event and the polar vortex.

> **Rebecca Lindsey, NOAA Climate (March 11, 2021b)**

Further analysis and professional opinion will no doubt be forthcoming as to the influence of global warming as a root cause of this extreme weather event. NOAA reports that there was warming in the upper stratosphere over the North Pole. While it seems counter intuitive that global warming could cause unusual cold like the February event, Amy Butler, a NOAA stratosphere expert explained.

> '. . .disruptions of the polar vortex occur when the vortex is bumped from below by large-scale atmospheric waves flowing around the troposphere. . .. the waves are always there, but anything that changes their strength or location—including changes in surface temperature and pressure that result from sea ice loss—can potentially influence the polar vortex. So the idea would be that even though you have an overall warming trend, you might see an increase in the severity of individual winter weather events in some locations.' At the surface, this stable stratospheric state is often associated with an even colder than usual Arctic, and milder-than-usual weather in the mid-latitudes.

> **Rebecca Lindsey, NOAA Climate (March 5, 2021a)**

In any case, the mid February event in Texas challenged the Texas electric grid managers like nothing else in the last 25 years—the time period since the State initiated the restructuring of the Texas electric utility system. The weather-induced fallout is discussed in some detail below. The failure of the electric, natural gas, and water systems added human pain and suffering to an already devastating year–long Covid–19 pandemic that has itself caused enormous human hardship, death, and economic loss. The winter event and utility systems failures, in addition to the pounding of economic activity across the board, also disrupted the ongoing attempts to produce, distribute, and administer Covid–19 vaccines. More than a week of delay ensued, preventing efforts to protect the population from severe illness and death, and importantly, to stop the spread of the virus. Ironically, the electric system failures in some cases made the job of treating Covid-19 patients at hospitals and other medical facilities more of a challenge than what already existed.

Key grid events and ERCOT responses as the Big Freeze unfolded and subsided

Electric Reliability Council of Texas (ERCOT) recorded a sequence of events and detailed ERCOT actions leading up to, and following, the Big Freeze. The events and timeline of actions are documented in a series of reports made available to the public following the event [Bill Magness (February 24, 2021) *Review of February 2021 Extreme Cold Weather Event*]. The grid manager issued extreme cold weather postings on the public website, made calls to market participants, and issued a news release on extreme weather expectations. Following these actions, ERCOT held a meeting with the Texas Energy Reliability Council on February 12, followed by a State Operations Center news conference and designating a Conservation Alert on February 13. On February 14 ERCOT issued a conservation appeal by a news release, performed a social media outreach, and held a media briefing.

But conservation from such appeals was not up to the task of allowing ERCOT to serve all loads as generators rapidly went off line. Under ERCOT protocols, the system became critical at 12:15 a.m. on February 15 when it entered Emergency Operations Level 1. The system reserves fell below 2300 MW. The unexpectedly severe weather resulted in generating units going off line one after another in the early morning hours of February 15. At 1:07 a.m. reserves fell below 1750 MW when ERCOT went to Emergency Operations Level 2. At 1:20 a.m. ERCOT entered Emergency Operations Level 3 when 10,800 MW of load were dropped by rotating outages. During this short period approximately 48.6% of ERCOT generation was forced out due to the severe weather (Fig. 2). Controlled outages had to be implemented immediately to avoid a total blackout of the system. Demand had to be reduced to match available supply.

The critical condition rapidly developed due to the severe multiple systems failures of generation from the combined sources of wind, gas, coal, and nuclear units going off line. A large portion of ERCOT generation capacity was forced out (Fig. 2). A further complication was that local utilities were limited in total load that could be shed and how they could rotate outages, which can usually be done so as not to leave individual customers without power for long periods of time. Under the large loss of generation, utilities were limited in how they could rotate outages because of the number of circuits with critical load which could not be shed.

As the early morning events unfolded on February 15, ERCOT had to act quickly in its role of maintaining system frequency. Over a short period of 20 minutes the system went from the norm of 60 Hz swiftly down to 59.302 Hz. As the frequency

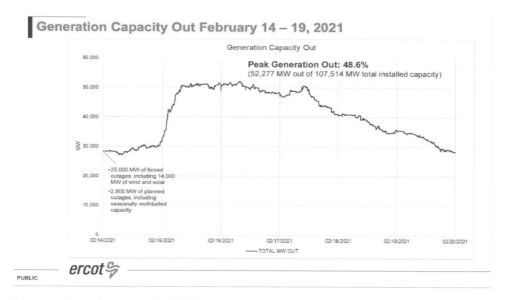

Figure 2 *Generation outage in ERCOT, http://www.ercot.com/content/wcm/key_documents_lists/225373/2.2_REVISED_ERCOT_Presentation.pdf.*

continued to fall ERCOT had to shed load to keep the system stable. Further reductions had to be implemented as additional generation went off line (Fig. 3). In short, the usual ability of ERCOT to keep the system stable by calling on spinning reserves, and incentives from high competitive prices in a tight market to balance supply and demand was not available. Only emergency actions on the demand side could keep the system stable and prevent a total blackout.

Not until February 18 did generation begin to come back on line when outages could be reduced. On Friday, February 19 ERCOT returned to Emergency Operation 2, followed shortly by Emergency Operations Level 1. At 10:35 a.m. the system was returned to normal operations.

Economic and human misery impacts of the Big Freeze

While it is clear that the economic impacts of the February extreme weather event cut across most domains of people, geography, and economic activity, it did not impact all sectors and groups equally. Disparities flowed from the extreme weather conditions independent of any utility system failures, but also from a failure to keep the lights on, from inoperable heating systems, frozen water systems (often due to loss of electric power), loss of natural gas for heating, and limited travel because of frozen roadways. The weather event and the associated utility service failures impacted all geographic regions of the State, most every sector of the economy, and curtailed activities of

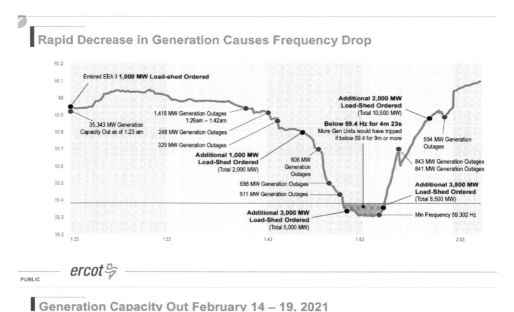

Figure 3 *Frequency drop due to rapid generation loss. ERCOT at http://www.ercot.com/content/wcm/key_documents_lists/225373/2.2_REVISED_ERCOT_Presentation.pdf.*

nearly every government agency, municipality, and school system. It also became an unexpected calamity, especially for poor families and other under-served population groups throughout Texas. The system disruptions were sometimes short-lived and in other cases lasted for days on end until power and gas services could be restored and water pipes repaired.

The weather conditions also made it nearly impossible to travel and left many people with an inability to move to public facilities for warmer housing and bathroom facilities. A number of deaths have been attributed to the multisystem failures although there is no official count since the reliable data source is the county coroner reports which will not be available for some time. Some newspaper accounts reported that 80 people have died as a result of the event. Small retail type businesses already shut down by Covid-19 restrictions had even less means to operate during the weather event. Some hospital systems had difficulty due to power interruptions even though they typically are equipped with backup generator capabilities. The event made obvious the State's vulnerabilities to interconnected and interdependent energy and water systems (electric, gas, water, and waste water utilities). Natural gas systems could not operate because of electric power outages; natural gas electric generators could not operate without gas.

The most obvious initial measure of economic impacts of the week-long disruption of the electric grid is the increased cost of electricity purchased by retail electric providers, and in some cases costs were passed on to end-use consumers. In some cases retail electric providers who are bound to fixed cost contracts with consumers have filed for bankruptcy. Perhaps a larger cost of the Big Freeze was the value of lost economic activity and from human misery and some deaths from the cold due to not having gas and electricity. In the longer term, the event caused damage to generation and distribution facilities that will need repairs or replacement. Estimates of economic impacts have been reported to amount to between $45 and $50 billion (*Joel Myers of AccuWeather*, February 17, 2021).

One of the most publicized negative outcomes of the weather event is a $16 billion electric energy cost attributed to performance failure of the grid manager (ERCOT) who reportedly failed to remove the administratively imposed electric price of $9000 MW/h for a 32 hour period beyond what was necessary under ERCOT rules. Without the $9000 cap operating in place of a competitive market price, the price for energy in the market would have decreased as forced curtailments were removed and generation came back on line. Note: the ERCOT market and ancillary services fundamentals operating during the event are explained in some detail in the next section below.

The ERCOT market structure essentials

ERCOT (the organization) is regulated by the Public Utility Commission of Texas (PUC) and the Texas Legislature. The Texas non-profit was established in 1970 with membership by seven market segments including (1) consumers (commercial, industrial and residential), (2) cooperatives, (3) independent generators, (4) independent retail electric providers, (5) investor-owned utilities, and (6) municipals. The original organization's purpose centered on coordination for reliability of the grid, but the organization was given a central role in the modern grid as the current Texas grid was restructured over the last 25 years. In 1999, the Texas Legislature restructured the Texas electric market and assigned ERCOT four primary responsibilities: provide system reliability, foster a competitive wholesale electricity market, foster a competitive retail market, and provide open access to transmission (ERCOT, 2019a, *State of the Grid Report*, available at http://www.ercot.com/).

The ERCOT electric grid is referred to in national reliability laws and regulations as the Texas Interconnection. It connects more than 46,500 miles of transmission lines and more than 650 power generation facilities which supplies power to more than 26 million Texas customers and represents 90% of the state's electric load. ERCOT is the first independent system operator (ISO) in the United States and one of nine independent ISOs in North America. ERCOT works with the Texas Reliability Entity (TRE), which is one of eight regional entities within the North American Electric

Reliability Corporation (NERC) that coordinate to improve reliability of the U.S. bulk power grid.

> *Market Participants in the ERCOT region are subject to both state and federal laws and regulations. Market Participants that own or operate facilities that are part of the Bulk Electric System, as defined in federal law, are subject to oversight by the Federal Energy Regulatory Commission (FERC), the North American Electric Reliability Corporation (NERC), and Texas Reliability Entity, Inc. (TRE).*
>
> **Compliance in ERCOT,** *available at http://www.ercot.com/mktrules/compliance.*

Texas began restructuring the electric industry in 1995, and over the next ten years completed the restructuring that included development of a competitive wholesale market, then a competitive retail market, leaving the transmission and distribution parts under the traditional rate of return regulatory systems. Municipal, co-op, and river authority systems were not restructured but are integrated into the larger system in selected ways, especially for reliability purposes. ERCOT carries out the functions listed above under the oversight of the PUC and the Legislature.

ERCOT has several major responsibilities for enabling the competitive wholesale and retail competitive markets and operating an ancillary service function that keeps the grid operating at or near 60 Hz; an absolute requirement for keeping the grid up and stable. ERCOT must, at all times (24/7/365), balance all consumer demand in the ERCOT region (load) and the power supplied by companies who generate electricity (generation) while maintaining system frequency of 60 Hz. ERCOT performs financial settlement for the competitive wholesale bulk power market and administers retail switching for nearly 8 million premises in competitive choice areas. Sophisticated tools are employed to support conservation and renewable energy involvements. ERCOT also supports new technology efforts such as accommodating and implementing rules for battery storage systems connected to the grid.

One of the most important functions of ERCOT is executing a series of emergency rules when the gird is under stress from rapid changes in demand and/or changes in supply, especially from generating units tripping off the grid. Executing these rules is essential, fundamentally because electricity cannot be stored on a shelf somewhere like food items in a grocery store or auto parts in an auto supply store. Electric supply must match demand within very narrow time limits for the grid to function, hence the requirement for 24 hour 365 day continual monitoring and adjusting of the generation and load sources, all within seconds, and at very near the 60 Hz standard for electric systems in the U.S. For example the wholesale market in ERCOT "clears" every 15 minutes so to support the trading of KWh between generators and retail companies. And the grid operator (ERCOT) must keep the grid up by continual monitoring and taking stability actions.

ERCOT is funded by a system administration fee to cover the cost of its operations. The fee is currently 55.5 cents/MWh. The average cost of ERCOT's services to residential households is $7 \, \text{year}^{-1}$. ERCOT does not set consumer electric rates.

Such rates are either set by the PUC, local municipalities and coops for regulated utilities, or by the companies that sell electricity at retail to end-users in the ERCOT competitive market. Transmission system costs by the "wires" companies (regulated by the PUC) are passed along to customers proportional to their use [Bill Magness, (February 24, 2021) *Review of February 2021 Extreme Cold Weather Event*] ERCOT Presentation available at http://www.ercot.com/content/wcm/key_documents_lists/225373/2.2_REVISED_ERCOT_Presentation.pdf.

The ERCOT geographical area covers most of the state's consumer base. Other customers are in the fringe areas (41 counties out of 254) served by neighboring Western Interconnection or Eastern Interconnection service areas (Fig. 4).

The ERCOT market has a diverse mix of generating power sources that in 2019 was 47% natural gas, 20% coal, 20% wind, 11% nuclear 1% solar, 0.2% hydro and 0.1% biomass.

The restructured Texas electric market has functioned remarkable well for fifteen years since full retail deregulation was completed in 2005—until the Big Freeze. An exception is noteworthy; the 2011 winter event was very cold but relatively short-lived making its impacts minor when compared to the Big Freeze. ERCOT went to EEA 3 (ERCOT level 3 of emergency procedures known as Energy Emergency Alerts) at 5:43 a.m. on February 2, 2011, when 1000 MW of load was shed. At 6:04 a.m. an additional 1000 MW was shed and the system frequency dropped to 59.576 Hz at 6:04. More load was shed until the system stabilized with the frequency returning by about 6:15 a.m., followed by load being partially restored beginning at 11:39 a.m. The system moved forward by restoring load as generation resources were restored, with the system fully operational again by 10:00 a.m. on February 3. The system frequency was restored to 60 Hz + by 6:30 a.m. As discussed elsewhere, a postmortem

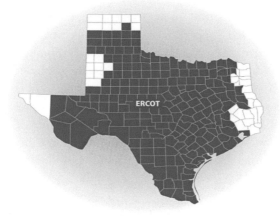

Figure 4 *ERCOT Geographical service area and other interconnection frenge areas. ERCOT available at http://www.ercot.com.*

analysis of the 2011 event called for improvements to be made to guard against a repeat of such system limits. ERCOT reported to the U.S. House Committee on Energy and Commerce on March 18, 2021, regarding the ERCOT area improvement activities. ERCOT reports as follows regarding Texas' follow up to the 2011 event:

> *This focus on generator weatherization appeared to have resulted in improvements in generator cold-weather performance, as demonstrated during subsequent cold-weather events. For example, on January 17, 2018, the ERCOT region experienced severe cold weather conditions that were very similar to those experienced in the 2011 event. Whereas ERCOT had lost 29,729 MW due to freezing weather conditions in 2011, ERCOT lost only 1,523 MW during that 2018 event and was able to serve system demand of 65,750 MW, which was, until recently, ERCOT's winter demand record, exceeding the next highest winter peak by over 6,000 MW.*
>
> *During this same event, areas outside of ERCOT in the South Central U.S. experienced outages to 183 generating units representing approximately 30,000 MW. This resulted in a separate inquiry into the cause for the outages and a 2019 report by FERC and NERC. The 2021 storm, however, was substantially more severe than these more recent severe events. Whereas during the 2011 event, low temperatures in the major load centers of Dallas, Houston, San Antonio, and Austin were 13 degrees, 21 degrees, 19 degrees, and 18 degrees, respectively, low temperatures for those same cities during the 2021 event were much colder: -2 degrees, 13 degrees, 12 degrees, and 6 degrees, respectively. As with the 2011 event, the ERCOT region also experienced high winds, ice, and freezing rain in 2021. Unlike the 2011 event, ERCOT also saw significant snowfall of several inches or more across most of the state in 2021. And the period below freezing during the 2021 event lasted much longer in most areas of the state in 2021 than in 2011. The strain on the ERCOT grid due to the combination of these conditions in February 2021 therefore far exceeded what ERCOT had experienced in 2011. While ERCOT had valid reasons to believe that its fleet was more capable of withstanding extreme conditions such as those in 2011, the 2021 event was far more severe, and the impact on generator availability was clearly that much more significant.*
>
> **ERCOT Letter Report to the U.S. House Committee on Energy & Commerce (March 18, 2021)**

The February 2021 event was more challenging than the 2011 event by far. Emergency operations Level 1 began at 12:15 a.m. on February 15, followed by Level 2 at 1:07 a.m. and Level 3 with rotating outages beginning at 1:20 a.m. when 10,800 MW of load was dropped. The system frequency moved from normal near 60 Hz at 1:30 a.m. to 59.302 Hz accompanied by multiple load sheds being executed until the level returned to 59.7 Hz, and then to near 60 Hz again by 2:00 a.m. A system Hz of below 59.30 would have triggered another frequency firm load shedding scheme which was narrowly avoided. The operator response brought the system to within 4 minutes of a total failure, which was avoided by the operator's swift actions to keep the system stable.

The generation taken off line because of the Big Freeze required a 20,000 MW peak load shed and the system was not returned to normal operations until 10:35 a.m. on February 19. So the 2021 Big Freeze lasted four days with sub-freezing temperatures all across the state with 48.6% of generation off line during the critical two-day period. Maximum generation capacity forced out in 2021 was 52,277 MW compared to 14,702 MW in 2011. The magnitude of generation failure in 2021 required a concentration of load reduction in residential circuits across the state as critical load had to be operational, therefore calling for targeted load reduction rather than the usual rolling blackouts.

Current ideas for reforming the Texas electric grid in response to the Big Freeze

The Big Freeze stands out and has called into question many of the assumptions and "as given" political judgments that have supported the ERCOT market operation for a decade and one half. A number of weaknesses are discussed next.

There are several weaknesses in the existing ERCOT market that the Big Freeze laid bare. The first is the unchallenged assumption that the past extreme conditions that need to be built in to the current response paradigm were valid going forward. An extension of that weakness is that the summer extreme weather events are the primary extremes to worry about in Texas. That assumption was "blown out of the water" when the peak demand in February 2021 more than matched the previous peak of 74,679 MW in the summer of 2019. The winter peak was 76,819 MW on February 14, 2021.

The second weakness is the "silo" structures in the key sectors. The Big Freeze exposed the weaknesses that assume electric systems (in the aggregate) can be planned for and rules made and implemented in isolation from that of gas, fuel oil and water utilities. Further, gas systems that have a much larger customer base than electric power plants and local gas utilities have not worried about the electric grid functionality—their primary customer base is industrial, commercial, and residential users, with Liquified natural gas (LNG) importance on the rise. Water systems primarily operated by municipal governments and special purpose water utility districts have not worried much about electric supplies to run water pumps. Finally, government regulatory and planning entities have been in silos associated with the three above sectors. These include the PUC, ERCOT, Texas Railroad Commission (TRRC) and Texas Commission on Environmental Quality (TCEQ). There has been special-purpose entities created in the past for dealing at a high level with public policy problems that require input and coordination among public agencies and private industry, joined with universities with R&D expertise. This principal is discussed in the forecasting section.

A third weakness is the practice of designating critical infrastructure for high priority systems to stay online in an emergency. The list of service providers for priority classification for uninterrupted power, natural gas, and water cannot currently be isolated for protection on an electrical circuit. Current practices and technology is confined to leaving an electrical circuit functional for people and businesses on the common circuit, while a critical infrastructure facility stays on as well. The need is for granularity in the ability to isolate individual entities for protection when everyone else on the circuit is curtailed.

While such critical infrastructure designations currently provide a source of comfort, a better solution begs for granularity that improves the efficiency of selective rolling blackouts. The entities designated as critical infrastructure is specified by the U.S. Cybersecurity & Infrastructure Security Agency (CISA). There are 16 critical infrastructure sectors whose assets, systems, and networks, whether physical or virtual, are considered so vital to the United States that their incapacitation or destruction would have a debilitating effect on security, national economic security, national public health or safety, or any combination thereof. Presidential *Policy Directive 21 (PPD-21): Critical Infrastructure Security and Resilience* advances a national policy to strengthen and maintain secure, functioning, and resilient critical infrastructure.

A fourth weakness of the current ERCOT grid is the continuity of institutional creations. Long term problems require long term commitments. A current example of an institution designed to survive the turn over from Presidential terms of office is the Federal Reserve whose members serve six staggered year terms.

A fifth weakness is the absence of equity among low income and marginalized populations who get caught up in difficult living arrangements that do not do well in extreme weather events. The obvious vulnerable conditions include poorly insulated housing and limited travel options to get to public facilities. Possible remedies discussed below include some form of public subsidies and rules to provide weatherization additions to housing to protect against extreme weather and insurance for equipment replacement, and/or direct payments for housing weatherization.

Where do we go from here?

The grid managers, their regulatory overseers, and political leaders have not been silent since the Big Freeze. Hearings at the PUC and the Legislature have collected information, taken testimony, executed preliminary executive actions, and introduced legislation to address the failures and limitations of the ERCOT market and its associated entities. Key leaders at ERCOT and the PUC have resigned or been fired and replacements have yet to be named. In the meantime, the processes which are quite adapted at keeping the systems up are ongoing. The paragraphs below summarize the key executive actions and draft legislation that has surfaced to address the system limitations so far.

A number of bills have been introduced in the Texas House of Representatives to address the apparent weaknesses of the ERCOT grid and its organizational structure. As usual there is not a uniform agreement on what the weaknesses are and less so on how to address them. No doubt there will be other legislative proposals, along with changes to the existing bills going forward. The current filings will see action, however, since the Governor has declared an emergency which allows the bills to be advanced out of line with the regular bill advancement process. The main ideas contained in the bills that have been introduced are summarized below.

House Bill 10
This bill would restructure the ERCOT board to have unaffiliated members appointed by the governor, lieutenant governor and speaker of the house. It would also require all board members to live within Texas and add one more consumer representative to the board.

House Bill 11
The legislation requires power generators and electricity suppliers to weather-proof their facilities.

House Bill 12
This bill would study the creation of a statewide alert system for weather disasters, similar to the Amber Alert for missing children.

House Bill 13
The bill would create a council made up of the PUC, Railroad Commission, ERCOT, and the Texas Department of Emergency Management to coordinate on energy during a disaster.

House Bill 14
The legislation would require the Railroad Commission to adopt rules requiring the weather-proofing of the natural gas infrastructure.

House Bill 16
The bill would ban variable rate electricity products for residential consumers.

House Bill 17
The legislation would ban political subdivisions and housing authorities from passing ordinances or rules that prohibit the connection of residential or commercial buildings to specific infrastructure based on the type or energy source it would provide to users. An example is banning an ordinance requiring new homes to be all electric.

The Big Freeze event is generating other ideas like that of Hunt Energy Network's new venture for adding 50 battery systems (500MW) on the ERCOT grid–battery systems located near the consumer end of the grid. The combined technology, institution, and market idea promises to take the Texas innovation prowess to a new level. This is one Big Freeze response promising to increase the resilience (stability and reliability) of the ERCOT grid (Alex Edwards, Dallas Innovates 2021 available at https://

dallasinnovates.com/hunt-energy-networks-new-venture-will-put-50-batteries-across-texas-giving-ercot-a-portfolio-of-energy-generation/

More time will be required to know what reforms actually get decided and implemented. In the meantime, it is worth reviewing the prospects for a more resilient and better managed electric grid that is predictably the outcome of the ongoing investigations and reform decisions. To identify such likely outcomes, the following discussion turns to a review of the current understanding of the energy sector innovation drivers, leading to predictions of the policy, institutional, market, technological and collaborative organizational changes that will follow in Texas.

As mentioned earlier, this Big Freeze event amounted to a system and cultural catastrophe metaphorically analogous to outcomes from "hitting a wall." The Texas restructured electric industry and the PUC/ERCOT institution with limited foresight did not see this event coming, and therefore was unable to prevent a catastrophic outcome. It did enable the grid managers to avoid a much worse outcome, to the credit of astute managers and protection systems that avoided a total collapse of the electric grid—an outcome that would not have been totally recovered for months.

The key drivers of innovation

Several discussion sections of the book describe a process where a catalyst for innovation comes from a "hitting a wall" kind of experience. I think Texas hit a wall with the February Big Freeze. Private discussions, public meetings, and Legislature committees have been trying to get a handle on the problem presenting itself from the Big Freeze. Leaders at the PUC and ERCOT have been fired. The Legislature has been holding hearings to clarify what just happened, to identify system and leadership failures, and to design reforms. The political process seems destined to find a villain and then design some reforms for the system "so this never happens again." The situation begs for innovation—innovation in markets, technology, public policy, institutions and collaborative organizations.

Mark Taylor has explored the perplexing and intriguing topic of identifying the drivers of national innovation in his book, *The Politics of Innovation*. He examined the innovation records of over 70 countries and found and examined the most innovative among them, as well as the laggers in innovation. The U.S. is among the list of the most innovative. Taylor's theme is a postulated theory he calls Creative Insecurity. Evidence from his examination of all these countries over long periods of time supports his creative insecurity theory of country-level innovation.

Taylor's creative insecurity theory states that there exists a national dynamic leading to innovation resulting from a combination of internal conflicts and strife on the one hand, but overarching external threats on the other. While Texas is not a country, there does seem to be parallels associated with the current Texas Big Freeze event and the unfolding processes. There certainly is internal conflict and strife now, especially at the political level.

There is also a bit of external threat, namely that some form of new federal regulatory action may interfere with the long standing commitment of Texas to keep the "Feds out of our business" as much as possible. This Texas independence streak has been part of the culture since the days when Texas was a nation before joining the Union in 1845. As a result, I predict a new round of innovation in the Texas energy sector. With that thought and brief historical reminder I now turn to the expected new innovations.

Texas new innovation direction projections resulting from the Big Freeze

Clearly the future extreme weather event prospects should be considered outside the extreme events in history, even the Big Freeze. Therefore, the **first innovation** to come from the follow up to the Big Freeze is **institutional innovation**. I predict that ERCOT, PUC, TRRC, TCEQ and TRE will jointly make two formal judgments each year concerning the prospects for an extreme weather event. Such judgments will be made as an outcome of the winter and summer forecasts discussed in the third innovation below. These judgments will take on a hearing and formal document format to be completed and published in the months leading up to the winter and summer seasons each year.

Much of our ability to prepare for extreme weather is up to the private sector. But the public sector entities have to provide early warning and timely public communication. Public communications using social media and news conferences will make the public aware of the group's judgments about the approaching winter and summer conditions. The group will use the best scientific work of the NOAA, National Weather Service, and university experts when reaching their judgments. This innovation calls for more astute weather predictions and public announcements.

While there cannot be much certainty in predicting a more severe winter event than the Big Freeze, Texas leadership will work the climate change topic until it helps enlighten us about whether the clear global warming trend will continue and feed the prospect that such warming could bring a more severe Big Freeze. Of course, the group will expand the work to include extreme heat event prospects and more severe hurricanes and storms. Extreme summer heat, hurricanes, and storms call for innovations that differ from severe winter events.

A **second innovation** to grow out of the Big Freeze event is a **market innovation** centered on the development and implementation of insurance remedies to cover the high level of uncertainty around such future weather events. Texas will explore a wide range of insurance options to cover a large set of activities not now covered in the private insurance markets. Of course, insurance remedies do not directly guard against the actual weather outcomes that destroy or make equipment obsolete, or prevent death and injury. Insurance just provides financial protection.

The private sector already uses hedging to provide insurance-like protections from extreme price movements in the ERCOT competitive electric markets. The agricultural sector uses crop insurance regularly, so expectations of coming extreme weather will get built into crop insurance. Home owners and commercial property insurance protections are commonplace for bank loans. Future insurance premiums will include rising risk levels from extreme weather. Premiums for future policies will adjust to cover future expectations. Lloyd's of London may offer ideas of how Texas might use unusual insurance policies designed to address winter and summer extreme weather events like the Big Freeze.

A **third innovation** involves an increased role for TRE as the existing **institution** that ought to be capable of solving the silo mentality challenge in Texas energy/utility related institutions. Electric industry users, owners, and operators within ERCOT are eligible for membership in TRE at no cost. Members include representation from one of six membership sectors, including cooperative utility, generation, load-serving, marketing, municipal utility, system coordination and planning, and transmission and distribution. Perhaps the natural gas and water system leaders will be added to TRE. Executive Board membership will be changed to provide long term stability by increasing the length of term and providing staggered terms.[1]

TRE is tasked with compliance, monitoring, and enforcement on the behalf of NERC to ensure bulk power system reliability. The PUC has authorized TRE to serve as its reliability monitor for the state of Texas and for ERCOT. TRE responded to NERC inquires following the 2011 event. A key question was: What does the ERCOT region do to benchmark its planning models against real system conditions, especially severe and unusual system conditions? When does the region benchmark its planning models? The answer was that the overall forecast methodology by ERCOT ISO has changed to include the extreme weather events of 2011 (both in the winter and summer). ERCOT ISO specifically publishes two seasonal assessments each year in summer and winter that aim to include impacts of more severe conditions. Models are updated to correct discrepancies discovered during operational studies and conditions, faults, and disturbances (Texas Reliability Entity response to NERC Request for Follow up Actions To 9/8/2011 SW Blackout — August 1, 2012). Models of gas and water system responses to extreme weather events will be added.

The **fourth innovation** is **technological improvements** to infrastructure, including weatherization. The U.S. and Texas economies are very dependent on infrastructure to function. A taskforce should be created to study and recommend needed changes to

---

[1] There is a historical example in the ten-year long follow up to the Arab Embargo, namely the Governor's Energy Advisory Council (GEAC) and follow up entities (including TEAC and TENRAC) in response to the Arab Embargo of 1973. Such a planning and coordination entity(s) need to be structured for the long term with a mission and staggered long-term director appointments, long enough to reach beyond the terms of office of political leaders (esp. Governor, Lt. Governor, and Senator). An example of long and staggered terms is the U.S. Federal Reserve.

the naming of critical infrastructure for priority service when considering weather impacts on electric, gas, and water systems. There are currently 16 sectors, energy being one of them, named as critical by the U.S. CISA. The energy infrastructure is divided into three interrelated segments: electricity, oil, and natural gas. Virtually all industry is dependent on electric power and fuels. The energy sector is vulnerable and is leading a significant voluntary effort to increase its planning and preparedness. The CISA institution will be the source for keeping this critical infrastructure designation process in place and Texas will continue to participate.

The top of the technology improvement list for preparedness is **weatherization** to improve the electric and gas systems resilience against extreme freeze events like that of February 2021. A number of approaches to the actual achievement of increased resilience through such technology driven improvements will be explored. One approach is to create a market price adder to market clearing prices in competitive market areas as a return for investment in weatherization, and direct reimbursement adders to provider's rate of return calculus in regulated sectors. Monitoring and inspections will be required. The difficulty is in knowing how much weatherization is justified to improve to probability that selected improvements will avoid the kind of generation trips and gas system freezes experienced in February.

A different rationale would be needed for gas and electric systems. Then there is the matter of technological improvements to home, commercial and industrial end users systems. There is an inescapable dynamic between what consumers believe the providers will do to "keep the lights on" and a decision to invest in onsite backup generators, for example. The matter of weatherization will require considerable work and political process to define and execute a weatherization program, but the process will be started soon.

The **fifth innovation** is decreasing the **equity inequality impacts** of extreme weather events among population groups for food, shelter and health care. This inequality will be addressed by including, but not limiting, the focus to the broad groups of rural and urban communities—rural communities are often at a disadvantage in policy decisions and implementation of change relative to the urban counterparts. A major focus will be increased incentives for weatherizing of housing and incentives for special clothing and food supplies for extreme weather events. These objectives can be best addressed by public entities joining with private humanitarian organizations supported by tax payer funded planning.

Another inequality laid bare by the Big Freeze was the unequal access to travel by low income and disadvantaged groups. Strategies will be developed to provide emergency services for disadvantaged groups to get to essential services and health care. Such strategies will be planned and implemented by refocusing the use of existing funding for public transportation entities.

CHAPTER 1

The dynamics of innovation and technology in a market economy

Contents

Introduction

The slow recovery of the US economy from the recession of 2008, and current evidence that the United States is losing the race for global innovation leadership is getting the attention of US political leaders, innovation leaders, and economic experts. Arguably innovation in the energy industries will greatly determine our success in rekindling and maintaining the US economic growth which is tightly dependent on dominance as the world innovation leader. An examination of the innovation successes in the Texas energy sector adds an important contribution to understanding and driving future progress through innovation.[1]

The 117-year Texas energy sector experience since Spindletop in 1901[2] is filled with innovation and technological advance that has been an interesting interplay of

---

[1] It is clear from the literature that innovation in the energy sector of the economy is a very important part of US innovation successes of the past, and will certainly be a major contributor to economic growth through innovation in the future. The Texas experience is a significant part of the story of the past, and the Lone Star State will be a key leader in the innovation of the future.

[2] This is a reference to the oil well completed at a place known as Spindletop Hill in the southern edge of Beaumont, Texas on January 10, 1901. The well that "blew-in" on a January morning shattered the pipe, spewed oil over 150 ft. out the top of the rig, and drenched the countryside at an estimated rate of 100,000 barrels of oil per day (Goodwyn, 1996). This event is the focus of Chapter 2, The Common International Energy Innovation Drivers, that follows.

Innovation Dynamics and Policy in the Energy Sector
DOI: https://doi.org/10.1016/B978-0-12-823813-4.00001-4

three forces that in large measure determine the role and importance of energy in the culture and the economy—policy, markets, and technology. This energy innovation triad has been and is ongoing not only in Texas, of course, but importantly here in the Lone Star State. Texas has a unique role in the energy market innovation dynamics that is the bedrock underlying the most important national and global challenges of our time—providing low-cost energy to industry and consumers everywhere while protecting the environment. Understanding the innovation process in the Texas energy experience can help us better prepare the state and nation for the challenges of the next century and beyond.

Texas has been the leading energy state in the United States since Spindletop. The Spindletop event set in motion a turbulent period of oil market innovation dynamics leading to the discovery of the East Texas Field in 1930. The wildcatters in East Texas flooded the world market with oil. The several events of this period and the two decades that followed set off a series of truly unique experiments in public sector intervention in private markets—creating an "institutional innovation." The East Texas Field event and the Texas Railroad Commission (TRC), empowered by the Texas Legislature, redefined the limits of property rights, created a constitutional crisis between state and national governments and introduced conditions ripe for innovation to flourish in the energy sector. While this institutional innovation clearly provided needed stability in a dysfunctional market place for about 40 years, its positive contribution soon turned negative and became an impediment to market innovation in Texas, opening the door to a Saudi Arabian lead Organization of the Petroleum Exporting Countries (OPEC) take-over of oil market dominance. But interestingly, the OPEC challenge to the Texas and US energy industry amounted to "hitting a wall" of sorts, which eventually led to the most important energy technological innovation in decades—hydraulic fracking. The OPEC challenge also led indirectly to a dynamic interplay of public policy, institutional change, and technological innovation to foster competitive energy markets for natural gas and electricity. Energy markets that leverage the internet, especially in the electric sector, are creating a new digital era that is still unfolding. Texas has been, and continues to be a leader in this new world of energy.

The East Texas Field event and all that followed put Texas in the forefront of energy sector development that continues today. These events set up the conditions for technological and market development throughout the energy sectors including oil and gas extraction, pipelines, refining, petrochemicals, electric utilities, and a renewable energy industry—and importantly created a culture of entrepreneurship that has benefited Texas and the national economies ever since. And, it has hastened a technological revolution in other sectors.

The process of innovation summarized in this chapter includes key topics addressed by experts when writing about this important but imperfectly understood subject. The following pages define innovation and summarize the increasing rate of growth in technological change, market development and policy implementation. Then the chapter defines the relationship of technological change and economic growth, explains the importance of human capital in driving innovation, explores the treatment of time in the law and economics, discusses the importance of property rights, explores government's role in advancing innovation, and identifies the fundamentals of economic models of the economy that quantify the influence of technological change on economic growth. Finally, the influence of politics on the Texas experience is explored. Just as war events and politically led initiatives like the "moon shot" of the 1960s create crises-driven innovation, so the new challenges of globalization and the internet raise new concerns about the ability of competitive markets alone to drive the innovation. There is a growing belief in the innovation economics profession that in this digital and internet-based technology era, global-level networking, is essential, and that free markets alone will not sustain US innovation leadership (Taylor, 2016).

More than ever the debate on the topic of innovation and market economics needs an understanding of when markets fail, and what public policies are needed to guide future market-based innovation. The topic is particularly difficult because there are no common anti-trust laws that span global private sector activity, and the information technology (IT) companies that drive much of today's innovation operate throughout the world. While the topic of market failure and related remedies have been an active part of economic theory and practice in natural resource economics for decades, the current challenges of the digital era inside and outside of the natural resource economics discipline seem front and center. We need to understand market power, market failure, and policy remedies that create economic efficiency which supports and drives innovation on a global scale (Arrow and Debreu, 1954).

The usual focus of experts writing on the topic of innovation is usually centered on the national government experience. Many key innovation and technology developments, however, have come to pass without a national directive or financial incentives from the federal government. In some cases state policy has driven innovation, with technological change following, fulfilling a purpose that was partly a market development and partly a broader public interest purpose. An example in the natural gas sector is an event that ended natural gas flaring (discussed in Chapter 7). In this case, a valuable natural resource was being "wasted" by burning the escaping natural gas at the well-head to produce crude oil that had a developed market and therefore a high value. Bill Murray, who was Chairman of the TRC at the time, recognized that the industry would only stop the practice of flaring natural gas if it could be shown that the technology existed to make reinjecting the gas into the oil reservoir an economic proposition. In a TRC-led effort a technology demonstration project succeeded and natural gas flaring was quickly brought under control.

Another example illustrates how market incentives led the triad when a technological challenge stood in the way of success. Spindletop, the subject of Chapter 3, opened the way for the great oil industry development in Texas. Spindletop came about only by the private sector raising enough venture capital to discover a large deposit of oil in a salt dome near Beaumont, Texas in 1901. But as explained later, the event would have failed except for the development of a down-hole technology to support drilling in soft horizons, thus allowing the driller to drill deep enough to reach the large oil reservoir. The first two of three wells had to be abandoned because the well structure fell-in on and trapped the drill stem. A technological innovation allowed this single event (the third well) to emerge, and the important side-effect was to define the structure of the oil industry operations—the roles of independent investors and operators, wildcatters, royalty-owners, rough-necks, roustabouts, land-men, and drillers that still exist today. It also set up an ongoing competition between major oil companies and independents which continues to enhance oil and gas industry development today.

The expectation of both US political leaders and economic experts for decades has been that the US capitalistic democracy will continue to provide world leadership in economic growth through technological, market (including business model changes), and institutional innovation—all supported and driven by public policy. The United States has clearly been the world's leader of innovation for many decades and the Texas energy contribution has been a significant part of it. Expectations that the US superiority will continue have been reinforced by experience on almost all fronts including energy technologies, internet-based communication, military systems, pharmaceutical and biomedical products, and other sectors where breakthroughs have changed the landscape.

The reinforcing evidence that the United States is the world innovation leader has been especially convincing since the end of World War II—until recently. Beginning with the 21st century the evidence of the US superiority in global competition for innovation leadership is less compelling, and it appears that economic growth from the advent of new technology has slowed. As global competition for innovation leadership increases many experts have begun to question the sustainability of US long-term economic growth that is increasingly dependent on maintaining world leadership through innovation of technology, competitive markets, and institutional change, all supported and driven by public policy. Having said that, however, there is an ongoing debate about whether the public policy change ought to be "getting government out of the way of private markets," or supporting government efforts to create the incentives, coordination of public and private interest, and funding priority research and development (R&D)—selecting the "winners" or just funding basic science.

Economists and other professionals are focusing more keenly than ever on understanding the innovation process, and importantly on developing better mechanisms, incentives, and public policies that can rekindle US innovation leadership. But a reading of the professional literature of the last decade or so yields two overarching conclusions. First, we do not adequately *understand the innovation process* so as to be confident that we can stimulate the "drivers" to innovation in a comprehensive way. Second, developing such an understanding requires being able to *measure innovation*, and *develop data sources that support ongoing research* efforts to advance technological, market, and institutional innovation. We are currently lacking on both counts.

Perhaps the recent development of university-based programs that are designed to shed light on the innovation process will lead the way to understanding innovation dynamics and turn what has historically been a hit and miss endeavor into a science itself. That is exactly what the MIT Lab for Innovation Science and Policy in Boston, Massachusetts, is trying to do. As stated on the Lab's webpage:

The MIT Lab for Innovation Science and Policy is an Institute-wide laboratory recently established to help develop the area of 'innovation science' — an emerging field that can be thought of as applying the scientific method to the practice of innovation.

Using a diversity of methods, the lab empirically investigates how innovation occurs, and pioneers more systematic assessments of possible interventions (such as policies, programs or incentives) to achieve desired innovation outcomes (such as the creation of innovation-driven enterprises, and in the longer run, job creation, economic and social impact, and a vibrant innovation economy).

The Lab for Innovation Science and Policy aims to become the place that policymakers, senior executives, and entrepreneurial leaders turn for evidence-based guidance on the design of innovation-focused job policies and programs in their organizations, local regions and nations.

The MIT Lab Areas of focus include:

Innovation Metrics: *The measurement, evaluation, and visualization of metrics for innovation, including those that trace the linkages among key ecosystem stakeholders (e.g. universities, corporations, government, entrepreneurs and risk capital providers).*

Innovation Policies: *Exploration of the impact of policies (e.g. taxation of early-stage investment, visa policies for entrepreneurs, intellectual property rules, broader legal frameworks for startups) on innovation-driven entrepreneurship and ecosystems.*

Innovation Programs: *Exploration of the impact of programs (e.g. accelerators, prizes, entrepreneurship education programs, etc.) on innovation-driven entrepreneurship.*

Innovation Boundaries: *Defining and understanding the factors that enable innovation practitioners to work most effectively across boundaries (e.g. universities with large corporations and with entrepreneurs, including 'good practice' on rules and policies for conflict of interest, IP ownership, licensing terms, etc.).*

Innovation Scale Up: *Working collaboratively to understand the role of manufacturing and production in the innovation process, particularly for capital-intensive innovations, across global supply chains, and how 'innovation diplomacy' can help with such global approaches to innovation.*

MIT Lab for Innovation Science and Policy (accessed April 22, 2018) https://innovation.mit.edu/lab-innovation-science-policy/

Except for the current concerns that challenge the excepted narrative, several key accomplishments during the last 75 years have convinced us, and the rest of the developed economies, that the United States leads in innovation. Perhaps the singular event of planning a moon landing in 1962 and landing on the moon in 1969 solidified the US performance as the world leader of innovation. John Kennedy's characterization of the moon shot challenge is a historic example of a visionary challenge that brought together the right policy, technology, and markets to achieve an extraordinary national purpose. Kennedy said:

We choose to go to the moon. We choose to go to the moon in this decade and do the other things, not because they are easy, but because they are hard, because that goal will serve to organize and measure the best of our energies and skills, because that challenge is one that we are willing to accept, one we are unwilling to postpone, and one which we intend to win, and the others, too.

ScienceFocus (accessed July 03, 2020)

The moon shot announced by President Kennedy in September 1962 is perhaps the most recognized technological and politically staged accomplishment of our generation. In retrospect the space program that achieved this extraordinary accomplishment was a high-risk endeavor that paid off, demonstrating to ourselves and the rest of the world that the United States knew better than anyone else how to bring together the forces of technological advance, public policy, and markets to achieve a national purpose. The United States has arguably maintained this reputation ever sense, but such a stature is now being questioned.

In the high-profile example of the moon shot, public policy and incremental additions to the technology base drove the effort and also produced unforeseen and unintended market developments like improved solar panels and battery storage that followed years later. In this case, policy (and institutional change) led markets and technology in the innovation process. Technology development and improvement of existing space technologies led the way to success in a highly focused objective. The market products that found their way into the market place came later and were never the primary purpose of the effort. The Hubble Space Telescope, NASA's Juno Spacecraft and the New Horizons project are spin-offs of the moon shot effort that have changed our knowledge of the universe and have perhaps become the one of the best examples of human capital investment that continues to generate new ideas.

Other examples come from our responses to unexpected challenges that amount to civilizations "hitting the wall." Wars are a source of innovation. While we would never propose going to war so as to generate innovation, such human experience has produced many important technologies. For example, the World War II effort following Pearl Harbor produced new aircraft, especially the B-27 and the B-29. Other parts of the war machinery were developed with a singular objective led by the federal government. In the end, the Manhattan Project that produced the atom bomb, which ended the War in the Pacific Theater came by a federal government—led effort with a single objective. The market outcomes that followed include nuclear power plants and modern refinery operations to mention two huge energy sector market outcomes of the war effort.

One of Bill Gates' current efforts also has a singular focus—produce one or more technologies that are of lower cost than fossil fuels, and which produce zero CO_2 emissions to give our world the ability to provide economic benefits to developed and underdeveloped countries alike and at the same time cease the CO_2 contribution to global warming (CNN's New Year's Day Interview with Bill Gates, 2013). The effort has similarities to the moon shot effort in that there is a clear, measurable objective. But there are two major differences. First, the moon shot was completely a US national effort while the Gates' low-cost energy and zero CO_2 efforts are multinational in scope. Second, the moon shot was completely a nationally funded effort. Gates proposes a multinational group of companies joined with multinational governments structured in private—public partnerships. Gates talked about what we can do to improve the lives of the poorest 2 billion people on the planet. Perhaps surprisingly, this talk was not on vaccines or seeds, but instead was focused on how energy impacts this population—and how we need "energy miracles," though innovation that will lead us to low-cost energy and zero carbon dioxide emissions (Gates, 2010).

A common approach of government for giving direction in response to a new shock to the economy and society is to set up a special commission to generate proposals for solutions, to recommend legislative and regulatory changes and to propose funding for commissions and agencies and, especially for R&D for the longer term. The most dramatic "shock" to the energy system, and thus the economy and society reviewed in the current writing, is the Arab Oil Embargo of 1973. The federal government in Washington DC and state governments across the nation created special commissions and often new government agencies to address the challenges flowing from the Embargo. Texas did so immediately.

Chapter 7, Panhandle Field and Natural Gas Flaring, summarizes the US and Texas creations of commissions and agencies created in response to the Embargo. The chapter reviews the structure and activities of these entities during the decades that followed. These commissions and agencies are examples of institutional innovations. An important part of Chapter 7, Panhandle Field and Natural Gas Flaring, is an assessment of the successes of these institutions by way of measures and indicators of their *impacts*. The commissions and agencies while being driven by duties defined in Executive Orders and statutes all had a fundamental institutional character—a means of generating new ideas. Measuring the effectiveness of such entities is a central issue in understanding how to generate institutional innovation in the future. Chapter 7, Panhandle Field and Natural Gas Flaring, discusses this challenge by way of examining the US and Texas experience following the Embargo. Governments both in Texas and the United States worked desperately for more than a decade to figure out how to respond to the economic and political upheaval brought on by the development of the "oil crisis" generated by the Embargo. The mix of energy policy and national defense dominated the debates both at home and abroad. The dynamics of energy, the Middle East and national defense still drives much of the ongoing debate. President Carter said United States faced a possible "national catastrophe" unless it responded with the "moral equivalent of war" (Carter, 1977).

Ideas often challenge the accepted narrative of the day, and new ideas usually find their expression in institutional change, policy revisions, and market development. Perhaps there is no better example than the experience during and following the Great Depression of the 1930s. The best known economist of the time, John Maynard Keynes, challenged the ideas of fellow economists, and had a great impact on not only the nation's rise out of the depression but his ideas resulted in institutional, policy, and market changes. In Keynes' best known book, *The General Theory of Employment, Interest and Money*, he wrote:

> *But apart from this contemporary mood, the ideas of economists and political philosophers, both when they are right and when they are wrong, are more powerful than is commonly understood. Indeed the world is ruled by little else. Practical men, who believe themselves to be quite exempt from any intellectual influences, are usually the slaves of some defunct economist. Madmen in authority, who hear voices in the air, are distilling their frenzy from some academic scriber of a few years back. I am sure that the power of vested interests is vastly exaggerated compared with the gradual encroachment of ideas.*
>
> **Keynes (1965, p. 383)**

Innovation defined

Understanding and measuring idea-driven innovation starts with a definition. While there have been several attempts to define innovation put forward in recent years there is a large measure of ambiguity surrounding the scope or domain of such a definition, much like the notion captured in the B.C. cartoon below.

Tuesday January 16, 2018

B.C. Reprinted by permission of John Hart Studios, Inc.

The ambiguity comes from recognizing that there are many influences that must be captured to do justice to this important concept. There is the change in technology part—machines, biotech, and software development—but there are also the important business model, institutional, and cultural development aspects, and conceptually, there is a process for achieving the capacity for, and the deliverance of, innovation. Understanding the "process" part is very important, not to mention the cultural aspects that encourage (or sometimes retard) innovation. These cultural aspects produce risk-taking entrepreneurs and an institution of venture capitalists in the modern economy. And, the process part includes both notions of large-scale "breakthroughs" from politically driven "moon shots" to incrementally small adjustments to improve existing tools, machines, and work place procedures. Innovation may also constitute a change in the business model. A characterization of innovation from a 2008 federal government Advisory Committee to the Secretary of Commerce provides a working definition as follows:

> *The design, investigation, development and/or implementation of new or altered products, services, processes, systems, organizational structures, or business models for the purpose of creating new value for customers and financial returns for the firm.*
> **Advisory Committee on Measuring Innovation in the 21st Century Economy (2008, p. 7)**

Innovation is not a new topic. The term has been used for centuries but it has been poorly defined and the process of innovation has not been well understood. Dispensing with the ambiguity of the concept has never been more important for the world's future and will most certainly influence our ability to drive innovation. There is a rising, overdue interest in innovation among leaders in both private and public sectors today because we all recognize that our future depends on our ability to innovate. Innovate we must because:

> *He that will not apply new remedies must expect new evils; for time is the greatest innovator.*
> **Francis Bacon**

The Germans have a term for describing the process of innovation called "Zeitgeist"—the spirit of the age or spirit of the time. Zeitgeist is the thought that a powerful force is embedded in the individuals of a society. The German word Zeitgeist, translated literally as "time mind" or "time spirit," is often attributed to the philosopher George Hegel. Another theory of the source of innovation is the so-called Great Man theory where societal changes come from the leadership of heroes and geniuses. The distinction between these two ideas seems to be primarily whether one believes that extraordinary leaders are the drivers of the innovation process, or whether the innovation just emerges from the many human interactions of the time. One wonders if the high-minded aggressive individual best represents the group of innovators of the age, or if more humble souls are the real source. Is it the humble, or the braggadocios among us who are the real drivers of innovation?

Then the matter gets more complicated when we recognize after achieving success with an innovation that fulfills our positive hopes there emerges a dark side. The topics of the good and bad of technological innovation have generated thoughtful comments by influential leaders and writers for centuries such as:

- *Technology made large populations possible; large populations now make technology indispensable.*
  - Joseph Wood Krutch
- *All this modern technology just makes people try to do everything at once.*
  - Bill Watterson
- *Just because something doesn't do what you planned it to do doesn't mean it's useless.*
  - Thomas Edison
- *One machine can do the work of fifty ordinary men. No machine can do the work of one extraordinary man.*
  - Elbert Hubbard
- *It has become appallingly obvious that our technology has exceeded our humanity.*
  - William Bruce Cameron
  - Albert Einstein
- *Which comes first, the chicken or the egg?*
  - Aristotle

The writings of innovation experts commonly recognize that there is a dynamic interplay of markets, technology, and policy within the context of a particular culture that drives innovation. Borrowing from other cultures also plays a part. Key ideas in the above statements recognize that technological innovation is time dependent and interactive (dynamic), has generated population growth, is a result of the pressure of population growth, generates schizoid personalities, displays surprising outcomes, is not a good substitute for clever minds, trumps the best intentions to enhance the human condition and challenges our ability to understand the triad's interdependencies. The history of technological innovation has clearly demonstrated that it has been our "salvation" in the modern world but innovation inevitably also has a "dark" side reflected in the age-old statements above.

The process of technological innovation seems basically described by Karl Marx's ideas of the dialectic. Marx believed that a dialectical mechanism exists in an economy and that each economic change is an outcome of a systemic process based on a current predominant mode of production where each advance brings up its opposite. Many of the Texas energy developments reviewed in this book appear to have followed such a dialectical process.

The clear answer to Aristotle's philosophical dilemma (which came first, the chicken or the egg) is "the prior egg." When thought of in the context of the dynamics of technological innovation "the prior egg" pretty much captures the element of time and dependence of a given technology on related technologies now or in the past. In a sense the dynamics of technological change cannot be understood except in the context of a system where time and interrelatedness with competing and complementary technologies, public policies, and markets are taken into account.

The Watterson and Einstein observations raise a fundamental question that defies a clear answer. Is the introduction of more technology improving the human condition? One way of trying to measure an improvement in the human condition is economic growth. But is it clear that the rapid growth in new technology is responsible for economic growth? Or is it the other way around? To answer that question one must understand the nature of the innovation process, be able to measure the advance in technology, and establish the relationship between technology and economic growth. And to make matters more complicated, determine the related environmental consequences not addressed by the markets which often threaten to negate the positive value of economic growth.

But technological innovation has become the driver of modern economies in virtually all facets of life. Innovation has had a major role in past energy sector contributions to our state and national wellbeing. The future economy, national and international politics, and more generally, our way of life, is surely to be driven by new ideas, technologies, and new ways of behaving that we can hardly imagine.

Technological innovation is in our state and national psyche. At every turn "technology" and "innovation" are the media headlines, and energy technology often dominates such stories. The media stories are about the latest stock prices of new technology companies, lawsuits over patent rights violations, Environmental Protection Agency regulatory rules that are based on adoption of the best available technology, and legislative debates concerning a government subsidy for clean energy development or dozens of related topics. But in one way or another much of the media attention today is about "technology" and "innovation" much of which is directly or indirectly about energy technology.

Technology is associated with products of hardware and/or software in the modern world. In the early 19th and 20th centuries new technology was primarily a machine like the steam engine, the gasoline engine, or the cotton gin. In the latter part of the

20th century and in all of the 21st century thus far, technology includes machines but also thousands of digital products of the computer and internet era. The new machines in factories, automobiles, trucks, and trains that provide transportation, and farm tractors and harvesters, for example, are integrated with and controlled by computer chips containing operating systems and algorithms for direction by people, and a new player—robots and self-driving cars, trucks, and tractors that perform human-like tasks.

While it is easy to see the contributions of energy sector technologies to our economy and way of life in the past, the future is not so obvious. Past examples abound. There is the substitution of automobiles for horses that eliminated 4 million pounds of manure from the streets of New York City in the late 1800s; air conditioning of buildings in the 1900s became the technology that made warm climates livable and redistributed population from the northern to the southern states of the United States; and the internet in the last half of the 20th century has made global communication a daily activity and mobilized global energy markets. Promises for the future include small solar units providing electricity to remote areas of the globe, widespread use of high performance electric vehicles and a greatly improved battery technology often referred to as the "holy grail" of the future electric grid, and a necessary companion of renewable, clean energy.

In contrast with the positive influences of past technologies the current challenge of public policy and/or private sector initiatives is to speed and enhance the innovation process for the future that are adequate to the challenges we now face. There is evidence that the past innovation process accelerated the rate of technological change at an increasing rate until about the turn of the century. But an apparent slow-down of innovation coupled with a set of unintended consequences make the future of net positive outcomes a challenge.

The new challenges come from both the success and failures of our chosen technology path. That is, while advances in technology have improved the human condition, technology often has a dark side. Clean nuclear power plants produce waste products that challenge our ability to isolate toxic waste from the environment. Computer systems can be hacked by intruders who wreak havoc on individuals, companies, and nations. Burning of fossil fuels emits CO_2 into the atmosphere exacerbating global warming. Further, the internet and related digital technologies have awakened the dynamic expectations of underdeveloped countries. Their citizens are more aware of the advantages of the developed world which give rise to unrest and political uprising, thus creating new challenges to world order.

The rate of growth of technological change

Is the rate of change in technology accelerating at an increasing rate as commonly believed? The support for the belief that acceleration has occurred during the last century is exemplified by the record of world-changing innovations since the late 1800s.

Electricity was introduced in 1873 but it took 46 years for the technology to achieve use by a mere one-fourth of the US population. The telephone was introduced in 1876 and it took 36 years for one-fourth of the population to have telephones. The radio was introduced in 1897 and took 31 years for such a penetration; television introduced in 1926 took 26 years; PC computers introduced in 1975, 16 years; mobile phones introduced in 1983, 13 years; and the Web introduced in 1991, 7 years (Kurzweil, 1993, p. 50).

Ray Kurzweil is perhaps the best known writer on the topic of exponentially accelerating technology changes. He captures his idea of acceleration in a thesis he calls the Law of Accelerating Returns. Kurzweil reviews a number of technology and knowledge domains which have shown a pattern of exponential growth, including computer technologies (where Moore's Law applies to integrated circuits), DNA sequencing, education and learning, biomedical, and other fields. His generalization of the Law of Accelerating Returns is:

- Evolution applies positive feedback in that the more capable methods resulting from one stage of evolutionary progress are used to create the next stage. As a result, the rate of progress of an evolutionary process increases exponentially over time. Over time, the "order" of the information embedded in the evolutionary process (i.e., the measure of how well the information fits a purpose, which in evolution is survival) increases.
- A correlate of the above observation is that the "returns" of an evolutionary process (e.g., the speed, cost-effectiveness, or overall "power" of a process) increase exponentially over time.
- In another positive feedback loop, as a particular evolutionary process (e.g., computation) becomes more effective (e.g., cost effective), greater resources are deployed toward the further progress of that process. This results in a second level of exponential growth (i.e., the rate of exponential growth itself grows exponentially).
- Biological evolution is one such evolutionary process.
- Technological evolution is another such evolutionary process. Indeed, the emergence of the first technology creating species resulted in the new evolutionary process of technology. Therefore technological evolution is an outgrowth of—and a continuation of—biological evolution.
- A specific paradigm (a method or approach to solving a problem, e.g., shrinking transistors on an integrated circuit as an approach to making more powerful computers) provides exponential growth until the method exhausts its potential. When this happens, a paradigm shift (i.e., a fundamental change in the approach) occurs, which enables exponential growth to continue (Kurzweil, 2001, p. 2).

But professionals do not always agree about the matter of exponential economic growth prospects from innovation and technology going forward, and question the conclusions from the recent experience. Such discussions are the focus of much writing by experts working in an organization called *Information Technology and Innovation Foundation* (ITIF). ITIF is a US-based nonprofit that focuses on understanding the complexities of the sources of innovation and advocates public policy to advance innovation (Information Technology & Innovation Foundation, 2016).

Generalizations about how to advance innovation are difficult. But knowing how seems critical today because knowing is required for maintaining the US lead in global innovation. Knowing is problematic because there are complexities of many types of technology that are fundamentally different, there is an inability to systematically measure technological change, and the overall process of change can be complex where discoveries and inventions interact and build off of other discoveries and inventions, only a small part of which trickle down to consumers in the form of new uses and cost reductions that are essential for market penetration.

Technological change is tough to measure. Accurate measurement is critical because technological change usually equals economic growth, although such economic growth may also produce undesirable social costs not measured by economic calculus. That is, there are often social costs of technological change that detract from the public good of economic growth. But an adequate understanding of both outcomes requires accurate measurement. The measurement challenge begins by recognizing that each type of technological change is different; some innovations improve quantity while others improve quality. And then analysts tend to be biased toward the present time period making it hard to look forward and account for new changes along the way.

There is concern that the US process for innovation is breaking down—not matching the performance of the past, and concern that the United States is losing our global advantage. The Bloomberg Innovation Index reported in January 2018 shows that the United States is no longer in the top 10 countries for innovation leadership— South Korea and Sweden are the number one and number two leaders.[3] Atkinson and Stephen Ezell's 2012 book explains that there is a fierce global race for innovation advantage that no longer means the United States will stay in front of the pack (Atkinson and Ezell, 2012). It may be that the failure to excel in innovation underlies the slow productivity growth that is now fueling political disagreements about

---

[3] The index scores countries using seven criteria, including R&D spending and concentration of high-tech public companies. The United States fell to 11th place from 9th mainly because of an eight-spot slump in the postsecondary, or tertiary, education-efficiency category, which includes the share of new science and engineering graduates in the labor force. Value-added manufacturing also declined. Improvement in the productivity score could not make up for the lost ground. Robert D. Atkinson states that he sees on evidence that this trend will not continue. Others are responding to the challenge with innovation policies like better R&D tax incentives, more government funding for research, and more funding for technology commercialization initiatives (Jamrisko and Lu, 2018; Atkinson and Ezell, 2012).

international trade policy, US technology and manufacturing company off-shore location phenomenon and related widening of income and wealth gaps in the United States. The annual total-factor productivity growth that averaged 1.6% from 1995 to 2005 has slowed to a disappointing 0.4% for 2005—15. This decline is the primary reason for the stagnation of wage rates and the widening of the income and wealth gap between the high-end and low-to-medium income classes in the United States. Productivity growth is fundamental to boosting the gross domestic product (GDP) growth rate back to that of better economic times. A 1.3% productivity growth rate, which was the average from 1948 to 2005—the postwar era until 2005—would support a 3.1% GDP growth rate, up from the recent 2.2%. Such a productivity rate increase is the challenge of a new round of innovation (Binder, 2017, p. A2).

The dependence of economic growth on technological change

Economists have a long history of efforts to understand and predict the determinants of economic growth both within countries and among developed and developing countries of the world. The theories and empirical work on the topic since the late 1700s have included the role of innovation or the rate of technological change.

The work of several well-known economists over the last two centuries have improved our understanding of the innovation process and innovations' influence on economic growth. While one can find some treatment of the topic going back as far as Karl Marx, the modern era contributions started with Joseph Schumpeter (February 8, 1883—January 8, 1950) who is best known for his theory called "creative destruction." The theory describes innovation as a process where entrepreneurs introduce new products or processes in the hope of realizing temporary monopoly profits as they dominate a market. Their success makes old technology obsolete, by destroying the rent (economic gain) derived from capital invested in the old technology (Schumpeter, 1942). The process also destroys jobs freeing labor to find new pursuits.

The era following Schumpeter saw a focus on modeling economic growth as influenced (driven) by technological change. Robert Solow and Trevor Swan developed what eventually became the main model used in growth economics in the 1950s. They once estimated that 80% of the long-term growth of US per capita income has been due to technological progress. The determination of national economic growth in the Solow/Swan model is driven by the rate of investment and the rate of technological progress, both of which are determined from outside the model (exogenous variables). The main value of the Solow/Swan model is that it predicts the pattern of economic growth once these two rates are specified, the patterns being primarily the varying rates of economic growth among nations. A weakness of the model is that in its original form it did not explain why capital investments in technology move among nations. Economic theory tells us that we should expect to see investment move to countries with low capital per unit of labor, where capital returns ought to be the greatest, but that has not been the case. Later versions of the model include variables that attract or detract capital investment (Solow, 1957, pp. 312–320).

The weaknesses of the Solow/Swan model of economic growth resulted in efforts to specify technological progress as internal to the model (endogenous). Leading these efforts were Paul Romer and Robert Lucas (Solow, 1956, pp. 65–94). These economists modeled economic growth by specifying a mathematical explanation of technological advancement. They included human capital measures, the skills that make workers more productive, in addition to physical capital. As a result the focus of the economics profession in recent times has been in explaining, and thus modeling economic growth to include investments in human capital (e.g., education) and physical capital as the source of economic growth.

The primary focus of the Solow/Swan, Romer/Lucas, and other economists' modeling of the time did not include the influence of an "ecosystem" or culture of entrepreneurship that are thought, in part, to drive innovation:

> *A successful strategy for American innovation must promote both technological-based and non-technological-based (i.e., institutional and organizational) innovation throughout all layers of an economy, including the private sector, government agencies, and non-profit organizations. In other words, the strategy should not only address innovation in government; rather, its chief aim should be to fundamentally change private sector activity and behaviors to spur greater levels of innovation.*
>
> **Ezell (2014, p. 1).**

Paul Romer's important contribution to economic thought has been called New Growth Theory. His work on this topic has become the defining understanding of how innovation occurs—understanding and quantifying the drivers of innovation. The older theory of economic growth recognized that growth was in large part due to capital investment and labor but did not integrate these variables with **new and better ideas** expressed as technological progress. So, ideas that drive innovation come from education and collaborative efforts of groups. But education can be quantified by such measures as a college degree or an on-the-job training course. Ideas may grow out of the education experience, but are distinctly different. The new understanding comes from recognizing, and incorporating human capital and new ideas into the theory of economic growth. The new theory makes a distinction between "ideas" and "things." Including things, like capital investment and education called human capital is easy enough to include and add to models of economic growth. But including ideas is different. A distinction between the two—human capital and new ideas—is that human capital can be owned/contained by an individual while ideas once developed are shared. Romer stated it this way:

> *Human capital is comparable to a thing. You have skills as a writer, for example, and somebody — reason — can use those skills. That's not something that we can clone and replicate. The formula for an AIDS drug, that's something you could send over the Internet or put on paper, and then everybody in the world could have access to it.*

This is a hard distinction for people to get used to, because there are so many tight interactions between human capital and ideas. For example, human capital is how we make ideas. It takes people, people's brains, inquisitive people, to go out and find ideas like new drugs for AIDS. Similarly, when we make human capital with kids in school, we use ideas like the Pythagorean theorem or the quadratic formula. So human capital makes ideas, and ideas help make human capital. But still, they're conceptually distinct.

Bailey (2001)

Romer is credited with the quote "A crisis is a terrible thing to waste." A crisis typically results in a refocusing of the research community and government leaders to correct the problem. Many new ideas come by this means.

The chief contribution of Paul Romer to the theory of economic growth is that the relationship of innovation and economic growth is an internal (endogenous) process. Innovation does not magically come into play from an outside force, but is internal to the systems of technology and economic behavior and should be modeled that way. Ideas are fundamental to innovation and ideas come principally from investment in various forms of human capital, but are not themselves human capital. Human capital is created—built up—by education of various types (a college degree, an on-the-job training course, etc.). But the key is that it is principally *human capital in research* that drives the production of new ideas. Romer has recently been awarded the Nobel Memorial Prize in economics for his work on this topic. He was joined in complementary work by William Nordhaus.

Ideas have the characteristic that once formulated they become available to everyone—they cannot be contained and owned by one person or a firm. A country's institutions translate the process of human capital creation that generates ideas into policy that helps drive economic growth. The practice of patent rights is the way the innovation process finds its way into markets that produce economic output and incomes for owners and employees. A patent basically is a legal means of allowing a monopoly to exist for a defined period of time. It becomes a promise that an investor with an idea growing out of a public and/or private research enterprise, engrossed by academic publications, peer review, collaborative engagement, and so on, gets a running head-start on actually generating a market product that will yield a favorable return on investment.

The implication of Romer's and Nordhaus' work is that a country can in fact internalize a process that invests in human capital (with a sustained amount of it devoted to research), and a set of related institutions that, on the whole, becomes an idea generator—ideas that become new technology that generates economic growth. There is an ongoing debate among professionals about the limits of the rate of growth that can be sustained in the long term. But a general consensus exists following Romer that we do not have to go forward expecting that ideas and new technology somehow just appear and get added to our investments of capital and labor, but the right policy and institutional arrangements can continue to generate economic growth in capitalistic democracies through an abundance of new ideas matched with labor and capital. There is an ongoing debate about whether authoritarian countries can successfully compete with capitalistic democracies to generate innovation, but it seems clear that capitalistic democracies win this debate since the strengths of capitalistic democracies flow from the benefits that transparency, decentralization, and diversity of populations provide for the generation of new ideas.

A major challenge in understanding and measuring the ongoing contribution of innovation to economic growth is the lack of important data for analysis. In the private economy where most of the important innovation occurs, the American system of private property allows individual firms to keep their advances secret for a long period of time. This is the focus of patent law and the protection of important competitive advantage. The importance of the lack of such data on both theory and measurement of the sources of innovation was recognized by a leading economist over 20 years ago (Griliches, 1994, pp. 1–23).

Leading theorists of innovation economics include both formal economists, as well as management theorists, technology policy experts, and others. The work focuses on innovation capacity to create more effective processes, products, and business models. While the theory is different for neoclassical economics that fails to recognize that innovation is internal to the economy, the purpose is the same—to understand how economic growth is driven by innovation and how public and private sector policies and R&D efforts yield important technological advances that boost economic growth.

The key to answering the question of technologies' contribution to economic growth is to understand the relationship to, and measuring the technology effect on productivity growth. The most referenced productivity growth measure is the value of product divided by a unit of labor, that is, labor productivity. A broader measure more appropriate for the technology issue is total-factor productivity, which measures the combined value of product effects of labor, capital, energy, and materials. Human capital is the store of capacity built through knowledge from various forms of education and training.

Economic growth is best measured by an increase in GDP (the value of all goods and services produced by an economy) after adjusting for inflation and on a per capita basis. This is the usual measure of economic improvement for a nation or the world economy at the macrolevel.[4] At the microlevel the measure is productivity growth for an industry or perhaps a firm, partially attributed to labor productivity (the value of product per unit of labor), an increase of which is an indication of whether wages should rise in a competitive market. The broader measure is total-factor productivity which is the value of product per unit of total inputs to production, usually measured

---

[4] An increase in economic growth caused by more efficient use of inputs (such as labor productivity, physical capital, energy, or materials) is referred to as intensive growth.

as a rate of change over a period of time. It is this measure that is the focus of contributions from technological innovation. Innovation is a label for a change in productive technology over a period of time. But the current understanding and the efforts of empirical studies still contain ambiguity and therefore public policy designed to drive economic growth through innovation remains a challenge.

But progress is being made. A worldwide group of innovation experts have developed a Global Innovation Index that quantifies the best nations for innovation progress and identifies all the various factors that influence innovation:

The Global Innovation Index (GII) 2015 covers 141economies around the world and uses 79 indicators across a range of themes … present us with a rich data set to identify and analyze global innovation trends … innovation-driven growth is no longer the prerogative of high-income countries alone.

Dutta et al. (2015, Ex Summary p. xvii)

Ray Kurzweil commented on the current state of affairs in quantifying the relationship between technological change and economic growth.

Current economic policy is based on outdated models which include energy prices, commodity prices, and capital investment in plant and equipment as key driving factors, but do not adequately model bandwidth, MIPs, megabytes, intellectual property, knowledge, and other increasingly vital (and increasingly increasing) constituents that are driving the economy.

Kurzweil (1993, p. 6)

Much of Kurzweil's concerns are dated given the contributions of Paul Romer, Peter Howitt, Robert Fri, Robert Solow, Trevor Swan and others of the 1990s and 2000s. In one way or another these writers have focused on understanding, measuring, and modeling the difficult contribution of human capital and the generation of new ideas—new ideas that flow from human capital devoted to research, government funding of R&D, and collaborative organizations.

Human capital and economic growth

Total-factor productivity analysis leads quickly to a focus on investment in human capital (education in various forms and health care) as the primary means of advancing productivity, and thus resulting in both a rise in wages and a higher return on capital investment by the firm, and ultimately to economic growth in the country.

The economic theory underlying the concept of human capital was made popular in the economics discipline in the 1960s by Professor Gary Becker at the University of Chicago. The theory, and economic analyses that have followed Becker's work, holds that the predominate contribution of capital investment that produces economic growth is attributable to the investment in human education in various forms, and to health care. The investment in higher education, for example, produces significant increases in wages and other forms of income in comparison to individuals with only a high school education. Education not only benefits the individual but also benefits the employer through higher returns, and ultimately the aggregate economy. According to Becker:

> *Education, training, and health are the most important investments in human capital. Many studies have shown that high school and college education in the United States greatly raise a person's income, even after netting out direct and indirect costs of schooling, and even after adjusting for the fact that people with more education tend to have higher IQs and better-educated, richer parents. Similar evidence covering many years is now available from more than a hundred countries with different cultures and economic systems. The earnings of more-educated people are almost always well above average, although the gains are generally larger in less-developed countries.*
>
> **Becker (2008, p. 1)**

Many others have written on this topic. For example, the writing of Robert Fri referenced earlier is about the role of knowledge in creating technological innovation in the energy sector. An example of empirical work that estimates the contribution of increasing levels of higher education above high school have been developed to encourage public sector support for investments in higher education. Each increment of achievement above high school demonstrably adds to the present value of life-time earnings (Holloway, 1995, p. vii). The referenced work and other studies like it follow the work of Becker referenced above. A recent study by Dale Jorgenson and others:

> *[R]eveals that replication of established technologies through growth of capital and labor inputs, recently through the growth of college-educated workers and investments in both IT and Non-IT capital, explains by far the largest proportion of US economic growth … Maintaining the gradual recovery from the Great Recession will require a revival of investment in IT equipment and software and Non-IT capital as well. Enhancing opportunities for employment is also essential, but this is likely to be most successful for college-educated workers.*
>
> **Jorgenson et al. (2014)**

Treatment of time in economics and the law

Innovation and technological change result from a process that is highly time dependent—changes do not happen in an instant. Economics and the law, among all the professions deal most explicitly with time in understanding, predicting, and guiding innovation and technological change. And, both disciplines deal with innovation that is not only about technology but also institutional change. Institutional innovation and cultural attitudes of the risk-takers are part of the larger innovation process.

The economics and law professions have well-developed, but very different structures for dealing with innovation dynamics (changes over time and interactions among economic sectors). And, each discipline depends on the other in important ways. Each one is in turn influenced by, and influences technological change. Both economics and the law struggle to understand the long term impacts of current technology decisions. The topic is discussed in detail by Kirk Smith in his 1977 PhD dissertation. Smith discusses three general categories of how the law deals with temporal aspects of technological change. The first principal is that the past is a valid measure by which to judge the future—these aspects of the law are backward looking, and remain a very powerful influence on decision-making through reliance on precedence from past court actions. The second is that the influences of the past on the future should be limited—a key example is the 99 year limitation on property leases. The third is that time has value—justice is not timeless and the reduction of uncertainty has merit in the law (Smith, 1977, pp. 223−260).

The law has two long-standing structures—common law and statutory law, each having a role in guiding/supporting and/or constraining innovation. Common law is fundamentally backward looking—courts are often bound by past decisions. This law structure depends on the experience of the past to guide current decisions about technology and the future. Common law is a process of imposing the traditions of the past on the future unless there is good reason to do otherwise, therefore this aspect of the law is conservative. The law in practice (from the common law perspective) puts the burden of proof on the party that seeks to change the status quo. Such a positioning of the burden of proof is codified, for example, in the National Environmental Policy Act that sets up the procedures for environmental impact analysis—a process where a technology proponent must show that the impact of the proposed technology is better than the available alternatives, including the "do nothing" case. The criminal law process in the United States likewise holds that the charged person is innocent until proven guilty—puts the burden of proof on the prosecution. Statutory law, however, is not predisposed to impose the traditions of the past on the future. Statues, which are bound by national and state constitutions, do allow legislation to define the time-dependent effects of technology to be set by the current political process. Thus patent law often defines the time period where the property right is protected from competitors—a legal status that is fundamental to investors who take risks by investing in a new technology.

Economic analyses has two basic purposes—one is descriptive and the other prescriptive. The descriptive purpose of economic analysis is simply to provide insight into the economic behavior of individuals, firms, industries, and nations. The prescriptive purpose is to determine the best decision given the decision-making entities' objectives (e.g., to maximize profits or the public good). Economics deals with technological innovation out of two structures—one regarding the economics of investment decisions by individual firms. The other is a broader treatment of public sector decision-making regarding investment of public funds in R&D, as well as various targeted incentives like tax credits, equipment depreciation, and resource depletion rates.

An analytical methodology for understanding and prescribing for time-dependent technological innovation is that of benefit—cost analysis (B/C Analysis). The key time considerations have mostly to do with the time period for which costs and benefits are expected to endure, and the fundamental aspects of discounting future values to a present value. Discounting is the recognition that a dollar tomorrow is worth less than a dollar today—if I borrow a dollar from you today to be repaid tomorrow I will need to pay you back not only the dollar borrowed but an interest payment (the opportunity cost—a value you forgo by not having the dollar employed in an alternative).

The application of B/C Analysis to individual firm investment decisions is fundamentally an analysis of the expected income and expenses resulting from a capital investment played out over the expected life-time of the equipment, resource, or human skill (human capital) using an appropriate discount rate to summarize the expected net income stream into a "present value." By specifying an appropriate discount rate, and considering the riskiness of the future events, one can compare various alternative investments to enlighten investment decisions. In contrast with the law's treatment of time, the future in economics is not directed by the past except as constrained by the legal and institutional structures of the law (which may well be so constrained as discussed above). The same analytical approach is useful for describing or prescribing decision-making by individual consumers or the combined investment focus of an industry. All of such technology investment decision-making by firms and individuals takes place in markets that function inside of the public policy and legal framework of the country, and sometimes inside of international law constraints.[5]

The application of economic analysis to public policy decisions may be more complex than that for individuals and firms. Investment decisions by individuals and firms regarding a new technology often have impacts on the nation or even the world where some benefits and costs are not incorporated into the calculus of competitive markets, for example, environmental impacts. Here the economics profession has developed a special discipline known as "welfare economics." The name is not to be confused with welfare programs of the government, though welfare economics may be useful in analyzing government welfare programs.

The principals of welfare economics and its application of B/C Analysis follow the writing and theories of such well-known economists as Kenneth Arrow and Joseph Stiglitz.[6] The application of B/C analysis to public policy decisions enlightens the

---

[5] There is also the matter of "black markets" sometimes dominate such as illegal drug markets. Black markets are not significant in US energy markets.

[6] Arrow is Joan Kenney Professor of Economics and Professor of Operations Research, Emeritus at Stanford University. Arrow won the Nobel Prize Nobel Memorial Prize in Economics with John Hicks in 1972. Stiglitz is a professor at Columbia University. He is a recipient of the Nobel Memorial Prize in Economic Sciences (2001).

public debate about important technologies such as the nuclear power waste disposal challenge, long-term CO_2 emission impacts on the environment, and other considerations of long-term effects of new technology. B/C Analysis, in addition to providing a rational way of comparing time-dependent benefits and costs, recognizes that public decisions get played out in the private sector markets and are greatly affected by other factors such as tax policy.

In short, both the theory and practice of law and economics as they deal with technology are supper important because they both impinge on and enlighten innovation and technological decision-making.

Property rights and innovation

There is nothing more fundamental to understanding the innovation process than the institution of property rights in the economy. Such property rights are sometimes set in the constitution of the United States and the various states, but also set in statutes, and perfected in the courts. Without clear specification and protection of property rights a competitive market system such as that of the United States could not function. Property rights are fundamental to the exchange of property and the enforcement of contracts. These determinations of property rights have evolved since the nation was formed and often were derived from English systems in the beginning.

The limits of property rights continually evolve, but importantly, such changes, by design, are made slowly such that private investor decisions can be made with a high degree of certainty. The courts are the remedy for resolving conflicts over property rights.

A good example of the evolving specification of property rights came to a head in Texas as oil and gas development progressed early in the 20th century. Crude oil and natural gas producers in the United States face a set of problems not faced by producers in most other countries. The problems stem from the combination of geologic and physical characteristics of crude oil and natural gas and the foundation of US capitalism, private property rights. Petroleum occurs in geologic deposits or pools at various depths beneath the land surface, but with no particular or standard relationship to the surface characteristics which form the basis for parcels of land traded in private markets. Since land became valuable for the first time, it was divided into standard geographical units. As crude oil and natural gas were discovered and also became valuable, the legal convention that developed was to assign property rights to the surface landowner. Thus the right to recover such minerals belongs to the surface owner, who has the right to remove all that is available beneath his or her land. The owner, of course, may also sell land and mineral rights together or separately.

Since crude oil and natural gas can flow within reservoirs from beneath one land surface property to another, it is the case that one landowner can pump a neighbor's oil and gas because it is impossible to distinguish one owner's oil and gas from another's. The basic rule of petroleum ownership and production then became the "rule of capture." To protect the mineral property, the neighbors must also pump or risk losing their asset.

The rule of capture is a doctrine based in English common law. Originally, the idea in England was that if a game animal or bird from one person's estate migrated to another, the owner of the latter had a perfect right to kill the game. American judges, when first confronted with the oil ownership problem, reasoned that this problem was fundamentally the same as the English game problem. Other property right—based conflicts, and the innovations that came to help resolve them are scattered throughout the following chapters.

Government's ability to advance innovation

The difficulty of getting to a consensus on government agency program and funding commitments that have continuity over time and become an effective driver of innovation is a long-standing challenge. The disagreements have grown out of a lack of understanding of how government can best help drive the innovation process. On the one hand there is always a belief that breakthroughs are the way to go—something heroic like the Manhattan Project or the moon shot. On the other hand there is support for funding a R&D program with continuity focused on a diverse set of technology areas at the research end of the research, development, and demonstration (RD&D) spectrum. For example, the current proclamation of the US Department of Energy (DOE) is:

As a science agency, the Energy Department plays an important role in the innovation economy. The Department catalyzes the transformative growth of basic and applied scientific research, the discovery and development of new clean energy technologies and prioritizes scientific innovation as a cornerstone of US economic prosperity.

Through initiatives like the Loan Programs Office and the Advanced Research Projects Agency-Energy (ARPA-E), the Department funds cutting-edge research and the deployment of innovative clean energy technologies. The Department also encourages collaboration and cooperation between industry, academia and government to create a vibrant scientific ecosystem.

In addition, the Energy Department's 17 National Laboratories are a system of intellectual assets unique among world scientific institutions and serve as regional engines of economic growth for states and communities across the country.

https://www.energy.gov/science-innovation

Political debate about the proper role of the federal government in driving innovation is often fueled by controversy from highly publicized government R&D failures. The best known are large demonstration project failures like Solyndra, the Silicon

Valley startup that collapsed in 2011, leaving taxpayers liable for $535 million in federal guarantees.

The ongoing debate that stems from a lack of a clear understanding of the best role for government is also complicated because there are three types of sponsors. There is the private sector acting alone, government as an innovator and user, and government/private partnerships. Innovation is driven through R&D programs, as well as government incentives for the private sector (especially tax policy and direct subsidies). For example, the focus of many state and local governments for promoting economic development in their economies during the last several decades has led them to set up incentive programs such as subsidies to attract the best scientific talent, and tax incentives for firms committing to bring jobs to the local economy. The evidence is that such programs are effective in attracting talent and adding jobs, but it is not clear that the benefits of the programs exceed the cost, or that they improve the innovation process. Moretti and Wilson reached the following conclusion after studying the successes of biotech economic development programs in the United States:

States forgo billions of dollars in tax revenue to attract economic activity—especially activity perceived to be high-tech or innovative—even though little is known about how these incentives affect the local economy. In this Letter, we examined how effective these policies are at attracting jobs in the biotech sector. We found that, after states adopted incentives, they experienced significant increases in the number of star scientists, the total number of biotech workers, and the number of establishments, but limited effects on salaries and patents. We also uncovered significant spillover effects from biotech incentives to employment in other sectors that provide services in the local economy such as retail and construction. In terms of policy implications, it is important to keep in mind that our finding that biotech subsidies are successful at attracting star scientists and at raising local biotech employment do not necessarily imply that the subsidies are economically justified. The economic benefits to a state of providing these incentives must be weighed against their fiscal costs—for instance, the loss of tax revenues and resulting loss of public services. Our research suggests that state incentives are successful at increasing the number of jobs inside the state. Nevertheless, our results do not suggest that the social benefit— either for that state or for the nation as a whole—is larger than the cost to taxpayers, nor that incentives for innovation are the most effective way to increase jobs in a state.

Moretti and Wilson (2014, pp. 20–38)

The ambiguity concerning the role of government has been an ongoing debate in the history of innovation in the energy sector for several reasons. First, government as a joint innovator and user complicates this discussion. Sometimes government R&D sponsors, especially the federal defense agencies, are also large users of new technology. The creator is also the user. Second, there is uncertainty about which stage or stages of the R&D process (basic research, development of prototypes, demonstration, and commercialization) is (are) the most effective for public sector investment of R&D dollars where the private sector is the user and where the tax payer is the source of the investment funds. Third, in recent history there have been combinations of government incentives (especially tax incentives) and R&D programs both intended to generate innovation in the private sector for a particular technology. Wind and solar power production are examples where a 30% market advantage against alternatives (through tax policy) and R&D programs to improve the performance of the technologies have coexisted and have complementary influences on innovation. Arguably both are not needed for effective innovation.

Robert Fri's view (referenced above) that "innovation is incremental, cumulative, and assimilative" helps to generate the right expectations; technological breakthroughs are rare. For example, Fri summarizes the results over a 25-year period of the DOE R&D program in fossil fuel and energy conservation technologies. Sixteen of twenty-nine technologies came to market while the rest did not. Of the successful ones only two represented a significant new direction for the technology/market combination. And, Fri concludes that successful technologies were on the whole less heroic efforts than was the case for unsuccessful ones, and in many cases represented assimilation of a technology from a different field. Further, the payoff for many of these incremental efforts has been very significant. For example, aero-derivative turbines (the heart of gas turbine combined cycle power plants), three-dimensional seismic technology (for oil and gas exploration), improved coal mining technology, and highly efficient gas furnaces arguably have had major impacts on the energy system. But these were not "moon shots"; they were the result of incremental efforts building off of existing technology, and sometimes assimilating improvements from other fields.

Revolution does happen, however. Perhaps best known is the internal combustion engine- driven automobile that replaced horse-drawn carriages, and ridded major cities of millions of pounds of horse manure (among many other benefits). Electricity replaced gas lighting. Hydraulic fracking is opening a new era of oil and gas recovery. A common term for such major innovations is "disruptive technologies." But even disruptive technologies are not radically new or different from a technology point of view. Rather, the radical transformation is perhaps primarily due to the performance of market incumbents (Bower and Christensen, 1995, pp. 43−53).

In short, the evidence seems to support the notion that government RD&D programs are most productive when focused on incremental changes and diverse program support at the early stage of the process rather than picking a technology and going for a "moon shot." There have been technology revolutions, however, and perhaps Bill Gates has a viable approach with a clear objective, and funding of several "paths" that all have some probability of becoming breakthroughs, and therefore a high aggregate probability of a single success or two.

The history of innovation in the Texas energy sector has some elements of "moon shots" but arguably the progress has been incremental, cumulative, and assimilative. The examination of the 12 Texas energy events in the following chapters explore this history and helps define a way forward that is the focus of Chapter 13.

Economic models of innovation in the energy sector

Bill Gates has a somewhat different definition of the stages of the innovation process which he believes we have to attain to meet both the economic growth objectives that sustain developed economies, lift growth in underdeveloped economies through cheap available energy and at the same time halt the CO_2 contribution to global warming. Gates' four stages are (1) basic research funding, (2) market incentives to reduce CO_2, (3) entrepreneurial opportunity, and (4) a rational regulatory framework.

Gates has organized a group of international investors and public entities he calls the *Breakthrough Energy Coalition* willing to put up billions of dollars into clean energy to simultaneously sustain developed-country economic growth, lift underdeveloped countries out of poverty and drive CO_2 emissions to zero by 2050. The objective is to support innovation from many ideas or pathways so that a winning technology or technologies will have lower cost than that of fossil fuels, is sustainable and emissions free. The fund will focus on electricity generation and storage, transportation, industrial uses, agriculture, and projects that make energy systems more efficient (Gates, 2010). Gates has decided that among the major world problems his Foundation has addressed, including health care and genetics, that the most important challenge to solve is energy—a new technology that is price competitive with fossil fuels, is easily transportable to poor countries of the world and which eventually drives CO_2 emissions to zero to slow or stop global warming. (Note: Gates seems to be focusing on a technological breakthrough rather than a series of incremental improvements but his focus is really a hybrid approach since he also advocates a focus on several technology paths so that the probability of success is improved.)

There is urgency about addressing innovation in the energy sector not present in earlier times. The urgency is present because of three overreaching conditions in the modern world recognized by Thomas Friedman. The theme of Friedman's book is the challenge of the current generation regarding whether our capacity to adapt is being outpaced by a "supernova," built from three ever faster things: technology, the global market and climate change. Friedman reminds us that 2007 was the year that, in the words of Netscape's founder, software began "eating the world." The arrival of the iPhone, Android, and Kindle all came in 1 year. Warming of the climate will cause massive movements of population from southern areas northward, stressing the abilities of host countries to assimilate them. Markets responding to globalization with the help of rapidly changing communication technology is increasingly allowing firms to operate from anywhere in the world and serve markets throughout the world. And Friedman points out that these three forces are interacting and are self-reinforcing the rates of change (Rose, 2016; Friedman, 2016).

Robert Fri discusses a typical demarcation of the stages of technological development in the energy sector that he labels a "model of innovation." Fri's focus is on the role of government in advancing technological innovation through Research, Development, Demonstration and Deployment and he describes the innovation process model thusly:

> The model of innovation that emerges from observation of successes and failures over the past few decades is thus a messy one. ... innovation is incremental, cumulative, and assimilative. Over time, the impact of this process can result in revolutionary change. In rare cases, disruptive technology offers new performance characteristics that can transform markets. The private sector is reasonably successful in dealing with this untidy process, although sometimes it cannot realize enough of the available economic benefit from innovation to warrant the costs of development. And in other cases, public benefits exist outside the market and so provide no economic incentive for innovation. Policy intervention must either bring the costs and risks of innovation into line with the available benefits, or must attach to the public benefit a value that gives the private sector an incentive to engage in the innovation process.
>
> *Fri (2003, p. 66)*

Lester and Hart define the innovation process a little differently including four basic stages in the progression from idea or concept to large-scale deployment in their recent book: (1) option creation, (2) demonstration, (3) early adoption, and (4) improvement-in-use. Their book is about energy innovation that is needed to maintain economic growth but addresses the world challenge of climate change. The authors lay out a 10-point program for a new American Energy Innovation System that urgently needs to be followed to meet the current challenges regarding energy for world economic growth with the eventual elimination of CO_2 emissions. Lester and Hart point out that the unlocking of America's innovation potential will not be accomplished in a centralized government program like the Manhattan Project. The task going forward is far more complex than building an atomic bomb and will not be done by the market alone. The 10 pieces of their new framework for energy sector innovations are:

1. **New Innovators** (new market entrants),
2. **Expanded Competition in Electric Power Markets** (the solution to zero CO_2 emissions is the electric sector),
3. **Smart Integrators** (industry actors whose job it is to manage the independent power producers, distributed generators, energy management service providers, customers, and many other players),
4. **An Invigorated Energy Efficiency Market Place** (a focus on the widespread adoption of building energy efficiency products and services),
5. **Regional Innovation Investment Boards or RIIBs** (a new group of institutions to identify and allocate funding for first-of-a-kind large-scale demonstration projects, postdemonstration projects, and early deployment programs),
6. **State Energy Innovation Trustees** (state-level entities to allocate funds to the RIIBs generated by a surcharge on all retail sales of electricity within the state),

7. **A Federal "Gatekeeper"** (entity to certify that projects and programs presented to RIIBs for funding have the potential to led to significant reductions in CO_2 emissions at a declining unit cost over time),

8. **Dynamic Pricing** (variable prices during a day that reduce peak loads, and provides incentives for participants in the electric market to respond to supply and demand conditions on the grid in the most efficient way),

9. **Open Grid Architecture and Customer Control** (open architecture for distributed generation and smart grid technologies that promote innovation "behind the meter"), and

10. **Breakthrough Innovations** (a larger and more diverse energy research structure to focus on creating new options for energy supply, delivery, and use with the potential to contribute on a large scale in the second half of the 21st century) (Lester and Hart, 2012, pp. 161−167).

The key to being able to enlighten the policy debate about the contribution of ideas through innovation depends to a large extent on being able to quantify the value of new ideas by modeling the economic process through econometric models. Romer's contribution provides the economic theory for completing this valuable work that requires explicit inclusion of an internal economic and institutional system representing idea generation innovation that is a common property resource while also including a measure of human capital through education. The key is to specify exact mathematical models that follow the theory and can be tested with actual observations from reliable data. Robert Fri has completed such work in the energy sector as described above. Other studies have formulated empirical studies of innovation contributions among countries as reported by specifying the various factors that are thought to drive innovation, aggregated into an index of innovation, and then comparing the index over a period of time among countries in the modern world.

The specification of an econometric model that represents one formulation of a Romer growth model is that by Chad Jones in 1996. The model emphasizes the importance of ideas and technology transfer in addition to capital accumulation to explain economic growth (Jones, 1996). Further, Hall and Jones determined that capital accumulation, productivity, and therefore output per worker among countries in large part are due to differences in institutions and government policies (Hall and Jones, 1999). Insights from the endogenous growth literature recognize that the use of an "idea" (on which much of the modern economies rely) by one person does not preclude, at a technology level, the simultaneous use by another person. This is the insight that has generated an enormous research effort and academic writing in the field of economic growth over the last 20 years. A number of studies are finding that research productivity is declining rapidly in recent years.

An interesting and perplexing finding by Jones and others is that ideas are becoming harder to find. A way of measuring such a concern is that data clearly show that research productivity is rapidly declining at the microlevel (for a particular research area) so we require a growing number of researchers to keep making the same rate of advancement (number of successful outcomes per unit of effort at the microlevel, like faster computer chips per number of researchers). One reason may be that the level of cumulative understanding a researcher may require to begin producing something new is growing rapidly. But for the economy as a whole this may not be true—not so because a particular area, like computer chips, may see declines in productivity that the effort generates (new chips per worker) but the effort spins-off ideas in a related area. So, measured at the macrolevel, overall productivity of research might stay constant, or increase. The policy issue this observation portends is whether we require increasing R&D funding just to maintain the same rate of economic growth, and more if we want to accelerate growth. Do we have to run faster and faster to maintain constant exponential growth we desire?

Overview of the Texas energy sector innovation experience

At times it has seemed that the sequence of the three members of the triad in energy innovation began with forward-looking policy that lead the other two influences, yet it is clear that new technology has at times been the leading driver, but always the markets of each era have provided the opportunity for, and validated the viability of the other two influences. Thus one must conclude that the innovation process is not a well-defined linear process but is complex in several ways. Further, the interplay process in Texas has had several important dimensions—dealing with national policy and participating in regional, national, and international markets inside and outside of her borders.

For more than a century, from the market free-for-alls following Spindletop in 1901 and the East Texas Field discovery in 1930, to today with a focus on the modern electric grid, Texas has dealt with all of the major energy issues of our time where innovation has unfolded in somewhat unpredictable ways. The last century has seen innovation in the control of well-spacing in the 1930s, the hot-oil crisis in 1931 (which led to enforcement of oil production controls), stopping natural gas flaring in the 1950s, dealing with the natural gas shortages of 1973—74, and responding to the global threat of the Arab Oil Embargo of 1973 accompanied by gasoline shortages of the 1970s. Innovation of a different kind developed to address the challenges from the imposition of federal oil and gas price controls and a Windfall Profits tax on crude oil. The innovations in nuclear technology generated the short-lived promise of nuclear power in the 1970s and 1980s, and a focus on high-level nuclear waste disposal. Other innovations led to restricting the use of natural gas as a boiler fuel for electric power generation, dealing with the massive importation of coal from Wyoming, handling the prospect of a US high-level nuclear waste disposal facility in the High Plains, and developing a low-level nuclear waste disposal facility in West Texas. More recently the deregulation of the electric industry—a national trend led by Texas—has opened a new path of innovation ranging from decentralized generation from solar and wind matched with consumer tools for managing their electric load to the unfolding of digital-based software for managing a much more complex electric grid.

The collapse of oil and gas prices in the 1980s generated the conditions for innovation of several types. The more recent rounds of innovation occurred following electric sector deregulation of the 1990s as mentioned above. Arguably the most important energy innovation of the century, hydraulic fracking, occurred importantly in Texas creating the current oil and gas "boom" that has upset the international geopolitical balance and sent the world oil price tumbling from $110 per barrel to $25 in 3 short years, only to rebound to about $50 in late 2016. As has often been the case the collapse of oil prices resulted in many bankruptcies, but also the resurgence of innovation to bring back the promise of profitable ventures in the fracking plays of Texas and elsewhere. Although it was not strictly an energy technology, the Superconducting-Super Collider project of the late 1980s had energy sector implications and was a great success story gone awry. It provided its own story of institutional and technological innovation in Texas.

The dynamics of the natural gas market intertwined with government policy (primarily regulation of well-head prices and eventual removal) has produced a new facet of the industry—that of a vibrant Natural Gas Liquid (LNG) market supported by LNG ports on all three of the US coastlines. Texas has played a major role in that development which now opens trade to both the import and export of natural gas.

Texas has seen the proliferation of wind turbines across the Texas landscape that is primarily an importation of technology from Europe, but follow-on institutional innovation has occurred in Texas that makes Texas the leading wind generation state of the nation. This technology importation, however, built on the Texas R&D efforts of the 1970 and 1980 that paved the way for massive deployment of wind turbines (wind farms) across West Texas beginning in the 1990s and continuing today. The promise of electricity storage in Texas for the electric grid has likewise seen innovation worth recognizing. While Texas is not the leader among the states in roof-top and utility-scale solar technology, development in Texas has seen selective innovation here. Finally, the Internet of Things is making an early presentation in Texas where innovation to fit the Texas market is occurring. So, indeed Texas has dealt with all of the major energy issues of our time and in doing so, advanced technology through innovation that has, and will continue to make major national and international contributions.

But as the events in the following chapters illustrate, innovation in the Texas energy sector is not an accident of nature or the simple fact of the geology of natural resource endowments of oil, natural gas, coal, uranium, good solar radiation, and wind. It has been also a deliberate concentrated effort of markets, institutions, and policy, and their connection to geographic location. Relocation of innovative firms has become a primary driver in government economic development programs important to extending the early achievements of combination of technology and geological location of resources.

The dynamics of the innovation process has evolved over time. This has been the case in general and in the energy sector, specifically. Geographic location that influences competitive advantage focuses on unique assets to spur innovation. The Austin Metroplex and the Woodlands of north Houston, for example, became hubs of innovation—a development somewhat like that of the Silicon Valley of California and the Research Triangle of North Carolina. Such centers of excellence combine talent, venture capital, government policy, government funding, and collaboration with universities to drive innovation. The combination of resources (supply) is met with companies and governments looking for innovation capacity to fulfill their program objectives (demand).

In Texas there are innovation centers trying to add university driven efforts to understand and promote innovation. For example, at Texas Tech there is the Innovation Hub at Research Park:

The Innovation Hub at Research Park *is Texas Tech's center for entrepreneurialism and innovation.*

The Hub connects entrepreneurs, with Texas Tech and Texas Tech University Health Sciences Center in Lubbock faculty and students to enable collaboration in launching new ventures that develop our intellectual property and to foster public-private partnerships between Texas Tech and industry that builds our technology base and the economic strength of our region as a public good.

The Hub is home to a number of programs and facilities to assist entrepreneurs bring their ideas to market.

https://www.depts.ttu.edu/research/researchers/researchguide/innovationhub.php/

Also, at the University of Texas there is *The Innovation Center*:

As a top-notch research institution devoted to changing the world, The University of Texas at Austin believes it's critical to foster the entrepreneurial spirit that runs deep in all of our colleges and schools. And that's why we've found dozens of ways to educate our student entrepreneurs, encourage local startup community engagement, support tech commercialization, and serve as an intellectual hub for pioneering research and enterprise. Our cutting-edge programs, resources, events and opportunities have helped to establish UT Austin as an essential engine of the booming innovation and knowledge economy, and we're proud to be at the forefront of the intersection between higher education, innovation and entrepreneurship.

University of Texas at Austin, Entrepreneurship & Innovation at https://www.utexas.edu/campus-life/entrepreneurship-and-innovation/

The UT *Bureau of Economic Geology*:

Established in 1909, the Bureau of Economic Geology is the oldest research unit at The University of Texas at Austin. The Bureau is the State Geological Survey of Texas and has been an integral part of the development of the state's economic success through the years. Our mission is to serve society by conducting objective, impactful, and integrated geoscience research on relevant energy, environmental, and economic issues. Our vision is to be a trusted scientific voice to academia, industry, government, and the public, all of whom we serve.

Bureau researchers spearhead basic and applied research projects around the world; among them, research in energy resources and economics, coastal and environmental studies, land resources and use, geologic and mineral mapping, hydrogeology, geochemistry, and subsurface nanotechnology. The Bureau provides technical, educational, advisory, and publicly accessible information via a myriad of media forms to Texas, the nation, and the world.

Talented people are the Bureau of Economic Geology's formula for success. Our staff of over 250 includes scientists, engineers, economists, and graduate students, representing 27 countries, often working in integrated, multi-disciplinary research teams. The Bureau's facilities and state-of-the-art equipment are world class, and include more than fifteen individual laboratories hosting researchers investigating everything from nanoparticles to shale porosity and permeability. The Bureau also maintains three major well core research and storage facilities, in Houston, Austin, and Midland—together believed to be the largest archive of subsurface rock material in the world, as well as an extensive wireline log library.

Bureau of Economic Geology (2017); http://www.beg.utexas.edu/

Also, at UT Austin there is the *IC² Institute*:

The IC²Institute was founded in 1977 as a "think and do" tank to test the belief of its founder, George Kozmetsky, that technological innovation can catalyze regional economic development through the active and directional collaboration among the university, government, and private sectors.

Since then, the Institute has researched the theory and practice of entrepreneurial wealth creation and has been instrumental in Austin's growth as an innovation and technology center and in the development of knowledge-based economies in over 40 countries.

Key Institute programs applying knowledge gained from ongoing research activities in Central Texas and throughout the world include the Austin Technology Incubator, the Global Commercialization Group and the Bureau of Business Research.

University of Texas at Austin, IC² Institute at http://ic².utexas.edu/about/

Also, at Texas A&M there is The *AM Innovation Center (AMIC)*:

The AMIC is a nonprofit 501c(3) corporation formed for the advancement of science, education, entrepreneurship and innovation. The AMIC works in collaboration with Texas A&M University System components headquartered in Brazos County, Texas, the Bryan-College Station community, and private industry … The AMIC was originally formed in 2007 as a collaborative program between the Texas A&M System's Office of Technology Commercialization, Texas A&M University, the Texas A&M University Health Science Center and the Research Valley Partnership. In 2014 the collaboration partners formed a stand-alone organization as a 501c(3) entity with initial support from the Texas Emerging Technology Fund, Texas A&M University, Texas A&M Engineering Experiment Station (TEES), the Texas State Energy Conservation Office, the United States Department of Energy and the private sector.

Texas A&M University, The AM Innovation Center (AMIC) at http://aminnovationcenter.com/ about/

And, *Mays Innovation Research Center*:

Drawing from academic disciplines across the Texas A&M campus, MIRC will examine the nature of innovation. Research at the center will focus on how innovation advances human potential; the essential conditions necessary for innovation to flourish; how innovation spreads; and the social, economic and legal frameworks necessary to support innovation. Many traditional university innovation centers focus on teaching the history, theory and practice of innovation. By contrast, MIRC is a research-oriented academic center that will engage in the study of innovation to advance knowledge in this important field.

Texas A&M University, Mays Business School Innovation Research Center (MIRC) at https://today. tamu.edu/2017/09/05/multimillion-dollar-gifts-establish-new-innovation-research-center-at-texas-am-mays-business-school/

Also, at Rice University there is the *Rice Alliance*:

The Rice Alliance for Technology and Entrepreneurship (Rice Alliance) is Rice University's nationally-recognized initiative devoted to the support of technology commercialization, entrepreneurship education, and the launch of technology companies.

One major focus of the Rice Alliance is The Rice Alliance Technology Venture Forums, some of the largest technology venture forums in North America where emerging technology companies showcase their new ventures in front of a diverse audience. The Rice Alliance holds three Technology Venture Forums a year - Energy and Clean Technology; Information Technology and Web; and Life Science.

For more information, see The Rice Alliance, Houston, Texas at www.alliance.rice.edu

Finally, at SMU is the *Center for Laser Aided Manufacturing*:

The Center for Laser Aided Manufacturing at SMU was created to develop a fundamental understanding of laser-aided intelligent manufacturing.

In August 2005, the National Science Foundation awarded SMU a grant to establish the Industry/University Cooperative Research Center (I/UCRC) for Lasers and Plasmas for Advanced Manufacturing. SMU is the fourth university site in the multi-institutional I/UCRC for Lasers and Plasmas for Advanced Manufacturing together with the University of Michigan, the University of Virginia, and the University of Illinois at Urbana-Champaign. In August 2010, NSF granted SMU an extension of the I/UCRC for the next five years.

The NSF's I/UCRCs program is structured to develop long-term partnerships among industry, academe, and government. The centers are catalyzed by a small investment from the NSF and are primarily supported by industry center members, with NSF taking a supporting role in their development and evolution. Each center is established to conduct research that is of interest to both the industry and the center. An I/UCRC contributes to the Nation's research infrastructure base and enhances the intellectual capacity of the engineering and science workforce through the integration of research and education.

www.nsf.gov/

The I/UCRC Program initially offers five-year awards to centers. This five-year period allows for the development of a strong partnership between the academic researchers and their industrial and government members. After five years, Centers that continue to meet the I/UCRC Program requirements may apply for a second five-year award. These awards allow Centers to continue to grow and diversify their non-NSF membership. After ten years, the Centers are expected to be fully supported by industrial, other Federal agency, and state and local government partners.

www.nsf.gov; SMU University, Center for Laser Aided manufacturing at https://www.smu.edu/ Lyle/Centers/

The following chapters explore the advent of new technology through an innovation process in the Texas energy market as they unfolded leading to an event, or as a response to an event. The 10 events began with Spindletop in 1901. This event greatly influenced the evolution of the energy sector in Texas over the next 115 years, which will no doubt continue into the future.

The pages that follow capture the key elements of each of the issues discussed in this opening chapter, and contain an assessment of the dynamics of an innovation process that has engaged the interplay of markets, policy, and technology in a still unfolding evolutionary process. A major part of the writing comes from my personal and professional experience in the Texas energy sector beginning with Arab Oil Embargo of 1973 until 2015 when I retired as President of a Texas electric sector nonprofit organization that was in the forefront of new technology development for the modern electric grid (Holloway, 2005—2015).

With each selected event in the following chapters the unique contribution is explored by identifying key parts of the innovation process. These key parts are described by Scott Berkun in a book entitled ***The Myths of Innovation*** (Berkun, 2007). Berkum details the combination of many factors that often characterize innovations. But he discusses the "seeds" of innovation thusly:

Invention, and innovation, have many parents: the Taj Mahal was built out of sorrow, the Babylonian Gardens were designed out of love, the Empire State Building was constructed for ego, and the Brooklyn Bridge was motivated by pride. Name an emotion, motivation, or situation, and you'll find an innovation somewhere that seeded.

Berkun (2007, p. 40)

Figure 1.1 *Texas energy share of gross domestic product.*
Data from U.S. Bureau of Economic Analysis.

From review of many innovation stories Berkun concludes that there are five categories that form a pattern: (1) hard work in a specific direction, (2) hard work with
direction change, (3) curiosity, (4) wealth and money, and (5) necessity. Most innovations involve many factors, so it is instructive to study the uniqueness of each (Berkun,
2007, pp. 41−42).

A way of stimulating the imagination about the role of innovation in the Texas
energy markets to set the stage for the following chapters is captured in the following
charts showing the energy sector share of GDP in Texas (Fig. 1.1) and separately in
the US economy (Fig. 1.2). GDP is the end product of innovation through new technology, markets, and policy. The share of GDP from the combined value of products
in oil, natural gas, electricity, pipelines, refined products and chemicals[7] in Texas over
the period of 1963 to date reached a peak of 26.8% in 1981 following the Iranian hostage crisis when oil prices reached $40 per barrel ($91 in 2015 price equivalents), and
26.5% in 2013 when the oil price reached $105 per barrel. The share dropped to a
low of 12.5% in 1998 when oil prices fell to $12.28 ($17.96 in 2015 price equivalents). The energy share of GDP for the United States peaked at 9.6% in 1981, and
8.0% in 2013 (Fig. 1.2) at these peak oil price periods.

The energy sector is a more important source of GDP in Texas relative to the
nation by a factor of three. By calculating the share of GDP originating from the

---

[7] This is a broad definition of the energy sector including the entire complex of interdependent industries
that have evolved from oil and gas products. The electric sector is included in the book as a key energy
source for the economy, although it evolved in its own dynamics that were not a direct outgrowth of
oil and gas production. But the electric sector is an integral part of energy for the economy in Texas
and throughout the developed economies of the world and is included in this book on energy sector
innovation while innovation in the refining, chemicals, and petrochemical industries are not included
here.

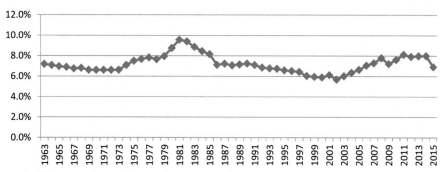

Figure 1.2 *US energy share of gross domestic product.*
Data from U.S. Bureau of Economic Analysis.

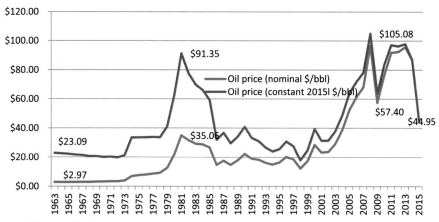

Figure 1.3 *Texas oil price.*
Based on Data for 1952–1977 from Stevens, M., Cummings, G., 1977. Texas Energy: A Twenty-Five Year History. Governor's Energy Advisory Council, Austin, TX; and data for 1978–2015 from Energy Information Administration, Annual Crude Oil Prices by State. <http://www.eia.gov/dnav/pet/hist/LeafHandler.ashx?n = pet&s = f0030483&f = a>.

energy sector in Texas we see that the contribution is both highly significant and highly variable over time, but the major source of the variation is due to the varying oil and gas prices, and since 2005, the influence of "fracking" technology. The contribution of energy sector GDP reflects the important variation in oil and gas prices during the period (Figs. 1.3 and 1.4), as well as the dynamic influence of technological innovation.

A major contributor to the GDP share variation is the price of oil and gas, but also hidden in the data is the influence of technological innovation. Innovation that developed fracking of oil and gas geological formations, for example, has contributed

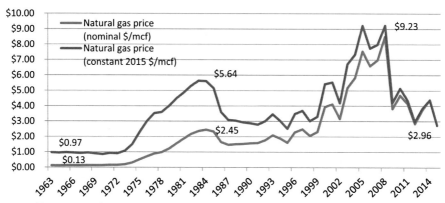

Figure 1.4 *Texas natural gas price.*
Based on Data for 1952—1977 from Stevens, M., Cummings, G., 1977. Texas Energy: A Twenty-Five Year History. Governor's Energy Advisory Council, Austin, TX; and data for 1978—2015 from Energy Information Administration, Annual Natural Gas Prices by State. <http://www.eia.gov/dnav/pet/hist/LeafHandler.ashx?n = pet&s = f0030483&f = a>.

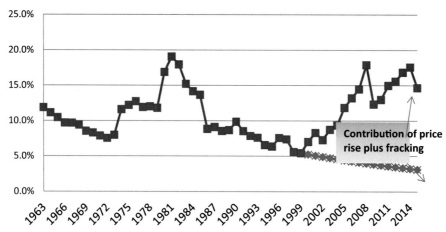

Figure 1.5 *US oil and gas share of gross domestic product.*
Data from U.S. Bureau of Economic Analysis and trend from 1990 to 1999.

greatly to the increased GDP share in Texas and the Nation since the year 2000. This important contribution began seriously 20 years earlier but did not become an important part of the market until 2005. Fig. 1.5 shows a rough estimate of the combined impact of oil and gas price rises and new production from fracking in both oil and gas reservoirs. The combined impacts of price and fracking technology returned the oil

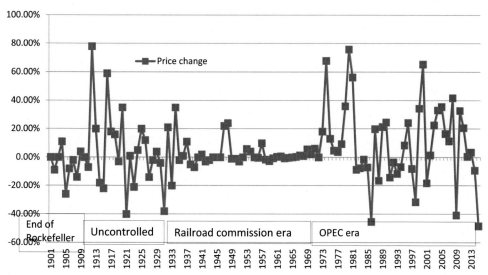

Figure 1.6 *Year-to-year percent change in Texas well-head price of oil.*
Data for 1902−1951 from Kohl, W.L., 1991. After the Oil Price Collapse: OPEC the Exporting States and the World Oil Market. The Johns Hopkins University Press, Baltimore, MD; data for 1952−1977 from Stevens, M., Cummings, G., 1977. Texas Energy: A Twenty-Five Year History. Governor's Energy Advisory Council, Austin, TX; and data for 1978−2015 from Energy Information Administration, Annual Crude Oil Prices by State. <http://www.eia.gov/dnav/pet/hist/LeafHandler.ashx?n = pet&s = f0030483&f = a>*.*

and gas sector share of Texas GDP almost to the 1981 level of 26.8%—a level that few experts expected as actual and forecasted reserves from old fields was in rapid decline (see illustration in Fig. 1.5).

Important contributions are imbedded in the GDP data from both technological advances, as well as institutional innovation. An example of an institutional innovation, joined with legislative policy is that of the TRC during the period from the discovery of the East Texas Field in 1930 to the Arab Oil Embargo in 1973 (Fig. 1.6). During this period the regulations of the TRC kept the world price of oil very stable compared with all of the rest of history since Spindletop in 1901. But did this period of price stability generate a positive influence on technological advance, or did it retard innovation? This topic and the events of this era is the focus of Chapter 3, A Game Change at Spindletop, and Chapter 4, From Chaos to Order in the East Texas Field. As a general matter the 12 chapters of the book explore the key influences of policy, markets, and technology developments through an examination of innovation's role in 12 key energy sector events beginning with Spindletop in 1901 and ending with "fracking" of oil and gas in 2005−15.

Innovation in the Texas energy sector through new technology has included, among other technologies, horizontal drilling, hydraulic fracking, control of natural gas flaring through the technology of reinjection, electric sector deregulation, the nation's leading wind power center, synchrophasor technology for the modern electric grid and many other innovations.

At this stage in the economics of innovation it seems clear that, in addition to good public policy to enhance competition in the private sector markets, timely innovation is essential to economic growth. Innovation is driven by factors that generate ideas, improvement in education of the workforce (human capital), investment in physical capital, and the optimization of the inputs to production (labor, capital, energy, and new technology). The innovation process is complex and difficult to predict, but we know that generation of new ideas, institutions, public policy, effective RD&D and competition in the market place are essential to future progress. The experience in the Texas energy markets reviewed in the following chapters is reviewed in the context of these key parameters of innovation.

Worked example

A paper by Ayres and Warr

> [T]ests several related hypothesis for explaining US economic growth since 1900. It begins from the belief that consumption of natural resources—especially energy (or, more precisely, exergy) has been, and still is, an important factor of production and driver of economic growth. However the major result of the paper is that it is not 'raw' energy (exergy) as an input, but exergy converted to useful (physical) work that—along with capital and (human) labor—really explains output and drives long-term economic growth. We develop a formal model (Resource-EXergy Service or REXS) based on these ideas. Using this model we demonstrate first that, if raw energy inputs are included with capital and labor in a Cobb–Douglas or any other production function satisfying the Euler (constant returns) condition, the 100-year growth history of the US cannot be explained without introducing an exogenous 'technical progress' multiplier (the Solow residual) to explain most of the growth. However, if we replace raw energy as an input by 'useful work' (the sum total of all types of physical work by animals, prime movers and heat transfer systems) as a factor of production, the historical growth path of the US is reproduced with high accuracy from 1900 until the mid 1970s, without any residual except during brief periods of economic dislocation, and with fairly high accuracy since then.
>
> *Ayres and Warr (2004, p. 1)*

The cited authors "model output growth accurately using a constant returns–to-scale (in all factors) Cobb–Douglas production function with useful work, rather than raw energy as a factor of production" (Warr and Ayres, 2006, p. 21). The authors conclude that much of long-run output growth has been driven by an incremental process of learning-by-doing in energy production and consumption technologies, with structural breaks corresponding to shocks that have generated invention-innovation breakthroughs and altered the long-run trends of factor substitution. The analysis estimates GDP in several mathematical formulations, and in the end, convincingly show that model specifications using energy service rather than energy inputs is the superior idea (see the following chart). In addition the authors illustrate that the models tested identify the disruptions that follow extreme events like the Great Depression, world wars, pandemics, and the Arab Oil Embargo of 1973. Two of their tested models are listed here.

$$\log(y) = \alpha * \log(k) + \beta \log(l) + \gamma \log(u)$$

where $y =$ GDP, $\alpha = 0.52$, $\beta = 0.04$, and $\gamma = 0.44$, $k =$ capital, $l =$ labor, and $u =$ useful work.

or

$$\log(y) = \alpha * \log(k) + \beta \log(l) + \gamma \log(b)$$

where $y =$ GDP, $\alpha = 0.56$, $\beta = -1.88$, and $\gamma = 2.32$, $k =$ capital, $l =$ labor, and $b =$ energy inputs.

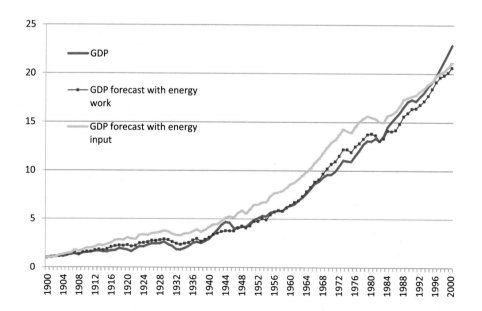

Year	GDP index	GDP forecast with energy work	GDP forecast with energy input	Year	GDP index	GDP forecast with energy work	GDP forecast with energy input	Year	GDP index	GDP forecast with energy work	GDP forecast with energy input
1900	1	1	1	1934	2	2.4728	3.5028	1968	9.3	9.6852	11.8304
1901	1.12	1.048	1.0576	1935	2.2	2.5528	3.518	1969	9.59	10.2124	12.3952
1902	1.13	1.1148	1.0788	1936	2.51	2.7996	3.7068	1970	9.6	10.6876	12.9036
1903	1.18	1.1544	1.2768	1937	2.64	2.976	3.9156	1971	9.92	10.9496	13.1768
1904	1.17	1.1736	1.3544	1938	2.51	2.6852	3.654	1972	10.46	11.5172	13.7364
1905	1.25	1.2884	1.4672	1939	2.72	2.882	3.8716	1973	11.06	12.186	14.296
1906	1.4	1.3776	1.5128	1940	2.95	3.0984	4.1744	1974	10.99	12.186	14.0804
1907	1.42	1.4752	1.8092	1941	3.43	3.3828	4.4296	1975	10.94	11.9092	13.922
1908	1.3	1.3708	1.6936	1942	3.87	3.5316	4.4492	1976	11.53	12.48	14.5352
1909	1.52	1.5472	1.8296	1943	4.38	3.7004	4.7716	1977	12.07	12.8084	15.0604
1910	1.56	1.6268	2.0088	1944	4.69	3.7824	5.1284	1978	12.72	13.2812	15.3804
1911	1.6	1.6176	2.0168	1945	4.61	3.7708	5.2568	1979	13.08	13.7496	15.6508
1912	1.69	1.7632	2.1264	1946	4.06	3.7684	5.1424	1980	13.04	13.8232	15.4856
1913	1.71	1.8416	2.3388	1947	4.03	4.1068	5.6004	1981	13.34	13.6688	15.3756
1914	1.63	1.7696	2.2376	1948	4.21	4.2648	5.8696	1982	13.05	13.0676	15.0244
1915	1.62	1.9	2.3948	1949	4.18	4.084	5.54	1983	13.57	13.338	14.998
1916	1.74	2.0992	2.5428	1950	4.55	4.2948	5.9828	1984	14.52	14.1368	15.6736
1917	1.76	2.2048	2.8104	1951	4.9	4.7464	6.4816	1985	15.04	14.062	15.8324
1918	1.97	2.2612	2.8804	1952	5.08	4.7588	6.4944	1986	15.5	14.1984	16.0872
1919	1.9	2.2448	2.8344	1953	5.31	5.0344	6.7304	1987	15.96	14.8404	16.6192
1920	1.82	2.3148	3.0744	1954	5.28	4.9164	6.7556	1988	16.57	15.6088	17.3268
1921	1.66	2.1612	2.9544	1955	5.65	5.4344	7.3596	1989	17.12	15.9584	17.5184
1922	1.92	2.2588	2.9124	1956	5.76	5.7388	7.6852	1990	17.33	16.3976	17.6824
1923	2.16	2.5076	3.3488	1957	5.87	5.8876	7.7844	1991	17.17	16.434	17.794
1924	2.15	2.5456	3.3868	1958	5.81	5.8216	7.92	1992	17.64	16.8076	18.1652
1925	2.33	2.6212	3.3384	1959	6.24	6.1536	8.2356	1993	18.05	17.1856	18.4792
1926	2.47	2.7608	3.5156	1960	6.39	6.4268	8.6032	1994	18.67	17.7216	18.8972
1927	2.47	2.7608	3.568	1961	6.54	6.7392	8.7808	1995	19.1	18.4336	19.1664
1928	2.48	2.8312	3.6468	1962	6.93	6.9128	9.1164	1996	19.76	19.1476	19.7804
1929	2.65	2.912	3.7864	1963	7.23	7.3388	9.5096	1997	20.54	19.6324	20.0652
1930	2.38	2.8444	3.6896	1964	7.65	7.8416	9.872	1998	21.33	19.8272	20.2616
1931	2.2	2.6416	3.4512	1965	8.14	8.3024	10.25	1999	22.13	20.1232	20.6436
1932	1.87	2.4744	3.3408	1966	8.67	8.846	10.8048	2000	22.93	20.648	21.1416
1933	1.84	2.3608	3.314	1967	8.89	9.08	11.2292				

The Warr and Ayers analysis was built on the work of others who have emphasized the use of endogenous models that treat innovation as endogenous rather than imposing representations of innovation from outside the model. Data for year 1990 through 2000 indexed to 1900 yields the model fit shown in the chart. The best fit is the model using energy services as an explanatory variable forecasting GDP rather than energy inputs (oil, gas, coal, and renewables). Energy services are based on the idea that the innovation focused analysis needs to use services provided by energy inputs since the productivity of energy inputs change with innovation in a number of ways. So, the energy services index measure is made to represent "useful work output" as a factor of production in specifying an aggregate economy production function. The analysis identifies where factor substitutions occur altering future GDP paths.

CHAPTER 2

The common international energy innovation drivers

Contents

Innovation Dynamics and Policy in the Energy Sector
DOI: https://doi.org/10.1016/B978-0-12-823813-4.00002-6

Any comprehensive comparison of countries in the world, even on the narrowest of targeted interests, is a daunting task. Such an international country-by-country comparison of the drivers of innovation and their successes in the energy sector is no exception. An understanding of the innovation process in a single country is ambiguous at best because, as addressed in Chapter 1, The Dynamics of Innovation and Technology in a Market Economy, there are several interdependent domains of concern—institutional, market, and technology components of innovation, all influenced by public policy in democratic nations or dictatorial directives in dictatorships. To understand the influence of the Texas energy sector innovation to the international community one must capture the dynamics of intercountry influence and do so over a period of time, for innovation is surely a dynamic process in both dimensions. My approach in this chapter is to explore the matter of the Texas international influence by examining the drivers of innovation while dealing with the Texas influence as though she was a separate country interacting with the rest of the developed world.

Capacity for and realization of innovation is part and parcel of the stages of economic growth that are captured by the global economic and political cycles as understood by experts who study Kondratieff cycles. The downsides of these long cycles are apparently driven by several influences in modern economies, including resource scarcity, over investment, and declining productivity as the payoffs from the current technological, institutional, and market structures peter out (the declining marginal product in economic theory). The upswings of the cycles appear to result from institutional, market, and public policy changes brought on by opportunity on the one hand, and the necessity to change on the other. The cycles seem to play out over 40–50 years that create prosperity in the upswings. Energy scarcity and rising resource prices and follow up episodes of increasing productivity are a fundamental part of these dynamic cycles.

Although the Kondratieff cycles seem to be global in nature it is also true that the drivers of innovation inherit in the upswings are heavily influenced by governmental structures and supporting cultural systems. This chapter identifies the relative strengths of capitalistic democracies compared to communist and socialist systems in promoting and achieving innovation in the energy sector. Major energy producing countries include all three major types of governmental systems. But the outcomes in the form of economic efficiency and prosperity are also greatly influenced by institutional arrangements such as "concessions" in the oil industry. These arrangements make clear economic distinctions between democracies and authoritarian systems complicated since the later systems manage to piggy-back off of the innovations of capitalistic democracies. Concessions allow private companies from capitalistic countries to operate in certain communist and socialistic countries to recover oil and gas and to operate various energy activities with benefits shared by the host government and the private companies.

Finally, the chapter details the importance of ownership of the energy resource in evolving economic efficiency of natural resource development and use. It also discusses the dynamics of intercountry transfer of technology that feeds the worldwide innovation process.

In addition to a focus on where the global economy is in the sixth Kondratieff cycle, a major focus of the chapter is the measurement of innovation leadership among countries, states within the United States and company leaders involved in the world economy. I have relied on two ongoing efforts that identify global innovation leaders among countries of the world. While there are several entities that provide insight into country and company rankings I have relied on the work of two particular organizations. These entities have, for more than a decade, developed and published quantitative measurements (indexes) that are useful in examining Texas and US energy sector innovation efforts in a global context. These index measures of the overall economy include major energy sources in the calculation of the annual indexes. These multiorganizational groups of experts produce and publish comprehensive global indexes that allow comparisons of innovation prowess among countries, among states of the United States, and among leading company innovation leaders.

One effort is known as the Global Innovation Index (GII). The GII is an annual ranking of countries. It is based on country capacity for, and success in innovation. The index is constructed from both objective and subjective data from multiple sources. The annual index is published by Cornell University, INSEAD, the World Intellectual Property Organization (WIPO), and several mother organizations and institutions (Dutta et al., 2015). The other is the Global Competitive Index (GCI), an ongoing effort that enlightens the understanding global competition and the influence of innovation. The GCI rates countries of the world based on an index of competitiveness performance. The index published annually in the Global Competitiveness Report by the World Economic Forum was first published in 2004 (Schwab, 2019).

Energy-intensive economies: Texas and US ranking among leading countries

This chapter identifies the leading world countries in energy production and consumption classified by the various energy sources including oil and natural gas, wind, solar, other renewables, and electricity. Comparisons are made among leading energy consuming countries, and those that produce the highest levels of CO_2 releases into the atmosphere. Energy production, consumption, and CO_2 emissions are all related to countries with high gross domestic product (GDP) and per capita income growth rates. Nations leading in renewable and nuclear are typically those with limited fossil fuel resources or easy access to fossil fuels. The part of the chapter focusing on dominant country energy supply, consumption, CO_2 emissions, and economic growth provides the base out of which innovation drivers are recognized. Innovation is examined as a function of public policy changes and market incentives in the presence of a countries' capacity on the one hand, and the requirement for change that follows an unforeseen crisis on the other.

The world supplies and reserves of fossil fuels in 20 countries currently make up 90 + % of the world's annual energy production and resources (capacity) but they are resident in a relatively few world economies. Twenty countries produce 90% of the world's crude oil, 89.4% of the world's natural gas, 97.9% of the coal, and earn 76% of GDP but have only 32.3% of the world population. Electricity production is, for the most part, also concentrated in the top 20 countries (80.4%) because the majority of the electric power is produced by burning oil, natural gas, and coal. The concentration of electric power production is not completely correlated with oil, gas, and coal production since electricity is importantly produced in part by hydro, wind, solar, and other minor fuels, thus the 80% estimate of the top 20 shown in Table 2.1, rather than the 90% that would

Table 2.1 Crude oil production as if Texas is a nation (2018).

Rank	Country	Volume (bbl/d)
1	Russia	10,759,000
2	Saudi Arabia	10,425,000
3	United States net of Texas	7,470,000
4	Iraq	4,613,000
5	Canada	4,264,000
6	Iran	4,251,000
7	China	3,773,000
8	Texas	3,492,000
9	United Arab Emirates	3,216,000
10	Kuwait	2,807,000
11	Brazil	2,587,000
12	Nigeria	1,989,000
13	Kazakhstan	1,856,000
14	Mexico	1,852,000
15	Angola	1,593,000
16	Norway	1,517,000
17	Venezuela	1,484,000
18	Qatar	1,464,000
19	Algeria	1,259,000
20	Libya	1,039,000
21	United Kingdom	1,000,000
	20 Country total (mbbl/d)	72.71
	World (mbbl/d)	80.77
	20 Country share	90.0%
	Texas share of the United States	31.9%
1	United States with Texas	10,962,000

Data from Crude Oil: World Fact Book (a), Central Intelligence Agency Library available at https://www.cia.gov/library/publications/resources/the-world-factbook/fields/261rank.html.

prevail if all electricity was produced by burning fossil fuels. As discussed in chapter 8, Upheaval in the energy markets: the Arab Oil Embargo and the Iranian Crisis of the book, the shift is underway in important ways toward renewable sources of electric power production, trending heavily toward wind and solar—hydropower sources depend on major river systems that have mostly been developed in the top 20 countries.

The leading three countries for oil production are the United States, Saudi Arabia, and Russia (Table 2.1). All three produced more than 10 million barrels per day in 2018 with the United States leading with 11 million barrels per day. (Note: Table 2.1 shows the US net of Texas production to rank third among the 20 countries to isolate the production ranking for Texas. Texas is shown as if a country. The total US production is reported at the bottom of the table.) Texas would rank eighth in the world with 3.5 million barrels per day in 2018. The calculations are as if Texas is included in the analysis as a separate country from the United States.

The elevation of the United States to number 1 above Russia and Saudi Arabia is the recent decade's contribution to United States and Texas production from the technological innovation known as hydraulic fracturing or "fracking." Fracking is a procedure where water and chemicals are forced under high pressure into the horizontally drilled oil well, which releases oil from "tight" rock formations. This technology originated in a natural gas formation in North Texas in the 1990s. Companies begin commercial operations in 2005 and the practice soon moved to oil formations focused heavily in the Permian Basin of West Texas. The years following have reversed the 40-year record of the United States being a net importer of oil, where Organization of the Petroleum Exporting Countries (OPEC) exercised market power to control world prices and increase their members' market share. As discussed in chapter 8, Upheaval in the energy markets: the Arab Oil Embargo and the Iranian Crisis this dependency on unstable Middle East countries for critical oil supplies has remained a threat to national security and drove every US President since Richard Nixon to declare "energy independence" as a national policy priority.

Natural gas production is also concentrated in a few countries where the top 20 countries make up 89.4% of world production with the United States, Russia, and Iran topping the list. The United States produced 773 billion cm^3 or 22.2% of world production in 2017. Texas ranked sixth among the countries of the world in 2017 with 25.3% of US production and 4.5% of world production (Table 2.2). The fracking technology located principally in the Permian Basin of West Texas has been the innovation that greatly increased the Texas and US world market share, adding to the national and Texas trade surpluses in recent years. Liquefied natural gas (LNG) import facilities in the Gulf Coast of Texas and Louisiana have in part changed from import to export facilities. This topic is discussed in more detail in chapter 10, Hydraulic fracturing: the Permian Basin challenges Organization of Petroleum Exporting Countries leadership of the book.

Table 2.2 Natural gas production as if Texas is a nation (2017).

Rank	Country	Volume (cm^3)
1	United States net of Texas	616,919,294,015
2	Russia	665,600,000,000
3	Iran	214,500,000,000
4	Qatar	166,400,000,000
5	Canada	159,100,000,000
6	Texas	155,880,705,985
7	China	145,900,000,000
8	Norway	123,900,000,000
9	Saudi Arabia	109,300,000,000
10	Australia	105,200,000,000
11	Algeria	93,500,000,000
12	Turkmenistan	77,450,000,000
13	Indonesia	72,090,000,000
14	Malaysia	69,490,000,000
15	United Arab Emirates	62,010,000,000
16	Uzbekistan	52,100,000,000
17	Egypt	50,860,000,000
18	Netherlands	45,330,000,000
19	Nigeria	44,480,000,000
20	United Kingdom	42,110,000,000
21	Argentina	40,920,000,000
	20 Country total (trillion cm^3)	3.11
	World (trillion cm^3)	3.481
	20 Country share	89.4%
	Texas share of the United States	25.3%
1	United States with Texas	772,800,000,000

Data from Natural Gas: World Fact Book (b), Central Intelligence Agency available at https://www.cia.gov/library/publications/resources/the-world-factbook/fields/269rank.html.

Low-cost natural gas, a relatively clean burning fuel, has allowed the electric power industry along with wind and solar power contributions to lower the CO_2 emissions from Texas and the nation. The increased natural gas supply has also provided additional feedstock for petrochemical plants concentrated in the Gulf Coast of Texas and Louisiana. LNG shipments through Texas and Louisiana ports have shipped product to several foreign countries including Brazil and multiple countries via LNG tankers from Spain, South Korea, India, and France. The shipments are flowing out the Sabine Pass export facility in Louisiana at the border with Texas at the mouth of the Sabine River in the Gulf of Mexico (Texas Railroad Commission, 2016). In addition Texas producers are exporting natural gas to Mexico through pipelines out of the Eagle Ford shale play in the Western Gulf Basin of south Texas (Texas Railroad Commission, 2013).

The current conditions in the Texas and US natural gas market is not the historical norm—quite the contrary. The Texas experience with bringing this important product to a fully functioning market has been a tortuous journey providing important lessons about the mix of competitive markets and government regulation. The natural gas market developed very much as a step child to the oil markets. In the beginning it was considered a waste product of oil production where the gas was flared at the well-head to produce crude oil (the two are joint-products in most underground reservoirs, although in some reservoirs there is only natural gas and associated natural gas liquids (NGL). As natural gas became the fuel of choice for electric power production, and in residential and commercial markets in Texas, the regulatory system failed to produce a well-functioning market. In the early 1970s a major producer in south Texas oversold his reserves leading to major shortages in the winter of 1974 resulting in the closing of schools and many commercial businesses in south Texas. The Texas Railroad Commission (the oil and gas regulator in Texas), the Texas legislature and the courts spent the next decade working out the proper mix of competitive markets and regulatory rules for the natural gas industry. The new policies were difficult to come. In addition to joint-product issues mentioned above, there were issues with pipeline development to move the gas to market, the technical challenges of "stripping" liquids from the raw product as it flowed from wells, and development of the NGL processing plants. The complexities of state and federal regulatory jurisdiction over parts of the industry added to the challenge.

A second complexity in the development of the natural gas market in Texas dealt with the mix of state and federal regulatory system that to this day governs the interstate and intrastate markets. Private companies need to ship natural gas to important markets in out-of-state locations like the Chicago. So there developed a dual system of intrastate pipelines; one system serving Texas consumers controlled by Texas governments, and interstate pipelines serving customers across state boundaries controlled by federal agencies (see chapter 7: Panhandle Field and natural gas flaring).

The increased production levels for both crude oil and natural gas in Texas, concentrated in the Permian Basin, has been responsible for expansion of pipeline capacity additions to move the products to refineries, power plants and petrochemical facilities in Texas and surrounding states. The over half century Texas experience with this valuable resource has been a challenge in natural resource development. In short, the Texas experience has helped inform the rest of the world about the value and complexities of developing and making the highest valued use of the relatively clean burning fuel, and valuable feedstock for petrochemical production.

Texas is well known for its contributions to US and world energy sector development from crude oil and natural gas. The state is less known for the production and use of coal. But there are actually three classes of coal produced in Texas; lignite, bituminous, and subbituminous (Railroad Commission of Texas, 2019). There have been minor levels of coal mining and use in Texas since the early 1900s, but as oil and gas became the preferred energy sources during the first half of the century, coal mining and use did not amount to much.

The history of Texas coal production and use began as early as the late 1800s before oil and natural gas provided better alternatives. Annual production of 36 million short tons in 2017 shows Texas ranked 12th in the world ("as if" Texas a nation) amounting to 6.0% of US coal production (Table 2.3). The original mining was via vertical shafts for underground access (Railroad Commission of Texas, 2020). But the large-scale mining of lignite began with the opening of the Sandow Power Plant near Rockdale, Texas, to provide electric power for a major Alcoa aluminum smelter. The diversification from a near total dependency on natural gas for electric power production in Texas began in 1954 when Alcoa Aluminum opened a major aluminum smelting plant in Rockdale, Texas 60 miles east of Austin. The lignite-based power plant

Table 2.3 Coal production as if Texas is a nation (2017).

Rank	Country	Volume (million short tons)
1	People's Republic of China	3,397,194
2	India	679,236
3	United States	602,728
4	Indonesia	494,709
5	Australia	442,748
6	Russia	312,811
7	South Africa	256,800
8	Kazakhstan	106,660
9	Colombia	90,549
10	Poland	65,975
11	Canada	51,500
12	Texas	36,382
13	Ukraine	24,281
14	Mexico	11,356
15	Czech Republic	5600
16	Germany	3836
17	United Kingdom	3041
18	Spain	2977
19	Turkey	2639
20	New Zealand	2599
21	Chile	2425
	20 Country total	6,596,046
	World	6,739,239
	20 Country share	97.9%
	Texas % of the United States	6.0

Data from Coal: World Fact Book (c), Central Intelligence Agency available at https://www.cia.gov/library/publications/resources/the-world-factbook/fields/252rank.html.

was opened and operated under an agreement with the largest Texas electric utility, Texas Power and Light.

The intensive electricity requirements for aluminum production lead Alcoa to open a major lignite mining operation and power plant near the aluminum plant. The Alcoa plant operated for 65 years until closing in 2018. Almost instantly after the aluminum plant closed a new kind of mining operation set up shop at the Alcoa site. The new enterprise is Bitmain Technologies who is opening a cryptocurrency mining operation at the site of a former Alcoa aluminum smelter. The cryptocurrency mining operations require the use large electric power operations to operate for high end computing hardware to verify and quantify the "blocks" in a blockchain. Blockchains log in an online ledger that keeps track of digital currency transactions (Payward, Inc., 2019). A new innovation has located in Texas and discussed in more detail in Chapter 13, The Road Ahead: Ideas Are Key.

Texas coal use, originally from the Rockdale mine, then from development of additional strip mines in northeast Texas, was supplemented by massive imports of western coal from Wyoming. This rising demand for coal to fire boilers of electric utilities across Texas was a response to the development of natural gas shortages. Shortages were generated primarily because of federal wellhead price controls (for natural gas delivered across state boundaries). The dynamic story of the Texas natural gas industry is summarized in Chapter 7, Panhandle Field and Natural Gas Flaring. A part of this Texas history also generated a great deal of interest in new technologies to make more environmentally friendly coal mining and use possible. Experiments with in situ gasification and fluidized bed gasification were carried out in the 1970s and 1980s, along with carbon capture and storage (CCS) research, development, and demonstration (RD&D) during the last decade. CCS could eliminate most of the CO_2 emissions from coal burning. There were also experiments in the 1980s with shipping coal from Wyoming to Texas (approximately 1000 miles) by pipeline. In short, The Texas experience with new technology to make use of low-cost coal, that is also environmentally friendly have been underway for about 50 years, lessons of interest to the international community of countries.

Texas coal production serves only Texas consumers of electricity. For the most part Texas does not export electricity or coal. The Texas electric grid (Electric Reliability Council of Texas or ERCOT) operates almost totally to serve the electric power needs of Texas. Exports are allowed only by (currently) three DC ties to send power for short periods out-of-state. There is a full treatment of the evolution of the ERCOT grid discussed in Chapter 9, Electric Industry Deregulation and Competitive Markets. The important innovations in this energy sector are discussed there. The Texas innovations in this sector, especially with the deregulation of the electric industry that began in the mid-1990s, provide important innovation experience like no other state in the United States.

Table 2.4 Electricity production as if Texas is a nation (2017).

Rank	Country	Electricity (kWh)
1	China	5,883,000,000,000
2	United States	3,617,648,000,000
3	India	1,386,000,000,000
4	Russia	1,031,000,000,000
5	Japan	989,300,000,000
6	Canada	649,600,000,000
7	Germany	612,800,000,000
8	Brazil	567,900,000,000
9	France	529,100,000,000
10	South Korea	526,000,000,000
11	Texas	477,352,000,000
12	Saudi Arabia	324,100,000,000
13	United Kingdom	318,200,000,000
14	Mexico	302,700,000,000
15	Italy	275,300,000,000
16	Iran	272,300,000,000
17	Turkey	261,900,000,000
18	Spain	258,600,000,000
19	Taiwan	246,100,000,000
20	Australia	243,000,000,000
21	Indonesia	235,400,000,000
	20 Country total	19.01
	World	23.65
	20 Country share	80.4%
	Texas % of the United States	11.7

Data from Electricity: IEA available at http://data.iea.org/.

Texas electric production ranks 11th in the world ("as if" Texas a nation) amounting to 2.0% of the world and 11.7% of the US electric production (Table 2.4). The fuel mix in the ERCOT that serves most of the state in 2019 was 47% natural gas and gas combined cycle, 20% coal, 20% wind, 11% nuclear, and other renewable, including solar, at 2%. As discussed in chapter 9, Electric Industry Deregulation and Competitive Markets, the fuel source for the Texas electric grid has evolved from near 100% natural gas in the 1970s to the current diversified mix present today by adding nuclear in the late 1970s and early 1980s, followed by coal in the 1980s and wind beginning in the 1990s. Solar generation on the ERCOT grid is currently rising rapidly, but is still a small share of the total capacity. The fuel mix in the other countries of the world is very different depending heavily on the endogenous resources within their boundary.

Table 2.5 Electricity from other renewable sources (2017).

All installed capacity (2016 million kW)	Rank	Country	Percentage of total installed capacity	Renewable kilowatts (million kW)—nonhydro
0.0006	1	Timor-Leste	100	0.0006
1.709	2	Luxembourg	67	1.15
14.34	3	Denmark	54	7.74
208.5	4	Germany	52	108.42
4.808	5	Uruguay	42	2.02
0.296	6	Eswatini	41	0.12
97.06	7	United Kingdom	39	37.85
21.56	8	Belgium	36	7.76
1.551	9	Nicaragua	35	0.54
20.56	10	Portugal	35	7.20
2.546	11	Honduras	34	0.87
9.945	12	Ireland	33	3.28
2.401	13	Kenya	33	0.79
82.887	14	Texas	33	27.52
114.2	15	Italy	32	36.54
105.9	16	Spain	32	33.89
40.29	17	Sweden	32	12.89
24.79	18	Austria	31	7.68
1.983	19	El Salvador	29	0.58
19.17	20	Greece	29	5.56
0.05	21	Samoa	29	0.01
774.5	20 Country % of capacity		39	302.42
6386.0		World	14	894.04

Data from Electricity from Renewables: World Fact Book (d), Central Intelligence Agency available at https://www.cia.gov/library/publications/resources/the-world-factbook/fields/265rank.html.

Electricity production from renewable energy (including wind but other than hydro) in the larger developed economies of the world range from 54% of capacity in Denmark, 52% in Germany, 39% in the United Kingdom, 36% in Belgium, 21% in Portugal, 33% in Ireland, and 32% in Italy, Spain, and Sweden (Table 2.5). Offshore wind and onshore solar make up most of the high concentrations of "other renewable sources" in the developed world outside the United States and Texas. Texas ranks 14th in the world for installed capacity of "other renewable sources," mostly due to wind energy capacity while the United States ranks 65th with 14% of capacity from "other renewable sources" (other than hydro). Most of the countries of the world have limited hydro. The United States is a long way down the ranking list at number 128 with 7.0% of capacity from hydro; Texas has less than 1% of grid capacity from hydro (ERCOT, Inc., 2019).

Texas produces the most wind power of any state in the United States (measured by electric power produced). Texas is best known for its oil production but she not only produces more petroleum than any other state, but also dwarfs other states in wind power production. In 2017 Texas wind turbines produced 21,044 MWs. Table 2.5 shows both the total installed capacity and the percent of total capacity that is "other renewable."

Wind capacity and generation is by far the lion's share of the "other renewable" worldwide. The largest shares of total capacity among the larger countries of the world are led by Denmark (54%), Germany (52%), and United Kingdom (39%). But Texas and Ireland at 33%, and Italy, Spain, and Sweden at 32% are not far behind (Table 2.5).

Refined petroleum products production in Texas is highly concentrated in the state along the Gulf Coast for two primary reasons. First, the natural resource base is nearby and spread onshore throughout the state as well as offshore, with easy pipeline capability to get raw products to processing plants with low-cost shipment. Second, refinery and petrochemical processing plants located along the Texas Gulf Coast have low-cost shipment to markets within the United States by major pipeline corridors, and by way of an intercoastal waterway to move products by barge. Also, there are several major coastal ship terminals that support very large oil tankers and LNG tankers to engage in international trade—both imports and exports. The Texas deepwater ports that support multiple commodity imports and exports include the Port of Corpus Christi, Port of Freeport, Port of Texas City, Port Lavaca, Port Sabine, Port Beaumont, and Port Orange. There are several new, mainly offshore terminals, currently proposed and pending regulatory approval. The new planned capacity is occurring mostly because of the large increase in oil and gas production in the Gulf Coast region, especially due to fracking in the Permian Basin of West Texas and Eastern New Mexico.

Texas (if as a nation) ranked fourth in the world for refined petroleum-product production in 2017 (see Table 2.6). The United States, with and without Texas included, ranked number 1 producing 14.5 million barrels per day, followed by China and Russia. The fracking technology in the United States has resulted in greatly increased exports of both crude oil and refined petroleum products. In short, the United States, and Texas within the United States, supplies a large portion of the world production. The United States produces 16.4% of world supplies while Texas alone produces 6.5%.

The growing concern about global warming places the primary source of controllable greenhouse gas emissions on the world consumption of fossil fuels, primarily refined petroleum products and coal. The world economic benefits from the development of fossil fuels, especially crude oil and natural gas refined products, however have freed the world from dependence on raw horse power and the resulting manure pickup and disposal challenges. Horses for work and travel, and the required manure cleanup and disposal was a great nuisance especially in large cities through the first decade of the 20th century. The automobile and petroleum-based, easily stored and transported—liquid fuels changed the world

Table 2.6 Refined petroleum products' production (2017).

Rank	Country	Volume (bbl/d)
1	United States less Texas	14,512,506
2	China	11,510,000
3	Russia	6,076,000
4	Texas	5,787,494
4	India	4,897,000
5	Japan	3,467,000
6	South Korea	3,302,000
7	Brazil	2,811,000
8	Saudi Arabia	2,476,000
9	Germany	2,158,000
10	Canada	2,009,000
11	Iran	1,764,000
12	Italy	1,607,000
13	Spain	1,361,000
14	Thailand	1,328,000
15	France	1,311,000
16	United Kingdom	1,290,000
17	Netherlands	1,282,000
18	Indonesia	950,000
19	United Arab Emirates	943,500
20	Venezuela	926,300
	20 Country total (Mil bbl/d)	71.8
	World (Mil bbl/d)	88.4
	20 Country share	81.2%
	Texas (Mil bbl/d)	5.8
	Texas 20 country share	8.1%
	Texas share of world	6.5%
	United States with Texas (bbl/d)	20,300,000

Data from Refined petroleum products' production: World Fact Book (e), Central Intelligence Agency Library available at https://www.cia.gov/library/publications/resources/the-world-factbook/fields/260rank.html.

enormously. Our modern cities are the result. Liquid fuels have transformed transportation of all types. And petroleum and natural gas-based products have supported the modern world's use of plastics, insecticides, and hundreds of other products of modern economies. But now, as in the past, "the solution becomes the problem." The challenge is whether the collective genus of the United States can solve this problem without "throwing the baby out with the bathwater." Texas will be a significant contributor of the solutions.

Information in Tables 2.7 and 2.8 sets up the fossil fuel use and global warming challenge. Texas, the United States and all modern economies of the world are recipients of the petroleum products "miracles" of the past century. The top 20 countries of the world consume 76.1% of the world total oil product consumption. Texas accounts

Table 2.7 Refined petroleum products' consumption (2016–17).

Rank	Country	Volume (bbl/d)
1	United States less Texas	16,027,880
2	China	12,470,000
3	India	4,521,000
4	Texas	3,932,120
5	Japan	3,894,000
6	Russia	3,650,000
7	Saudi Arabia	3,287,000
8	Brazil	2,956,000
9	South Korea	2,584,000
10	Germany	2,460,000
11	Canada	2,445,000
12	Mexico	1,984,000
13	Iran	1,804,000
14	France	1,705,000
15	Indonesia	1,601,000
16	United Kingdom	1,584,000
17	Thailand	1,326,000
18	Singapore	1,322,000
19	Spain	1,296,000
20	Italy	1,236,000
21	Australia	1,175,000
	20 Country total (bbl/d)	73.3
	World (mil bbl/d)	96.26
	20 Country share	76.1%
	Texas	3.9
	Texas share of top 20	5.4%
	Texas share of world	4.1%

Data from Refined petroleum products' consumption: World Fact Book, Central Intelligence Agency Library available at https://www.cia.gov/library/publications/resources/the-world-factbook/fields/266rank.html.

for 4.1% of world consumption and 5.4% of the top 20 countries. The top 20 countries make up 97.9% of coal consumption (Table 2.3) and 76.1% of the oil product consumption (Table 2.7). The top three countries (China, India, and the United States make up 69.4% of the world's coal consumption and 38.4% of the oil products consumption. Texas consumes 0.5% of world coal consumption and 4.1% of the world oil product consumption.

CO_2 emissions are directly linked to coal plus oil product consumption. China, India, and the United States are to top fossil fuel consumers and therefore the top CO_2 emitters. The three countries emit 60.9% of the CO_2. The top 20 countries account for 90.3% of the CO_2 emissions. So, the challenge is clear. The burden of reducing CO_2

Table 2.8 Carbon dioxide emissions from consumption of energy (2017).

Rank	Country	Amount (MT)
1	China	11,670,000,000
2	United States less Texas	4,584,579,570
3	India	2,383,000,000
4	Russia	1,847,000,000
5	Japan	1,268,000,000
6	Germany	847,600,000
7	South Korea	778,400,000
8	Texas	657,420,430
9	Saudi Arabia	657,100,000
10	Canada	640,600,000
11	Iran	638,300,000
12	South Africa	572,300,000
13	Indonesia	540,700,000
14	Brazil	513,800,000
15	Mexico	454,100,000
16	Australia	439,100,000
17	United Kingdom	424,000,000
18	Turkey	379,500,000
19	Poland	359,000,000
20	Thailand	355,000,000
21	Italy	351,000,000
	20 Country total (MT)	30.4
	World (bil MT)	33.62
	20 Country share	90.3%
	Texas share of top 20	2.2%
	Texas share of world	2.0%

Data from Carbon dioxide emissions: World Fact Book, Central Intelligence Agency Library available at https://www. cia.gov/library/publications/resources/the-world-factbook/fields/274rank.html.

emissions from fossil fuel use falls heavily on the top 3 countries, and almost completely on the top 20. Texas is a large part of the US source of CO_2 (12.5%), 19.7% of the oil product consumption, and 6% of the US coal consumption.

The identification of the large fossil fuel consumers and the associated CO_2 emitters of course is not the whole of the story of the global warming challenge. Greenhouse gases include other gases and other sources of emissions such as livestock farming and ranching operations, residential and commercial buildings, manufacturing plants, and many others. These gases and their sources are all on the emission side of the equation. Offsetting the emissions there are also the various systems that absorb greenhouse emissions, and technologies that sequester the emissions and remove them from harm. These topics are discussed in Chapter 12, Capture and Global Warming: The Technology and Regulation Debate.

The international energy and associated organizations: vehicles for innovation

During the move toward globalization of the last several decades international communities of professionals, trade associations with particular industry clientele, and environmental interest groups have established many international organizations. These organizations, both nonprofit and for profit entities have become major means of communicating innovation breakthroughs across the globe. Some have public policy goals in mind while others are organizations of professionals promoting learning among the group from work around the globe. Still others, like OPEC have been able to exercise market power through production controls exercised by the cartel's production limits. The following list contains the main entities that are most associated with the innovation topic of this book. Others not considered highly important for the current writing are listed at the end of the section.

OPEC

The Organization of the Petroleum Exporting Countries (OPEC) is an intergovernmental organization formed in September, 1960. Original members included Iran, Iraq, Kuwait, Saudi Arabia and Venezuela. Membership grew over the years adding Qatar in 1961, Indonesia in 1962, Libya in 1962, United Arab Emirates in 1967, Algeria in 1969, Nigeria in 1971, Ecuador in 1973 Angola in 2007, Gabon in 1975, Equatorial Guinea in 2017 and Congo in 2018. However, the membership often changed as a number of the smaller countries have left after joining or been suspended. In some cases countries have joined again.

The stated objective of the organization is to provide stabile and fair prices and to coordinate policy among producing counties. Also, the purpose is provide a regular and economic supply to consumers and to provide a fair return on capital investment.
Organization of the Petroleum Exporting Countries (OPEC) organizational documents available at https://www.opec.org/opec_web/en/about_us/24.htm.

International Association for Energy Economics

The International Association for Energy Economics (IAEE) is a global non-profit organization that strives to provide an interdisciplinary forum for the exchange of ideas, experience and issues among professionals interested in the economic analysis of energy resources. The Association is based in the United States, where it was founded in 1977, but includes members from over 100 nations. We maintain national affiliate organizations and conduct conferences and meetings in many parts of the world.

Although founded coincident with the global oil crisis of the 1970s, the scope and interests of the IAEE have grown significantly over the years, in ways that parallel the development of alternative sources of energy, the evolution of markets and regulatory structures, major advances in technology, growing concern about energy's impact on climate change and the environment, and ultimately the need to identify sustainable paths of economic growth.

Our members come from academia, the business community, all levels of government, as well as non-governmental organizations worldwide. We seek members who are interested in energy economics and those who shape opinions and prepare for events which affect the energy industry. We aim to advance the knowledge, understanding and application of economics across all aspects and forms of energy, and to foster communication among a diverse community of concerned professionals.

International Association for Energy Economics (IAEE) organizational documents available at https://www.iaee.org/.

World Coal Association

The World Coal Association is a global industry association formed of major international coal producers and stakeholders. The WCA works to demonstrate and gain acceptance for the fundamental role coal plays in achieving a sustainable and lower carbon energy future. Membership is open to companies and not-for-profits with a stake in the future of coal from anywhere in the world, with member companies represented at Chief Executive or Chairman level.

World Coal Association (WCA) organizational documents available at https://www. worldcoal.org/.

World Petroleum Council

WPC conducts the triennial World Petroleum Congress, covering all aspects of the industry including management of the industry and its social, economic and environmental impact. Accordingly the WPC does not have a formal position on issues but does act as a forum to bring together in dialogue the various sectors of society that have views on specific issues. WPC is a non-advocacy, non-political organization and has accreditation as a Non-Governmental Organization (NGO) from the UN. The WPC is dedicated to the application of scientific advances in the oil and gas industries, to technology transfer and to the use of the world's petroleum (oil and gas) resources for the benefit of all.

Headquartered in London, since 1933, the World Petroleum Council includes 65 member countries from around the world representing over 96% of global oil and gas production and consumption. WPC membership is unique as it includes both OPEC and Non-OPEC countries with representation of National Oil Companies (NOC's) as well as Independent Oil Companies (IOC's). Each country has a National Committee made up from representatives of the oil and gas industry, academia and research institutions and government departments. Governing body is the Council consisting of representation from each of the country national committees.

World Petroleum Council (WPC) organizational documents available at https://www.world-petroleum.org/.

International Renewable Energy Agency (IRENA)

The first intergovernmental organization designed to facilitate cooperation, advance knowledge, and promote the adoption and sustainable use of renewable energy. "IRENA promotes the widespread adoption and sustainable use of all forms of renewable energy, including bioenergy, geothermal, hydropower, ocean, solar and wind energy in the pursuit of sustainable development, energy access, energy security and low-carbon economic growth and prosperity."

IRENA was officially founded in Bonn, Germany, on 26 January 2009. The organization currently has 160 members with headquarters in Masdar City, Abu Dhabi, United Arab Emirates. The organization has an IRENA Innovation and Technology Center In Bonn, Germany and the Office of the Permanent Observer to the United Nations in New York, NY, USA.

International Renewable Energy Agency (IRENA) organizational documents available at https://web.archive.org/web/20190406045612 or https://irena.org/aboutirena.

Global Wind Energy Council

The Global Wind Energy Council (GWEC) was established in 2005 to provide a credible and representative forum for the entire wind energy sector at an international level. GWEC's mission is to ensure that wind power is established as one of the world's leading energy sources, providing substantial environmental and economic benefits. GWEC is a member-based organization that represents the entire wind energy sector. The members of GWEC represent over 1,500 companies, organizations and institutions in more than 80 countries, including manufacturers, developers, component suppliers, research institutes, national wind and renewables associations, electricity providers, finance and insurance companies. Working with the UNFCCC, REN21, the International Energy Agency (IEA), international financial institutions, The Climate Group, the World Bank Group, the IPCC and the International Renewable Energy Agency (IRENA), GWEC represents the global wind industry to show how far we've come, but also to advocate new policies to help wind power reach its full potential in as wide a variety of markets as possible, with a focus on emerging markets. The GWEC Secretariat is located in Brussels, Belgium with an Asia office is located in Singapore.

Global Wind Energy Council (GWEC) Organizational Documents available at https://gwec.net/.

International Solar Energy Society

The International Solar Energy Society (ISES) is a global organization formed in 1954 for promoting the development and utilization of renewable energy. ISES is a UN-accredited NGO headquartered in Freiburg im Breisgau, Germany. "SES aims to bring recent developments in solar energy, both in research and applications, to the attention of decision makers and the general public, to increase the understanding and use of this non-polluting resource in everyday life through its events and publications, ISES promotes research, development, and use of technologies which are directly or indirectly fuelled by the sun. These technologies can provide sustainable solutions for the supply of energy in both industrialized and developing countries. In spite of a historic focus on direct solar energy, ISES today is involved in all renewable energy fields. ISES is the largest international solar organization, with extensive membership worldwide. ISES has members in more than 110 countries, and Global contacts and partners in over 50 countries with thousands of associate members, and almost 100 company and

institutional members throughout the world. Young ISES, a network of students and young professional ISES members, is now connecting young solar professionals worldwide."
International Solar Energy Society (ISES) organizational documents available at https://www.ises.org.

International Atomic Energy Agency (IAEA)

The IAEA is the world's centre for cooperation in the nuclear field. It was set up as the world's "Atoms for Peace" organization in 1957 within the United Nations family. The Agency works with its Member States and multiple partners worldwide to promote the safe, secure and peaceful use of nuclear technologies.

The IAEA Secretariat is headquartered at the Vienna International Centre in Vienna, Austria. Operational liaison and regional offices are located in Geneva, Switzerland, New York, USA, Toronto, Canada and Tokyo, Japan. The IAEA runs or supports research centers and scientific laboratories in Vienna and Seibersdorf, Austria, Monaco and Trieste, Italy.
International Atomic Energy Agency (IAEA) organizational documents available at https://www. iaea.org/.

Other international organizations

Energy and Climate Partnership of the Americas

European Renewable Energy Council International Centre for Hydrogen Energy Technologies

OREDA

Organization of Arab Petroleum Exporting Countries

World Wind Energy Association

World Council for Renewable Energy

World Wind Energy Association

Energy Technology Data Exchange

European Network of Transmission System Operators for Electricity

European Network of Transmission System Operators for Gas (ENTSOG) is an association of Europe's transmission system operators (TSOs)

Interamerican Association for Environmental Defense

International Association of Oil & Gas Producers

International Energy Agency Energy in Buildings and Communities Programme

International Energy Forum

International Fuel Gas Code

International Gas Union

International Partnership for Energy Efficiency Cooperation

International Symposium on Alcohol Fuels

International Solar Alliance

International Geothermal Association
International Hydropower Association
International Renewable Energy Alliance
International Energy Agency
World Energy Council
World LPG Association
Organizations available at https://en.m.wikipedia.org/wiki/Category:International_renewable_energy_organizations

Current status of the global economy in the sixth Kondratieff cycle

There are two interesting and useful efforts of analysis that I have relied on heavily for this chapter. One such effort is the work of Walt Rostow and others following him (in time if not in influence). Rostow spent a significant part of his career at the University of Texas analyzing and writing about the long cycles of economic growth across the major countries of the developed world. His analysis to an extent followed the work of an 18th century Russian economist, N.D. Kondratieff who first recognized long-term economic cycles but was unable to adequately explain the underlying causes of the cycles. Rostow's effort was largely an attempt to understand these cycles of the developed world economic history across two centuries since Kondratieff's 18th century generation by exploring and analyzing the dynamics of national and international economic growth (Rostow, 1978a,b). I believe Rostow was primarily concerned with what appeared at the time to be the coming era of rising scarcity of various foodstuffs and raw materials (including energy), and with the associated character of leading sectors in the major world economies. In short, he was heavily focused on this scarcity condition that existed in the 1970s following the Arab Embargo of 1973, and the expectation that such was destined to continue for decades ahead, leading eventually to the upward movement of growth and prosperity associated with a new Kondratieff cycle. Such analytical results formed the background for Rostow's public policy proposals to drive innovation.

The Kondratieff cycles contributions to understanding the global ebb and flow of prosperity are detailed by Hans-Jörg Naumer (hjn), Dennis Nacken (dn), and Stefan Scheurer (st) of Allianz Global Investors who relied on L.A. Nefiodow's analysis. Nefiodow identified six cycles beginning in 1780 ending in 2010 (below):

1st Kondratieff 1780−1830—Steam engine.
2nd Kondratieff 1830−1880—Railway and steel.
3rd Kondratieff 1880−1930—Electrification and chemicals.
4th Kondratieff 1930−1970—Automobiles and petrochemicals.
5th Kondratieff 1970−2010—Information technology and communications technology.
6th Kondratieff 2010-20XX—Environment technology? Nano-/Biotechnology? Health care
(Nefiodow, L.A., 2006. "Der Sechste Kondratieff" ["The sixth Kondratieff"], 6th edition; Table: Allianz Global Investors Capital Market Analysis)

The Nefiodow analysis identifies the major technological innovations that developed on the heels of global upheavals centered in the United States and Europe during the last two and one-half centuries. The sixth Kondratieff cycle is now unfolding; the authors expect environmental technologies (including green energy), nanotech/biotechnology, and health care to be the main sources for the economic growth explosion of the current unfolding cycle ending perhaps in the 2050 time frame.

These 40- to 50-year cycles, driven jointly by crisis (followed by) technological change and other innovation have resulted in transformation of entire societies where old industries disappeared replaced by new ones; new professions developed, and corporate structures and associated cultures changed as well. Corporate entities and their economic returns from investments drove rising equity markets. In short, extended periods of high economic growth and prosperity followed crisis. One hopes that the coronavirus disease (COVID-19) pandemic of 2020 that has become the new crisis will drive transformative growth in the latter half of the sixth Kondratieff cycle. Fig. 2.1 illustrates the third through fifth Kondratieff cycles since the Great Depression in the 1929−39 decade, and the beginning of the sixth Kondratieff following the Financial Crisis of 2007−09. Rapid economic growth and equity investment returns in the stock market followed each of the three crisis events.

As Hans-Jörg Naumer et al. completed their paper in 2010 it was a bit uncertain whether the Financial Crisis was to mark the beginning of the sixth Kondratieff

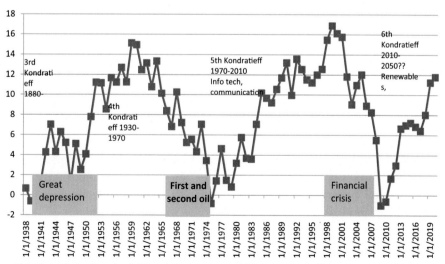

Figure 2.1 Data from Kondratieff cycle-long waves of prosperity rolling 10-year yield on the S&P 500 (annualized % returns). *Nefiodow, L.A., 2006. "Der Sechste Kondratieff" ["The sixth Kondratieff"], 6th edition; Table: Allianz Global Investors Capital Market Analysis.*

cycle. Now in early 2020 it seems clear that it did. Following recovery from the Crisis excess capital came into the markets, the advent of the internet as a resource in global trade began to support internet-based information exchange and global trade to great effect. The internet and related digital technology has generated a major structural change in global trade supported by free-trade policy—trade that has exploded on the global stage. While free-trade structures and agreements are not perfect arrangements like one encounters in economic textbooks, the movement in the direction of tariff-free trades has seen enormous advancement during the last century.

Government policy repaired the US and global banking system that was the main source of financial system failure in 2008−09. Investment banks, encouraged by banking deregulation of the 1990s created incentives for excessive market risks taken on by the large banks. Breaking with the banking system policies of the past, the 1990s changes opened up participants in the banking industry that became overextended with risky loans, especially in the housing industry. Essentially, banks became speculative investors, not just lenders as was the historical practice throughout with one exception: the savings and loan crisis of the 1970s when these entities took investment interests in the housing industry while also serving their traditional lending role.

The economic downturn of 2008−09 also pushed several large automobile companies to the brink of bankruptcy requiring a federally funded bailout that left the federal government owning equity shares in the firms. In the end, however, two large auto companies repaid the federal funding and along with it the government returned company equity to private owners. Economic growth returned at a modest pace following 2010 through a somewhat subdued recovery through 2017. But the economic GDP growth rates in the United States, Europe, and Asia have been robust for 2017−2019.

Then came the COVID-19 epidemic that has become the next crisis to challenge the world's innovation capability. A new book may need to be written to addresses the innovation challenge that now threatens world economic progress. Not only do we face an innovation challenge in the medical industry, the COVID-19 crisis has spilled over into the energy markets by greatly impacting the demand for energy of all types, but importantly, the oil market.

The recent dynamics of innovation (fracking) in the oil and gas markets has allowed the United States to become "energy independent" for the first time since the early 1970s before the Arab oil embargo. Due to the development of fracking technology, a product of US innovation prowess, the technology removed the 50-year concern of being a net importer of oil, heavily dependent on OPEC and the unstable Middle East. Returning to energy independence has been the policy proclamation of every US president since 1973, a condition that has finally come to fruition. Failure to achieve energy independence "dogged" the United States since the embargo of 1973 because of the national security threat that OPEC dominated international market dependence entailed. But that concern disappeared as oil recovery innovation from fracking raised US oil production up from 5.0 mmbd in 2008 to 12.2 mmbd in 2019.

As of early 2020 innovation challenges abound for the oil and gas industry and the US economy that depends on abundant energy supplies. The US oil industry has been met with a drastic decrease in world oil demand due to a major economic decline resulting from the COVID-19 epidemic. This continuing shock to the demand for energy developed at the same time that Russia and Saudi Arabia continue a fight over world oil market share. The combination of demand decline and fracking driven supply surplus is forcing oil prices down to historic lows. This "perfect storm" drove oil futures market prices to negative values in mid-April of 2020.

The COVID-19 epidemic seems likely to result in important social and institutional change, not to mention changes in technologies and market structure and, importantly an onslaught of public policy changes. The almost complete surprise and worldwide spread of the virus will no doubt cause governments and individual citizens and their families to rethink their job and leisure activities that will change how we work and play. Private companies and governments will rethink how they protect employees and how they join with others in both public and private cooperative arrangements to more adequately counter future virus epidemic threats. Finally, creative entrepreneurs will evolve new businesses and business models to deal with fallout of the virus epidemic. They will develop and market innovative institutional and market products to better handle future would-be pandemics. This is the "stuff" of the sixth Kondratieff cycle—changes begun following the Financial Crisis now reinforced by the COVID-19 pandemic.

So, what might we expect in the decade ahead as the world finishes the sixth Kondratieff? Certainly the ongoing trend toward globalization with its impact on demographics is a major driver, even though the current nationalist backlash has temporarily slowed the trend. The forces of globalization and accompanying demographics are powerful forces not likely to be reversed. Technological change of the internet and the digital economy is a genie that can't be put back in the bottle. The manufacturing industries that have freely moved around the developed and developing countries have in recent decades not included the service industries constrained to operate primarily within national boundaries. But the internet has to a significant degree removed that constraint.

Governmental structure influence on energy sector innovation

Authors of college textbooks typically contrast the major economic/political systems in the world economy pointing out the fundamental differences regarding the alternative economic/political systems are (1) major stated objectives, (2) the organizational structures, and (3) the principal philosophies that guide them. While such descriptions of economic and political systems differ among economists and political scientists there are commonly recognized traits of basic political/economic systems that arguably influence the rate of energy sector innovation. The three subsections below describe the three most basic economic/political systems in the world today and lays out the various strengths and weaknesses of each concerning energy sector innovation. The fundamentals of economic systems which are common to all three systems include (1) who owns the factors of production, (2) how products and services are produced and distributed, and (3) the extent to which government entities directly engage in one or both of these basic functions.

Capitalism

Economies similar to the United States foster and protect the institution of private property ownership of the factors of production. Private sector companies own and exploit natural resource deposits, own land and plant and maintain ownership of products produced until title is transferred to buyers in competitive markets (Spaulding, 2020). This is the case in the United States and many European countries, as well as key examples in Asia. But as is immediately recognized, this fundamental characteristic is not universally followed in all sectors. For example, while the United States is recognized as the leading capitalistic democracy in the world, local, regional, and national governments often own and operate electric utilities in the United States. But it is also true that most, or at least the largest utilities, are privately owned enterprises. The Bonneville Power Administration, the Tennessee Valley Authority, and municipally owned utilities are prime examples of government ownership while (the former) Texas Utilities, Inc. in Texas has been among the largest privately owned electric utility in the United States. Private member-owned cooperatives are also common in rural areas of the United States.

The second feature of capitalist systems is that resources, products, and services are allocated through decentralized market systems where buyers and sellers bargain over price and other terms of trade. Oil and gas reserves and products, for example, are almost universally traded in competitive markets in the United States. Consumers aggressively shop among alternative providers for food, fiber, and fuel, and producers try their best to attract consumers and thereby increase market share, all the while focusing on earning a profit. Services are traded similarly, including labor markets where the standard is that competition determines wages and other benefits from employers, while employers are constantly searching and bargaining for skilled employees. Such are the characteristics of democratic systems that champion freedom—freedom to take risks, freedom to set out in new directions, freedom to run for public office, freedom to choose leaders, and above all, freedom to think for one's self—ideas not dictated by a supreme leader.

The primary function of government in capitalistic systems is to promote and provide legal structures for competitive markets, to keep inflation under control and aggressively protect property rights. Governments also have other essential functions including maintenance of security, and law and order, maintaining a standing army and operating a court system to administer justice.

On the innovation front, governments in a capitalistic system have a fundamental but shared role in promoting innovation through research and development (R&D) planning and funding. Such a function is mostly unique to capitalistic systems and not present in the other two economic/government systems discussed here. The shared roles of the private and public sectors in R&D planning and funding is a key issue of this chapter of the book. The joint funding provides important support especially during the early R&D stage where high uncertainty of long-term outcomes limits solo private sector investment. Later stages—RD&D and RD&DC (research, development, demonstration, and commercialization) joined with patent protection is more easily supported by private-only funding. Examples of key innovation outcomes are contained in the following chapters 3 through 12.

Communism

Communism is an economic system with a heavy authoritarian center where the government owns the major factors of production and also decides the allocation and distribution of products and services produced (Spaulding, 2020). Karl Marx and his associate Frederick Engels in the 18th century envisioned a system that removed class distinctions between the rich and the poor. The rich in capitalist systems, they believed, owned all of the property and the poor only had their labor—an unstable system where citizens would eventually revolt and overthrow the government. Under the communism system the government owns the means of production, employees live under the direction of party-appointed planners who set prices and wages, determine output targets and organize to serve party preferences.

A centrally planned economy is the hallmark of communist countries, but they have a real challenge directing the many enterprises that grow in number and complexity as the production process evolves from basic resources to manufacturing then to distribution through wholesale and then retail stages. This complexity leads the government to focus on basic resource sectors like oil, natural gas and coal production and the following refining and manufacturing sectors. The central government tends to focus on the early stage sectors and on the military that allows them to maintain political power. Communism envisions the use of military force and other violent means to achieve control.

One might think that the narrow dictatorial edict of communist systems to produce basic resources like crude oil, and to direct the downstream refining and manufacturing sectors would be more efficient than the same industries in capitalist systems with the messy conglomeration of companies opting for market share. But such communist system outcomes are not supported by the evidence, especially when it comes to the matter of innovation. For example, the worldwide discovery and development of oil and gas resources came primarily from the US and British companies starting in the late 1800s with developments first at home, then spreading abroad. A seven company consortium known as the "seven sisters" expanded their reach abroad mainly in the Middle East where vast deposits of oil were discovered. Royal Dutch Shell, Standard Oil of California, Gulf Oil, Texaco, Standard Oil of New Jersey (later Exxon), Standard Oil of New York, and Anglo-Iranian Oil Company (now BP) controlled 85% of the world's oil reserves (Spaulding, 2020).

The three separate Standard Oil companies grew out of the US antitrust actions of the late 1890s that finally broke up Standard Oil Co. Inc. in 1911. The decade long court battle determined that John D. Rockefeller and Standard Oil was a monopoly dominating the oil industry, eliminating competition and setting anticompetitive prices. But the short version of this story is that technological, institutional, and market innovation that developed the oil and gas industry came from competitive investments of private companies in the United States, not from a communist dictator system. The same is true of the most recent innovations of enhanced oil recovery technologies and fracking that found their footing in the capitalistic system of the United States—strongly from Texas and not from Russia or China. These innovation developments are discussed in more depth in Chapter 6, West Texas and the Permian Basin Early Innovations, and Chapter 10, Hydraulic Fracturing: The Permian Basin Challenges OPEC Leadership.

Socialism

Socialism is generally understood to be an economic/political system that lies somewhere in a middle ground between capitalism and communism (Spaulding, 2020). The focus is on the equitable redistribution of wealth achieved by communal ownership of most means of production but allowing small nonessential enterprises to be privately owned. Although some ideas of socialism allow military force and other violent means to achieve ends, most advocates do not, relying instead on political parties to wield enough power to deliver socialism ends. Trade unions representing the interests of labor are the modern institution for such political movements.

The modern form of socialism relies on prominent political parties like the British Labour Party to deliver equity and welfare for the masses but not with a focus on class-based politics and communal ownership of the means of production. The key focus is to achieve ownership or at least control of businesses by their employees. The essential idea of modern socialism is an economic/political system that is a blend of competitive markets at the retail level with social ownership and control of property and primary means of production.

Measuring the common factors for driving change

Recent years have seen an explosion of interest in gaining better insight into the definition, causes, and the process of innovation. The increased interests have arisen because everyone recognizes that now, more than any time in history the continued improvement of future outcomes of the economy and well-being of people across the globe depends on the pace and substance of innovation. Since so much depends on the next round of innovation we need desperately to know how to make it happen—how to "drive" it forward, and how to correct the mistakes of the past innovation round that invariably produces "dark," unintended outcomes.

A number of economists and political science professionals have written about the innovation topic—the topic was a major focus of Chapter 1, The Dynamics of Innovation and Technology in a Market Economy. I will not repeat that account here but turn to the topics of Texas and US global competition enhanced by innovation and the attempts to measure and rank countries for innovation leadership. This requires distinguishing inputs from outputs, and importantly identifying whether inputs are drivers of innovation, or whether they only support and enhance the process. If that is the case, what then are the drivers? One of the clearest is major unexpected events like epidemics, war, and environmental and weather catastrophes. But there are certainly other recognized challenges that seem to defy solutions that are not catastrophes but excite the curiosity of humans, like what would it be like to walk on the moon.

When dealing with this innovation topic there are several distinctions that need to be made and concepts that need explanation. First, there is the matter of the definition of innovation. The term is used freely by writers but not always defined. Since there are many sides to this "diamond" that holds our future in its grasp one needs to clarify exactly what is being discussed. Just the mention of innovation immediately brings up the notion that it has to do with technology, markets, institutions, public policy, and economics. It is difficult to discuss innovation without delving into each of these subjects. Chapter 1, The Dynamics of Innovation and Technology in a Market Economy, cited a definition of innovation that will do for now:

> *The design, investigation, development and/or implementation of new or altered products, services, processes, systems, organizational structures, or business models for the purpose of creating new value for customers and financial returns for the firm.*
> **Advisory Committee on Measuring Innovation in the 21st Century Economy (2008, p. 7)**

Following the acceptance of a definition of innovation the discussion turns immediately to how to measure it. One is tempted to follow Albert Einstein when he said,

"Not everything that can be counted counts, and not everything that counts can be counted." But to concede that not everything that counts can be counted will not do! How can we make important public policy to guide innovation if we can't measure it? How can the culture's attitudes and perceptions that guide political support in a capitalistic democracy be reasonable if we can't measure the phenomena we try to influence?

The innovation topic has been a part of classical economics for decades. But recently there has developed a new subdiscipline called Innovation Economics. The professionals who participate in this new discipline have been critical of modern "neo-classical" economists because they (innovation economists) fundamentally do not accept the claims of the traditional economists who believe that free markets automatically generate innovation.

Rejection of the "free market automatic-generation notion" is very much the beginning of the innovation economist's work. Much of this professional work on the topic of innovation now focuses heavily on measuring, tracking, and public policy to "drive" innovation. Mark Taylor (2016) makes a convincing case that the usual list of favorable inputs to innovation do not necessarily qualify as "drivers" but they are certainly needed for support of innovation, however exactly innovation is defined.

STEM (science, technology, engineering, and mathematics) is always a part of the discussion, and a requirement for maintaining and supporting innovative ideas. Taylor believes that the true primary driver is "**Creative Insecurity**" and that innovation is not the same thing as diffusion. Creative Insecurity is "the condition of feeling more threatened by external hazards than by domestic rivals" (Taylor, 2016, p. 13). External hazards may be war, but there are many other hazards like extreme weather, health pandemics, cybersecurity attacks, global warming, nuclear waste disposal and economic cartels. So, Taylor's writing, although it includes the innovations generated by war, such is not his central focus.

Taylor identifies what he calls the five pillars of innovation: intellectual property rights, research subsidies, education, research universities, and trade policy. When they function properly, markets appear to be excellent producers of innovation (Taylor, 2016, p. 276). But there is often market failure which is not corrected by free markets or public policy.

A factor that is different from the past is that in our current world systems scientific and technical knowledge, investment capital, equipment and supplies, and STEM workers can now diffuse globally with relative ease. The inputs to innovation can travel cheaply. Therefore location should not matter so much as it did in the past. Nations of all types should be rapidly converging toward the S&T frontier. But that is not convincingly the case.

Politics that support and appose innovation tool creation and use explains why many nations innovate, or fail to do so. Distributional politics, self-interests of industries and social groups often override the public interests. There is no particular national innovation system, type of government, or variety of capitalism that strongly correlates with successful innovation (Taylor, 2016, p. 276). In short, Taylor makes the case that even when creative insecurity would be expected to generate exceptional innovation rates, such does not always happen.

No doubt the debate over the causes of innovation will continue among professionals. In the meantime, those who follow the innovation economics ideas focus heavily on measurement and rankings among countries. A number of groups now produce indexes that rank countries as to their innovation prowess. Two following sections summarize recent rankings that place the United States in a list of the top 20 countries out of the approximately 125 nations. One index is known as the Global Competitiveness Index and the other the GII.

Global Competitiveness Index

The World Economic Forum has published an annual Global Competitiveness Report every year since 2004. The report summarizes the results of ranking 141 economies amounting to 99% of world GDP. The rankings are made by way of comprehensive index that compares the countries competitiveness using a combination of assessments from about 103 individual indicators and qualitative judgments of 13,500 business leaders from 141 countries. The latter is derived from responses to an Executive Opinion Survey. The quantitative data is drawn from publicly available sources such as the United Nations.

The GCI separates countries into three specific stages: factor-driven, efficiency-driven, and innovation-driven, each implying a growing degree of complexity in the operation of the economy. The indicators are organized around what is called 12 pillars. The pillars are: institutions, infrastructure, information and communications technology (ICT) adoption, macroeconomic stability, health, skills, product market, labor market, financial system, market size, business dynamism, and innovation capability. The conceptual framework for the index is measuring the drivers of total factor productivity (TFP). This is the part of economic growth not explained by the factors of production—labor and capital. TFP growth is calculated by subtracting growth rates of labor and capital inputs from the growth rate of output in the economy, sector by sector. The calculated index by country is an average of the scores of the 12 pillars (see Tables 2.9 and 2.10). GCI and individual pillar and scores are expressed on a 0−100 scale and are interpreted as "progress scores," indicating how close a country is to the ideal state.

The 2018 report gave the United States a score of 85.6 out of 100 (rank 1 in the world). The ranking was reflective of several good scores among the 12 pillars. Most relevant to the topic of energy innovation is the measures of business dynamism and

Table 2.9 Global Competitiveness Index (2019).

Rank	Country	Score
1	Singapore	84.8
2	United States	83.7
3	Hong Kong	83.1
4	Netherlands	82.4
5	Switzerland	82.3
6	Japan	82.3
7	Germany	81.8
8	Sweden	81.2
9	United Kingdom	81.2
10	Denmark	81.2
11	Finland	80.2
12	Taiwan	80.2
13	South Korea	79.6
14	Canada	79.6
15	France	78.8
16	Australia	78.7
17	Norway	78.1
18	Luxembourg	77.0
19	New Zealand	76.7
20	Israel	76.7
21	Austria	76.6
22	Belgium	76.4
23	Spain	75.3
24	Ireland	75.1
25	United Arab Emirates	75.0
26	Iceland	74.7
27	Malaysia	74.6
28	China	73.9
29	Qatar	72.9
30	Italy	71.5

Data from Schwab, K. (Ed.), 2019. The Global Competitiveness Report 2019. World Economic Forum, Geneva, Switzerland.

innovation capacity. The United States ranked well—86.5 (rank 1) for business dynamism and 86.5 (rank 2) for innovation capability. Individual measures within business dynamism ranked the United States rank 1 for "insolvency regulatory framework," and "embracing disruptive ideas." Individual measures within the innovation capability pillar gave rank 1 for "state of cluster development" and "multistakeholder collaboration," and also rank 1 each for "scientific publications," "quality of research institutions," and "buyer sophistication." For the labor market pillar, the United States

Table 2.10 Global Competitiveness Index (2018).

Rank	Country	Score
1	United States	85.6
2	Singapore	83.5
3	Germany	82.8
4	Switzerland	82.6
5	Japan	82.5
6	Netherlands	82.4
7	Hong Kong	82.3
8	United Kingdom	82
9	Sweden	81.7
10	Denmark	80.6
11	Finland	80.3
12	Canada	79.9
13	Taiwan	79.3
14	Australia	78.9
15	South Korea	78.8
16	Norway	78.2
17	France	78
18	New Zealand	77.5
19	Luxembourg	76.6
20	Israel	76.6
21	Belgium	76.6
22	Austria	76.3
23	Ireland	75.7
24	Iceland	74.5
25	Malaysia	74.4
26	Spain	74.2
27	United Arab Emirates	73.4
28	China	72.6
29	Czech Republic	71.2
30	Qatar	71

Data from Schwab, K. (Ed.), 2019. The Global Competitiveness Report 2019. World Economic Forum, Geneva, Switzerland.

ranked 1 for "pay and productivity" (the best rank). This US innovation ecosystem pillar (business dynamism and innovation capacity) got an overall score of 84.1 in the 2019 report.

The conceptual framework includes the "productivity frontier"—the maximum efficiency the economy could obtain—is given a score of 100. The measured performance of each country is then some share of 100. In 2019 the highest score was 84.8 for Singapore, followed by the United States with a score of 83.7 (see Table 2.9). The other highest ranking countries include most of the West European economies (the Netherlands, Switzerland, Germany, Sweden, United Kingdom, Denmark, and Finland, plus Canada) and five Asian countries (Singapore, Hong Kong, Japan, Taiwan, and South Korea).

The 2018 report contains rankings that differ somewhat from the 2019 report but for the most part the same countries dominate the highest rankings. The United States ranked first followed by Singapore. Others in the top 30 from included Germany, Switzerland, the Netherlands, Sweden, Finland (all Western Europe), and Canada in North America. Asian countries in the top 30 included Japan, Hong Kong, Taiwan, and South Korea (Table 2.10).

The competitiveness index is more than a ranking methodology. It also indicates the improvements that might be made that would push the economic performance out toward the score of 100. A rising score for the world averages, and for individual countries, would indicate an improvement in economic efficiency in the use of world resources and improvements beyond factor productivity.

An examination of performance by country results in a ranking among countries on each of the 12 pillars, and in the aggregate. The innovation capability is the part of the score that ties this index closest to the following index system called the GII. For example, the rankings are favorable for the United States on several individual indicators. The 2019 report shows growing efficiency of electric vehicles, increased capacity and electric power generation. But the United States lags several European countries for incorporating renewable resources, an outcome heavily influenced by the historical, and current access to plentiful supplies of fossil fuels. The world remains 80% dependent on nonrenewable energy sources.

Global Innovation Index

The second effort I have relied on beyond the Global Competitiveness Index and the Kondratieff cycle literature is a contemporary analysis and framework called the GII developed and updated annually by a group at Cornell SC Johnson College of Business, jointly with INSTEAD, WIPO, Confederation of Indian Industry, and others. The annual index ranks innovation contributions from approximately 130 countries. The effort attempts to measure an economy's innovation performance. The first section of this chapter made key comparisons of major energy-intensive countries with the United States and with Texas based on production and capacity. The Innovation Index analysis provides a framework for evaluating Texas and US contributions to global innovation (see Tables 2.11 and 2.12).

The central themes of professionals who study the factors that affect or drive innovation contain what may be called a matrix of innovation drivers or **inputs** to the innovation process.

Table 2.11 Global Innovation Index rank by country (2019).

Rank	Country	Score
1	Switzerland	67.2
2	Sweden	63.7
3	United States (the)	61.7
4	Netherlands (the)	61.4
5	United Kingdom (the)	61.3
6	Finland	59.8
7	Denmark	58.4
8	Singapore	58.4
9	Germany	58.2
10	Israel	57.4
11	Republic of Korea (the)	56.6
12	Ireland	56.1
13	Hong Kong, China	55.5
14	China	54.8
15	Japan	54.7
16	France	54.2
17	Canada	53.9
18	Luxembourg	53.5
19	Norway	51.9
20	Iceland	51.5
21	Austria	50.9
22	Australia	50.3
23	Belgium	50.2
24	Estonia	50
25	New Zealand	49.6
26	Czech Republic (the)	49.4
27	Malta	49
28	Cyprus	48.3
29	Spain	47.9
30	Italy	46.3

Data from Cornell SC Johnson College of Business, INSTEAD, WIPO, CII, BSDASSAULT SYSTEMS, SEBRE, and CNI, The Global Innovation Index (GII), 2018 (Dataset/Material); Cornell SC Johnson College of Business, INSTEAD, WIPO, CII, BSDASSAULT SYSTEMS, SEBRE, and CNI, The Global Innovation Index (GII), 2019 (Dataset/Material).

The broad areas are:
1. Institutions **(political environment, regulatory environment, and business environment)**.
2. Human capital and research **(education, tertiary education, and R&D)**.
3. Infrastructure **(ICTs, general infrastructure, and ecological sustainability)**.
4. Market sophistication **(credit, investment, trade, competition, and market scale)**.

Table 2.12 Global Innovation Index rank by country (2018).

Rank	Country	Score
1	Switzerland	68.4
2	Netherlands	63.3
3	Sweden	63.1
4	United Kingdom	60.1
5	Singapore	59.8
6	United States	59.8
7	Finland	59.6
8	Denmark	58.4
9	Germany	58
10	Ireland	57.2
11	Israel	56.8
12	Republic of Korea	56.6
13	Japan	55
14	Hong Kong (China)	54.6
15	Luxembourg	54.5
16	France	54.4
17	China	53.1
18	Canada	53
19	Norway	52.6
20	Australia	52
21	Austria	51.3
22	New Zealand	51.3
23	Iceland	51.2
24	Estonia	50.5
25	Belgium	50.5
26	Malta	50.3
27	Czech Republic	48.7
28	Spain	48.7
29	Cyprus	47.8
30	Slovenia	46.9

Data from Cornell SC Johnson College of Business, INSTEAD, WIPO, CII, BSDASSAULT SYSTEMS, SEBRE, and CNI, The Global Innovation Index (GII), 2018 (Dataset/Material); Cornell SC Johnson College of Business, INSTEAD, WIPO, CII, BSDASSAULT SYSTEMS, SEBRE, and CNI, The Global Innovation Index (GII), 2019 (Dataset/Material).

5. Business sophistication **(knowledge workers, innovation linkages, and knowledge absorption)**.

The GII analysis on the input side results in the **Innovation Input Sub-Index**.

The other key measurement is the outputs of the innovation process. The measures of **Innovation Output Sub-Index** developed for the GII include:

1. **Knowledge and Technology** (knowledge creation, knowledge impact, and knowledge diffusion)

2. **Creative Outputs** (intangible assets, creative goods and services and online creativity)

The analysis on the output side results in the **Innovation Output Sub-Index**.

The overall **GII score** of innovation leadership is the simple average of the Input and Output Sub-Index scores. Tables 2.11 and 2.12 show the GII rank and scores for the top 30 countries for 2018 and 2019.

The innovation index measurement by the Cornell group also includes a calculation of the **Innovation Efficiency Ratio** which is the ratio of the Output Sub-Index score to the Input Sub-Index score. It shows how much innovation output a given country is getting for its inputs.

There are several factors that determine the global leadership in innovation—all detailed in the GII annual reports. Cornell SC Johnson College of Business (2018, 2019). The United States and Texas rank well among the more important factors of R&D expenditures, patent applications, scientific, and technical publications. Also very important are the measures of market sophistication, innovation clusters and knowledge impact. The United States also leads in computer software spending and rankings of high and medium tech manufactures. The country excels in quality of universities, venture capital deals, and state of cluster development.

The United States has the largest number of innovation clusters in the world. She is the location of the globe's top innovation clusters including Silicon Valley in California, Chapel Hill in North Carolina, and Austin in Texas. If these clusters in the United States were countries, they could rank uniformly high on the innovation index scale rankings. Most top science and technology clusters in the world are in the United States, China, and Germany; Brazil, India, and Iran also make the top 100 list.

The dynamics of innovation progress in the world that is unique to our current time period is just how easy it has become to innovate. A major part of this development is the speed with which systems are spread around the world. Globalization since World War II has removed many barriers to innovation. Scientific and technical knowledge, investment capital, equipment and supplies, and professional STEM workers can move around the world easily and inexpensively. Hence location does not matter so much as it has historically. As a result countries are more easily moving toward the productivity frontier.

The top 10 countries for the GII innovation rankings for 2018 and 2019 include Switzerland, Sweden, the United States, the Netherlands, the United Kingdom, Finland, Denmark, Singapore, Germany, and Israel (2019). The same countries occupy the top 10 in 2018 except Ireland came in 10th, one place ahead of Israel (Tables 2.11 and 2.12). The western democratic capitalistic countries clearly outpace the communist countries of Russia and China.

The United States and China, however, are the largest world contributors of absolute, innovation inputs and outputs, including R&D expenditures and patent applications. But the United States and other western countries out rank China by a large measure. The United States moved up from sixth in 2018 to third in 2019. China, who ranked 14th in 2019, gained ground in the GII ranking, moving up to the 17th place in 2018 (Tables 2.11 and 2.12). China's weight in both the input and output sides of the innovation process, however, is huge. In absolute terms, China's number of patent applications by origin and scientific and technical publications, as well as its number of researchers is the highest in the world. China ranked fifth in knowledge and technology outputs in the 2019 report. China ranks well in knowledge and technology outputs (fourth in 2019). She ranked 22nd in knowledge diffusion in 2019.

Fundamentals of resource ownership

The western styled capitalistic democracies, most of whom have followed the US example since World War II, depend heavily on the institution of private property ownership as a cornerstone of free market operations, as well as basic personal freedoms beyond economic markets. But the focus here in this chapter is with the functioning of markets, especially energy markets and the ways property rights guide and control such markets. Without clear specification and legal protection of property rights a competitive market system such as that of the United States could not function. This is generally the case but importantly a fundamental principle for guiding the production, transport, distribution, and use of energy.

The concepts of property and ownership are based on a bundle of legal rights recognized and enforced by society—a primary function of democratic systems. From a legal sense, a property right is the right to a physical object or the right to enjoy the benefits from the exercise of a specific right. The property is the legal right to use, enjoy, sell, mortgage, or rent the physical object. Patents are fundamental to property rights and the functioning of markets. The owner may sell the patent, license others to produce the patented article or he may produce it himself. The process for filing a patent application and approval is a basic function of the federal government. Property rights are fundamental to the exchange of property and the enforcement of contracts (Lusk, 1963).

In depth discussion of property rights and their functioning in energy markets requires one to deal with the notion of possession, the distinction between real and personal property, tangible and intangible property, and importantly, public and private property. Energy markets begin primarily with real property, such as oil and gas deposits that involve a unique legal system for defining the right to control access to and produce the resource and deliver it to market. Solar energy is a different matter since the property rights that insure access is not about ownership of a portion of the sun, but access to the sun rays via the use of solar collectors placed on land or buildings. The legal systems are about protecting access. Solar access laws provide protection for solar customers by prohibiting or limiting private restrictions on solar energy installations.

The determinations of property rights have evolved since the Nation was formed, and often were derived from English systems in the beginning. Property is a fundamental part of the ideas built into the US constitution. The ideas contained in several Articles and Amendments to the Constitution have undergone gradual change over the 240 + years since the constitution was written but underlie both personal freedoms that can't be taken away by governments (or be infringed upon by others, including the government), guarantees that protect both individuals and business participants. But it is also the case that property rights are not unlimited—governments can and do limit the exercise of privileges flowing from property rights.

The limits of property rights continually evolve, but importantly such changes, by design are made slowly such that private investor decisions can be made with a high degree of certainty. A major change in the exercise of energy property rights in modern times that did not command attention during the adoption of the US Constitution has to do with waste products. Major concerns now days are the matters of pollution from energy use, and disposal of waste products from the production and use of energy. The most problematic are high-level nuclear waste disposal that pose a threat to life forms of all kinds, and greenhouse gas emissions that produce global warming.

Democratic capitalistic systems like that in the United States face an ongoing process of refining the limits of property rights for individuals and firms, while protecting the public interest. The benefits and costs that go on outside of competitive markets are among the most important of current endeavors. All three branches of government have a role—the legislative branch to revise and enact new statutes, the executive branch to implement laws and regulations and the courts to adjudicate. A question of concern in this writing is whether the rather messy processes of democratic capitalistic systems are up to the task of resolving such conflicts, and whether they can do so better than communist and socialism systems.

Concessions and their role in international energy markets

The early development of oil (and less so natural gas) took place inside the United States and the British Empire. The capitalist countries had the power of private enterprise motivation, a developing, supportive legal and regulatory system and government's tax revenue prospects to drive investment and development. The dictatorships, especially in the Middle East and Venezuela, had neither the exploration and drilling capacity or the financial means to develop their oil and gas resources, which were government owned. So, the large US and British companies came to their rescue. The mutual interests produced an institution called a "concession." Schlumberger defines a concession in the oil and gas business as:

> *Governments around the world (especially sovereign powers) provide "concessions" to private companies. These governments have land ownership overlying oil and gas or other mineral deposits. Grants or permits are made to allow private companies to explore and recover oil, gas, and other minerals. These "grants" typically carry a fee or other compensation payment in exchange for access. The compensation may take the form of a bonus, license fee and royalty or even the sharing of production. These grants are usually provided for a limited period of time.*
>
> **Schlumberger (2020; https://www.glossary.oilfield.slb.com/en/Terms/c/concession.aspx/).**

Concessions allowed British companies (primarily Shell and BP) to compete with Standard Oil of the United States to develop and market oil products in about every part of the global as the markets for fuel oil, and eventually gasoline developed with the advent of the automobile, and the transition of coal based rail transportation to oil. Oil deposits were well known and being exploited by Russia in the Baku area while British companies used concessions in several locations in the East Indies and eventually in Persia to develop oil reserves. Standard Oil focused on developments inside the United States in the last decade of the 19th century before going abroad. The company expanded its holdings in the United States to include all parts of the industry from crude oil production to refining, pipelining, wholesale, and retail. The company so dominated the industry with the formation and operation of a trust arrangement that it was eventually determined that the company was exercising monopoly power in violation of US antitrust laws. The courts got involved, and in the end the Supreme Court broke up the company in 1911. The antitrust basis was restraint of trade under the Sherman Act.

The Standard Oil Company dissolution resulted in the creation of several separate companies. There was Standard Oil of New Jersey (that eventually became known as Exxon), Standard Oil of New York (which became Mobil), Standard Oil of California (which became Chevron), Standard Oil of Ohio (which became Sohio, then the American arm of BP), Standard Oil of Indiana (which became Amoco), Continental Oil (which became Conoco), and Atlantic (which became a part of ARCO and eventually of Sun; Yergin, 1991, pp. 97–110).

The often repeated theme of this book is the recognition that sudden changes—"hitting the wall" and less dramatic, but fundamental, often external changes generate innovation. That was true in this case. The break up of Standard Oil resulted in the several new competing companies that actually increased the aggregate stock value of the combined set of companies by a factor of two, and set them off into somewhat different directions. They developed somewhat different business models as well as technological innovation. Standard Oil of Indiana moved forward after the break up and developed a breakthrough in refining known as "cracking." This is a heat and pressure technology designed to produce gasoline and diesel for the rapidly developing auto industry of the 1920s. The process also just about doubled the quantity of usable gasoline produced from a barrel of crude oil (Yergin, 1991, p. 111).

The World War I experience ran up the demand for oil, and immediately following the end of the war, the growing automobile industry increased the demand for gasoline and therefore crude oil. In the meantime the outlook for the US oil supply was growing increasingly pessimistic. The United States became fearful that projected declines in US oil reserves in the face of rising postwar demand, would drive the United States to become more dependent on foreign, unstable supplies. The US Bureau of Mines in 1919 predicted that within the next 2—5 years the US supplies of oil would begin to decline. The support for such a belief came as actual shortages of fuel oil occurred during the winter of 1919—20. This concern sparked the beginning of a near century-long concern of US industry leaders and politicians about the national security exposure from reliance on unstable foreign oil sources. This ongoing declaration for US energy independence as it has been called, has only recently dissipated with the advent of fracking and the resulting positive net export position for oil that continues today.

The development of US dependence on unstable foreign oil followed the rush of British and US company competition to use the concession institution to exploit the oil reserves of the East Indies and the Middle East during the two decades following the end of World War I. The developed world in Europe and North America needed the oil to satisfy growing demand, provide oil supplies in times of war and satisfy anticipated shortages due to a slowdown of oil development inside their own territories. Concessions opened the gates to East Indies and Middle East oil development pursued vigorously by the big British companies (Shell and BP) and the United States by the several American companies formed after the break up of Standard Oil Company in 1911. Concessions became the means for development of oil resources within the dictatorship countries.

While concessions remained a necessary arrangement of contracting that gave individual companies a degree of investment confidence, the situation began to change by the formation of cartels to exercise market power. In 1951 Iran nationalized its oil industry previously owned by Anglo-Iranian Oil Company (now BP) and then imposed an international embargo. Anglo-Iranian Oil Company joined with Royal Dutch Shell, and American companies, Gulf Oil, Standard Oil of California, Standard Oil of New Jersey, Standard Oil of New York, and Texaco to form a consortium that became the "Consortium for Iran" in an effort to bring Iranian oil back into the world market. The consortium became known as the "Seven Sisters" a group of companies that controlled most of Middle East oil production through the 1960s.

The dynamics of seeking control of the oil industry eventually allowed the totalitarian governments to organize the OPEC in the 1960s with an early attempt to take major market control in 1967. This cartel originally included Saudi Arabia, Iran, Iraq, Kuwait, and Venezuela. The membership has changed over the years since the beginning in 1960 and now includes 13 countries adding Algeria, Angola, Congo, Equatorial Guinea, Gahon, Libya, Nigeria, and United Arab Emirates to the group.

The power of OPEC, led by Saudi Arabia, to maintain market share and control price prevailed from the Embargo of 1973 until the fracking boom in the United States beginning in 2005. The market power of OPEC has finally been successfully challenged by innovation from the US Concessions have largely been outdated as the totalitarian countries have come to nationalize the oil industry within their boundaries and rely on OPEC to maintain stability and market influence. This has not been Russia's behavior as she has mostly stood alone. However, Russia has sometimes joined with OPEC in the latest attempts to control the world market in the face of United States increased production.

US companies with strong ties to Texas: spreading innovation throughout the globe

The United States has been a world innovation leader for over 100 years, and Texas has been a major part of that experience, especially in the energy sector. The US innovation has included machines, tools, manufacturing processes, transportation systems—airplanes, railroad infrastructure and locomotives, and the automobile with a national highway system—military equipment and weapons, electricity sector development, including generators, transmission lines, electric motors, light bulbs and all the rest, including city lights that make earth's inhabitation visible in the night sky from a satellite or orbiting spaceship. And more recently, innovation leadership has produced the internet, computers, and rocket systems capable of sending men to the moon and by doing so, changed our understanding of the universe. These things are all examples of technological innovation. Importantly the US innovation of modern corporations has become the institutional mechanism for raising investment capital, managing complex systems and sharing of the benefits and risks of new ventures. This business structure innovation has been supplemented by modern venture capital institutions, when matched with the internet, information technology (IT), ICT, and now artificial intelligence, have dramatically changed our way of life. And there is more to come.

The American innovation contributions have found their way into the world economy by two primary means. First, global trade policy has become less and less restrictive over the last century allowing trade among countries to more easily share the benefits of technological innovation through open markets. Second, American companies, taking advantage of freer trade have set up company branches across the globe. Thus the innovation prowess of the United States and Texas has spread economic benefits across the globe, and importantly, free-trade and capitalistic companies have often lowered the risk of war.

In short, the world has never produced a country in all of human history that has exceeded the contemporary US innovation experience that has, above all else, helped lift most of the world's people out of hunger, poverty, and misery. Such an outcome is from a system where the government is the servant of the people rather than the other way around. Kingdoms and dictatorships that constitute top-down societies have dominated most of human history. Capitalist democracies following the US example have upended that dominance, and a large part of the reason has been US innovation prowess. Such has been the outcome of capitalistic democracies that upended Thomas Malthus' belief that the inescapable destiny of the world was such that the "misery index" would remain high as the natural exponential population growth would continually outstrip the food supply. A major part of this modern "miracle" has been the development and evolution of energy systems where the United States has been the clear leader, and where Texas has been a "giant" contributor to the energy story.

The United States and Texas innovation contributions over the last 125 years have created enormous global benefits. The process of spreading innovation around the globe is a complex process but it seems clear that three factors have had a highly positive impact. First, US leadership in international trade has encouraged US companies to go abroad taking their technologies and market knowledge with them. Second, US government participation in international institutions, especially the United Nations, World Trade Organization, International Monetary Fund, World Bank, Organization for Economic Co-operation and Development, and North American Treaty Organization has been exceptional and provided leadership to promote peace and prosperity through innovation. Third, the US participation in extreme events (war, pandemics, and extreme weather) and national initiatives like the space program have been a powerful forces driving innovation. But fundamental to all three influences are "ideas" and "freedoms" that flourish in capitalistic democracies.

Individual US energy companies have had a major role in driving energy sector innovation over the last century. Freedom to go abroad under advancing free-trade policies have been key. These company activities abroad have provided benefits to the countries but also to the companies who have learned valuable lessons from their experiences. The experience of several key energy companies exemplify the United States and Texas contribution to the benefits of energy innovation. The following companies have been exemplary.

Avangrid

AVANGRID is a US company with important assets in Texas and 21 other states valued at $32 billion. The company holds assets and lines of business that includes networks (Avangrid Networks) and renewables (Avangrid Renewables), employs 6500 people and serves 3.25 customers in New York and New England. The networks branch owns eight electric and natural gas utilities while the renewables branch owns and operates 6.6 GW of electric capacity. Avangrid Renewables participates in or owns two major wind farm projects in Texas including the Texas Gulf Coast Wind Farm (a 286 MW Karankawa Wind Farm) via a power purchase agreement (PPA) and it owns and operates a 605 MW South Texas Coast Complex wind farm and the 120 MW Barton Chapel Wind Farm in Mid-West Texas. The company is committed to renewable development throughout the country as evidenced by its investments and through support of the United Nation's Sustainable Development Goals.

Avangrid Renewables available at www.avangrid.com.

Chevron Corporation

Chevron Corporation's global headquarters are located in San Ramon, California. The company is the second-largest integrated energy company headquartered in the United States. Through their subsidiaries and affiliates, Chevron produces crude oil, natural gas and many other essential products. Chevron is one of the several companies formed under the direction of the U.S. Supreme Court case that broke up Standard Oil of New Jersey in 1911.

Chevron products are sold in the more than 7,800 Chevron® and Texaco® retail stations across the United States. The company is also a major supplier of aviation fuel in the U.S. Chevron's four U.S. refineries have the combined capacity to process 932,000 barrels of oil per day.

Chevron was one of the top producers of net oil-equivalent in California during 2018, and one of the largest net acreage leaseholders and producers in the Permian Basin. Production by Chevron in the Permian Basin dates back to 1920 as the industry first learned of and began primary production, followed in later years by water flooding and then CO_2 recovery techniques. The company currently has 500,000 acres leased in the Midland Basin and 1.2 million acres in the Delaware Basin (both within the Permian Basin). The company uses multiwell pads to drill a number of horizontal wells that are completed using hydraulic fracking technology.

Chevron is one of the leading producers in Deepwater Gulf of Mexico where in 2018 she produced an average daily yield of 186,000 barrels per day, in addition to 105 million cubic feet of natural gas and 13,000 barrels of NGL from seven fields. The wells are in water ranging from 3,500 to 7,000 feet and a reservoir depth of 30,000 feet.

Chevron has a shipping company called Chevron Shipping Company LLC based in San Ramon, California. The company has a fleet of both U.S.-flagged vessels and foreign-flagged vessels. The U.S.—flagged vessels are primarily shipping refined products in the coastal waters of the U.S. The foreign vessels transport crude oil, LNG, refined products and feedstocks to and from a number of locations around the globe.

Chevron is also in the power generation business via a subsidiary, Chevron Power and Energy Management Company. This subsidiary looks after Chevron's gas-fired and renewable power generation assets. The company also provides engineering and support services for Chevron's entities worldwide. The power plants provide electricity and steam for support of enhanced recovery projects. Chevron's renewable operations include wind, solar and geothermal assets in Arizona, California and Texas.

Chevron is also one of the world's top producers of petrochemicals via a Chevron Phillips Chemical Company located in The Woodlands, Texas, a 50% joint venture with Chevron Phillips 66. The company produces olefins, polyolefins, aromatics, styrenics and specialty chemicals. These are basic petrochemical products used to make consumer and industrial products. The company began operation of a new 1.5 million metric ton per year of ethylene from its ethane cracker plant in Cedar Bayou, Texas in 2018.

Current technology projects will produce returns worldwide as the projects reach commercial maturity. Chevron owns proprietary technologies in exploration, resource characterization, well drilling and completion in operations at the company's extensive unconventional resources focused on Permian Basin challenges. Second, Chevron has improved seafloor boosting pumps that have increased production by approximately 20,000 barrels of crude oil per day being applied in the Gulf Coast. Third, Chevron is leveraging advanced digital technologies that employ wireless connectivity, sensors, plant and process data analytics, and mobile worker solutions to improve safety, enhance equipment monitoring and reduce maintenance costs.

Chevron is investing in future innovation as well, as evidenced by $350 million in venture capital investments in 2018, and she has set up a new Future Energy Fund targeting emissions reduction with a $100 million commitment. The company has made investments in ChargePoint, an electric vehicle charging technology scaled for fleets, and Natron Energy, a company developing stationary energy storage systems at electric vehicle charging stations.

Chevron's work can be found worldwide in Angola, Indonesia, Republic of Congo, Argentina, Iraq, Russia, Australia, Kazakhstan, Saudi Arabia, Azerbaijan, Kuwait, Singapore, Bangladesh, Malaysia, South Korea, Brazil, Myanmar, Thailand, Cambodia, Netherlands, United Kingdom, Canada, Nigeria, China, Philippines, Venezuela and Colombia.
Chevron Corporation available at https://www.chevron.com/worldwide/united-states and https://www.chevron.com/worldwide (September 3, 2020).

CMS Energy

CMS is moving aggressively, changing the way they generate electricity so they can improve Michigan's future. By the year 2040 the plan is to eliminate the use of coal to generate electricity and reduce carbon emissions by 90%. The company is pushing hard toward another 2040 goal, where 40% of our electric generation will come from renewable sources and energy storage. It's a balanced energy approach to secure a more of a sustainable future.

CMS is switching away from coal as the primary fuel for power generation. Coal plants have historically been the backbone of our energy supply. For decades, CMS and others counted on a dozen companies around the state to produce a consistent flow of electricity to fuel homes and Michigan's economy. These companies have earned numerous awards recognizing reliability, safety, and environmental stewardship. But the world is changing. More and more people want their energy to come from renewable sources. So CMS is shifting to cleaner energy sources as part of our commitment to build a sustainable future for people and the environment.

CMS' remaining five coal plants have been equipped with sophisticated air quality technology and which uses low-sulfur coal to reduce emissions. The plants are significantly cleaner. Having a diverse fuel mix is important to keep electric bills lower and energy value high.

CMS Energy available at https://www.cmsenergy.com/home/default.aspx.

ExxonMobil

ExxonMobil is currently among the largest energy companies in the world with operations across the globe. The full history of the company is far too voluminous to include here so the focus is on recent developments in innovation and new technologies. But it is worthwhile to include a brief historical account, beginning 135 years ago. The company began as a marketer of kerosene in the late 1800s, eventually becoming part of the Standard Oil complex that was broken up forming 34 different entities when the long lasting antitrust case ended in the US Supreme court ruling breaking up Standard Oil Company of New York and Standard Oil Company of New Jersey. A major outgrowth of the 34 entities was the formation of Humble Oil and Refining Company of Texas in 1919. In 1972 Jersey Standard changed its name to Exxon Corporation. In the years following the company became Exxon and then on the hills of a merger in 1999 became ExxonMobil.

Today ExxonMobil explores for oil and gas on six continents, operates refining facilities and markets final products in most countries of the world by operating in over 50 countries in North America, Europe, European Union, the Middle East, and North Africa. Faced with realities of today the company conducts R&D of new technologies to meet the dual challenges of fueling global economies and addressing the risks of climate change.

ExxonMobil's downstream enterprises include refineries integrated with lubricants and chemicals. The products are derived from crude oil and other feedstocks. The final products are provided to customers around the world through a network of manufacturing plants, transportation systems and distribution centers. The company's major products include asphalt, aviation fuels and lubricants, base stocks, commercial vehicle lubricants, crude oil sales, industrial lubricants, LNG, marine fuels and lubricants, passenger vehicle lubricants, retail fuels, waxes, white oils, and wholesale fuels.

A significant line of business comes from natural gas development and marketing that supports production of NGLs and electric power. There is a full natural gas value chain from raw natural gas in the field to the stripping of NGLs from the raw gas stream, pipelining of both basics, followed by market products produced and processing plants, refineries, and chemical plants. ExxonMobil participates in most of these business enterprises.

While there are new products coming from the value chains of both crude oil and natural gas there are several new and emerging technologies that promise to, or already have, changed the energy landscape. Foremost among the recent technology developments is the fracking technology most prominently in evidence in the Permian Basin of West Texas and Eastern New Mexico. The development and implementation of this technology has changed the geopolitical landscape since 2005. Because of this technological breakthrough the United States has become a net exporter of crude oil for the first time since 1970, due to the rise of US oil production from a low of 5 million barrels per day in 2005, to a peak (following fracking deployment) of 13 million barrels per day in 2019. Exxon has been a major player in deploying fracking in the Permian Basin through its XTO Energy, formally Cross Timbers Oil Company subsidiary.

But ExxonMobil is contributing technology advances in many other areas. Their technology effort currently underway includes drones and robots, digital manufacturing, biofuels, and energy efficiency technologies. The company is working on technologies to reduce greenhouse gas emissions. The company sponsors an online resource called Energy Factor. This resource provides information on the ExxonMobil cutting-edge technology and innovation efforts. Current study efforts include applications to reduce carbon in the atmosphere and from power plants and industrial facilities; a technology for "soaking up" carbon dioxide that essentially vacuums the air, adding a chemical called amines that remove CO_2.

ExxonMobil available at https://energyfactor.exxonmobil.com/about-us/.

Fairmount Santrol (now Covia)

Fairmount Santrol merged with Unimin Corp in 2018 and took on a new name called Covia Holding Company. Covia is located in Independence, Ohio.

The company offers a variety of solutions for drilling, cementing, stimulation and production. Their production capacity and supply chain infrastructure enable the company to be flexible and responsive to changing requirements. Reservoirs vary in composition, and the right proppant solution will help get more out of a well. Covia's wide selection of proppants include Northern White fracturing sand, regional substrates, flowback-preventing curable resin-coated proppants and crush-resistant precured resin-coated sands alongside many other enhanced well solutions.

Covia is a premier provider of industrial materials and proppant solutions serving the industrial and energy industries. The company has an extensive network of plants and operating terminals located across North America. The company has more than 2,000 customers across North America.

Covia Holding Company available at https://www.coviacorp.com/energy/permian/ EnergyCS@coviacorp.com (September 1, 2020)

First Solar

First Solar is an international company with advanced thin film technology connected to the electric grid. The company has the distinction of being first to be certified by VDE Institute. The firm has several gigawatts connected to the grid. The technology allows customers to benefit from advanced control features that help stabilize the grid.

The testing during certification involves electrical and mechanical safety, system accurate energy yield, proper system operation and independent verification for investors and other stakeholders. The carbon emission benefits accrue because thin film photovoltaic (PV) technology has approximately half the carbon footprint of conventional crystalline silicon PV modules.

First Solar's major innovation is community solar where increased economies of scale that reduce cost, create 50% less carbon emissions than residential solar and streamlines grid integration. The company now provides more than 1 MW of solar to the community of Adkins, Texas, east of Austin.

First Solar's headquarters is located in Tempe, Arizona with North America offices in Houston, New Jersey, San Francisco and Ciudad, Mexico. South American offices are in Andar, Brazil; European and African offices are in Brussels, Dubai and Frankfurt; and Asia-Pacific offices are in Singapore, Sydney, and Tokyo. India offices are in Mumbai and New Delhi. The company has manufacturing plants in the United States, Malaysia, and Vietnam.

First Solar available at http://www.firstsolar.com/PV-Plants/Community-Solar.

Halliburton Company

Halliburton which was founded in 1919 in Duncan, Oklahoma currently has its headquarters in Houston, Texas and in the United Arab Emirates. The company has 50,000 employees located in more than 80 countries. The employees form a diversified labor force representing 140 nationalities.

The several company divisions provide several products and services to the energy industry. The company primary function is to provide services to the energy industry, helping companies maximize value from beginning to end of an oil and gas reservoir over the life cycle of the deposit. Halliburton's expertise ranges from locating hydrocarbons and managing geological data, to drilling and formation evaluation, well construction and completion, and optimizing production throughout the life of the asset.

Like many other companies that grew up with the industry during the 20th century, Halliburton began small establishing its expertise with its first research laboratories where the company tested cement mixes and began offering acidizing services to break down the resistance of limestone formations and increase the production of oil and gas. From there the company performed its first offshore cementing job on a drilling rig in the Creole Field in the Gulf of Mexico. Having established a firm reputation from this early work, the firm so became the world's most extensive offshore service.

From this early service industry experience the company soon expanded its US operation to include Canada, Europe, Venezuela, Columbia, Ecuador, Peru, and the Middle East in 1946. The company later began performing services for the Arabian-American Oil Company which is the forerunner of Saudi Aramco. Following the Middle East experience the company expanded into Argentina and Germany. The company provided all of the equipment for the first multiwall platform offshore China in 1984.

Halliburton now has 14 product service lines operating in two divisions. One division performs drilling and evaluation services while the other division offers services in well completion and production. The consulting and project management service spread across the two divisions that drives the ability to integrate the several service areas.

Halliburton Company available at https://www.halliburton.com/en-US/about-us/history-of-halliburton-of-halliburton.html?node-id = hgeyxt5y

Hess Corporation

The Hess Corporation began much like other small energy companies during the development of the industry in the early part of the 20th century. Leon Hess began the company with an oil delivery truck in 1933 near Asbury Park, New Jersey. The simple focus broadened in the years that followed adding a distribution and storage capacity focusing on providing truck fleets for delivery services for large companies. By 1948 Hess had added a 10,000 ton tanker to the operation. In 1961 the company added a refinery located in Port Reading, New Jersey, followed with gasoline stations, then in 1962 expanded with a merger that allowed the formation of Hess Oil & Chemical Corporation in Cleveland, Ohio.

A second merger in 1969 allowed the company to begin operations in Prudhoe Bay. The company name was changed to Amerada Hess. The company completed its first wildcat in 1969. This beginning soon expanded to the Gulf of Mexico and the North Sea. In 1981 the company began acquisitions in North Dakota that put the company in a position to begin fracking operations in the Bakken of North Dakota. The fracking experience soon put the company into position to carry out fracking operations in Utica Shale formation in Ohio via an acquisition and joint venture. Hess later developed a number of offshore projects under the name, Hess Midstream Partners.

Hess Corporation LP available at https://www.hess.com/company/hess-history

Marathon Petroleum Corporation

Marathon Petroleum Corporation was founded in Northwest Ohio in 1887 as the Ohio Oil Company. The company constructed its first oil pipeline in 1906, from Martinsville, Illinois, to Preble, Indiana. In 1924 the company purchased Lincoln Oil Refining Company that included

a refinery in Robinson, Illinois, and 17 Linco brand service stations in Robinson and Terre Haute, Indiana. Six years later, Ohio Oil purchased Transcontinental Oil. The deal brought oil and natural gas wells, three refineries, bulk storage plants and filling stations. It also included the Marathon product name. In 1959 the company purchased Aurora Gasoline Company, which included 680 service stations and a refinery in Detroit.

In 1962 the Ohio Oil Company officially changed its corporate name to Marathon Oil Company in conjunction with the company's 75th anniversary. That same year, it purchased the Plymouth Oil Company and a refinery in Texas City, Texas, which took the company into the wholesale gasoline business and opened a presence in the Texas market. In 1977, Marathon acquired its refinery in Garyville, Louisiana. In 1990 the company moved its corporate headquarters to Houston, while its refining operations remained based in Findlay, Ohio. In 1998, Marathon formed a joint venture with Ashland Petroleum, which added three more refineries and an inland barge fleet. On July 1, 2011, Marathon Petroleum Corporation (MPC) became a stand-alone refining, marketing and transportation company, headquartered in Findlay. MPC formed a midstream master limited partnership, MPLX LP, in 2012, of which MPC is the general partner. MPC purchased the Galveston Bay refinery in Texas City, Texas, in 2013, and in 2014 acquired Hess retail operations, adding 1,256 stores in 16 states. In 2015, MPLX LP acquired MarkWest Energy Partners L.P., expanding MPLX's midstream business into natural gas gathering and processing.

In 2018, Marathon Petroleum Corporation acquired Andeavor, which extended its operations nationwide. Included in the acquisition were 10 refineries in eight states: Washington, North Dakota, Texas, New Mexico, Alaska, California, Utah, and Minnesota. The acquisition also added approximately 3,300 retail stations and the general partner of Andeavor Logistics. In 2019 MPLX LP acquired Andeavor Logistics LP, creating a leading, large-scale, diversified midstream company.

Marathon Petroleum Corporation available at https://www.marathonpetroleum.com/ and license agreement January 18, 2021.

Occidental Petroleum Corporation (Oxy)

Occidental Petroleum Corporation or Oxy is the largest onshore oil producer in the United States but also has major operations in the Middle East, Latin America, and Africa. The company is also a major offshore producer in the Gulf of Mexico. The company has a number of subsidiaries including Oxy Low Carbon Ventures, LLC, and OxyChem. OxyChem is the producer of building blocks that put the company in a position to grow the business to focus on leading edge technologies that produce growth while also reducing emissions.

Oxy is the number one producer in the Permian Basin, Niobarra–DJ, and Uinta Basins. The company is also number one in CO_2 projects, number four producer in the Gulf of Mexico and the number one independent producer in Columbia. In addition to oil and gas production the company is number three in production of polyvinyl chloride (PVC), chlorine, and caustic soda.

Occidental Petroleum Corporation available at https://www.oxy.com/Pages/default.aspx

Phillips 66

Phillips Petroleum Company (now Phillips 66) is a US-based diversified energy manufacturing and logistics company. Phillips was formed in 1917 with headquarters in Bartlesville, Oklahoma. The company merged with Conoco in 2002 becoming ConocoPhillips. The developed over the years now includes refining, midstream, chemicals and marketing, and specialties. The midstream business includes transporting and storing of NGLs, crude oil, feedstocks and refined and specialty products, and fractionating NGLs.

Phillips 66 now has more than 14,000 employees worldwide. Phillips 66 now has its headquarters in Houston, Texas. The company recently completed a major midstream and export project on the Gulf Coast. The company is international currently with 13 refineries in the United States and Europe. The refining capacity worldwide is 2.2 million barrels of crude oil per day.

Phillips 66 available at https://www.phillips66.com/

Schlumberger

New technology in the oil and gas industry has been a fundamental driver of efficiency in the industry from the beginning. Schlumberger has been a major contributor of innovation and technology improvement since the early days of oil discovery and production. The company has been a leader in technology for reservoir characterization, drilling, production, and processing to the oil and gas industry since it began in 1920. The early contributions of Schlumberger included technology for mapping subsurface geologic structures that revealed underground features, such as bed boundaries and the direction of formation layer dips. This capability is fundamental for providing information for locating traps that contain oil and gas. The current day versions of this fundamental technology requires use of a digital foundation via the cloud, big data, user experience and cybersecurity to enhance efficiency and reliability for the customer. Schlumberger provides that capability. The firm now employees 103,000 people representing 170 nationalities with products, sales, and services in more than 120 countries.

Schlumberger's interest in innovation and new technology extends to the educational domain. The company has formed a non-profit organization to promote higher education through the Schlumberger Foundation. The Foundation purpose is to support science and technology education. The Foundation supports funding for women from emerging economies to pursue Ph. D. and postdoctoral studies in STEM fields at leading foreign universities.

Schlumberger available at https://www.slb.com/who-we-are/technology-development and license agreement January 18, 2021.

Sempra Energy

One of the most dynamic energy companies serving the US energy sector is Sempra Energy. The company with headquarters in San Diego California was formed in 1998 by the merger of Pacific Enterprises and Enova Corporation. The result of the merger was to create the largest regulated utility customer base in the United States. With operations in company employed 20,000 people and served 40 million consumers worldwide including an expansion into South America in 1999 with the acquisition of Chilquinta Energía in Chile and Luz Del Sur in Peru. Investments in Mexico followed in 2014. Sempra Energy's Mexican subsidiary initiated an initial public offering establishing a natural gas segment of pipelines, natural gas distribution and LNG, and a power generation segment of wind and solar facilities.

A major expansion of Sempra's holdings took place in 2018 with the acquisition of Energy Future Holdings Corp of Dallas, Texas. The merger included a $9.45 billion investment of 80% interest in Oncor Electric Delivery Company LLC, the largest of the Texas electric utilities. This acquisition created the largest US customer base in the nation.

Sempra expanded in other ways as well. International trade in LNG ramped up brought on in major part by the fracking addition to US oil and gas supplies. In May 2008 Energía Costa Azul in Baja, California, began commercial operations, making it the first LNG receipt terminal on the West Coast at a location about 15 miles north of Ensenada. The terminal can process up to 1 billion cubic ft. of natural gas per day.

In 2013 Sempra, joined with Mitsubishi, established a joint venture to set up the financing and construction of a LNG export facility at Cameron LNG import terminal in Hackberry, Louisiana. Sempra's Mexican sub subsidiary signed long-term contracts with Valero Energy Corp to create storage capacity for the liquid fuels marine terminal. The terminal was to be constructed in Veracruz with island storage facilities located in Puebla and Mexico City.

Sempra has also recently carried out a major program under incentives provided by the California Public Utilities Commission to install thousands of electric vehicle charging stations at businesses and multifamily communities. The program is supported by state subsidies for electric rates.

Sempra Energy available at https://www.sempra.com/history#time-line-slide-out-1374

The major companies reviewed above provide important perspectives about the extent of energy sector innovation at home and abroad. While US government policy supporting the sharing of technology abroad are important, it seems clear that positive innovation outcomes carried abroad by our American companies is likely the most important way for innovation to spread across the globe. All of the companies reviewed are doing business in the energy sector in the United States with activities in and/or headquarters in Texas. The focus of the companies universally has been to take their US-based experience and put it to work in countries around the world, thus spreading the innovations to other countries.

Fundamental to the creation of shared innovation outcomes is the process that generates ideas under the banner of freedom—freedom to go where "no man has gone," freedom to venture investment under protection of a legal system that allows for equitable treatment through the bankruptcy courts if the venture fails, and freedom to talk your neighbors into joining your dream. Spindletop discussed in chapter 3, A game change at Spindletop followed that American model. Texas at Austin, Bureau of Economic Geology, thematic map SM 10, map scale 1 inch = 100 miles.

CHAPTER 3

A game change at Spindletop

Chapter outline

The event

The modern era of oil as the world's dominant energy source began with the discovery of crude oil in Pennsylvania in 1859, but events that launched this industry into the 20th century economy began with a single well drilled at a place known as Spindletop Hill in the southern edge of Beaumont, Texas on January 10, 1901. The well was the by far the greatest producer the world had ever known. Spindletop is a salt dome oil field along the Gulf Coast of Texas. The well that "blew in" on a January morning "shattered the pipe, spewed oil over 150 ft. out the top of the rig, drenching the countryside at an estimated rate of 100,000 barrels per day" (Goodwyn, 1996, p. 19). Nearly a million barrels of oil was lost before the well was brought under control by a device known as a "Christmas Tree." The photo in Fig. 3.1 of the well is notorious.[1]

Spindletop was the outcome of the effort of an entrepreneur named Patillo Higgins of Beaumont, Texas. The success of Higgins' well at Spindletop was like a "shot heard around the world." Men, equipment and dollars flowed into Beaumont much like that of a gold rush in California or Alaska in the late 1800s. By 1902, 285 wells were in operation at Spindletop (Wooster and Sanders, 2009; Handbook of Texas Online). Beaumont grew into a boom town overnight.

But a lot of things had to fall into place before this new industry catapulted into the modern world by Spindletop could be fully functional. There were pipelines to

---

[1] Image is available on the Paleontological Research Institution web site. The site credits the American Petroleum Institute; however, a Texas roadside historical marker credits John Trost (June 24, 1868–August 4, 1944). Original photo by John Trost. w-en:Image:Lucas_gusher.jpg by Nv8200p, 4/12/2006 16:29 (UTC), Public Domain, https://commons.wikimedia.org/w/index.php?curid = 3425193/.

Innovation Dynamics and Policy in the Energy Sector
DOI: https://doi.org/10.1016/B978-0-12-823813-4.00003-8

Figure 3.1 Spindletop: Lucas Gusher. *Courtesy the American Petroleum Institute.*

build, refineries to turn raw crude into usable products, retail outlets (service stations) to move product to market, and so on. So it is understandable that things did not really get moving until the East Texas Field was discovered in 1930 north of Spindletop at Henderson in deep East Texas.

The effort that finally paid off at Spindletop began with the vision of Higgins who through pure grit pursued his hunch that oil was below. Prior oil discoveries typically began because oil could be seen seeping to the surface but Higgins was sure that oil lay far below the surface at the spot known as Big Hill south of town. He found a spot on Big Hill where natural gas seeped to the surface so he was sure that oil lay below in the salt dome. He began promoting the idea of discovering oil among the town's leading citizens and got local financial backing and formed a company known as Gladys Oil Company. The plan was to drill 1000 ft. deep to where Higgins believed he would strike oil in the salt dome below. The company began drilling the first well in 1893, but it failed, as did two more in the months ahead.

Higgins lost the support of the locals to continue, so he turned to a former navy man named Captain Anthony Lucas, who was an engineer and had knowledge of salt mines in Louisiana where he had encountered a measure of sulfur and oil associated with the salt. Lucus invested in a lease at Big Hill and began drilling only to encounter the same soft-sand formation challenges that Higgins reported. By this time there were four failed wells at Big Hill and both men were out of money to continue.

Neither Higgins nor Lucas was willing to give up the fight. Lucas attempted to gain financial backing from Standard Oil, but failed. Standard's experts were not convinced that oil was present in the salt dome. Lucas, armed with a favorable scientific report from a University of Texas geologist, then gained the interest of an investor and drilling company named Guffey and Galey with oil finding experience in Pennsylvania and Kansas. After extensive negotiations Guffey and Galey reached a deal with Lucas, who in turn kept Higgins in the agreement with a small share of the play if they succeeded. Being the experienced oil finders, and given the dire straits of Lucas, Guffey and Galey took seven-eighths of the deal, but both Higgins and Lucas kept a stake in their dream. Guffey and Galey agreed to fund the work with $300,000 to drill a single well, but should it fail the venture would end. Having the agreement with Lucas in place, Guffey and Galey then turned to Andrew Mellon, a New York banker to provide the funding (Goodwyn, 1996, p. 16).

The drilling proceeded but ran into several challenges that might easily have rendered the well a failure but for the energy and thoughtfulness the last ditch effort brought to the team. They figured out how to drill successfully through the soft sand that stopped the earlier wells, and then they figured out how to drill through large strata of loose rock and gravel and finally hit pay-dirt at 1020 ft. To the amazement of the play developers, and to the world in the days to follow, the well blew in by first cascading thousands of pounds of drill pipe out the top of the derrick, followed by a volcano of mud, rocks, gas, and oil sent hundreds of feet into the air. The oil drenched the countryside in a black mist that continued unabated for days before it could be brought under control. The flow rate was perhaps at 100,000 barrels per day and continued for 9 days flooding the surface with 600,000 to 1 million barrels of oil. This well was by every measure the largest oil well the world had seen and became known as Colonel Guffey's great gusher in Texas (Goodwyn, 1996, p. 19).

The rush of investors, drillers, landmen, and others brought drilling on postage-stamp sized leases. Every well drilled on the salt dome produced oil. It made the California gold rush look like a picnic. The discovery brought sightseers, fortune seekers, and oil field workers. Land that had sold for $10 per acre now went for $900,000 per acre. Beaumont's population grew over night from 10,000 to 50,000 and upward of 10,000 people moved in to tents on the hill. The legal system of fundamental property rights tied to the land area, the worldwide stories of the discovery, and the flow of financing to all parts of the enterprise brought production up so fast that the reservoir pressure declined and Spindletop petered out within 2 years. But in the meantime the production from the area flooded the oil market. Soon after Spindletop blew the price in some boom towns dropped to as low as 3 cents a barrel while water sold for 5 cents a glass (Goodwyn, 1996, p. 23).

Spindletop was a "flash in the pan" in the crude oil world but the event found its way into the history books for its influence on the industry development. The innovation contributions made their mark and when combined with related events during the next 30 years transformed the way oil was discovered and produced. The technological advances developed by the Higgins/Captain Lucas team described in the following section amounted to innovations that would help bring the petroleum era on to the world stage, and by 1930 the Spindletop innovation achievements made Texas the leading producer in the United States, a title she continues to hold today. In addition to the technical innovations, the other innovations that grew out of Spindletop included business model and institutional changes and began the recognition that policy changes were essential to the stability and long term economic contributions of the energy sector, including but not limited to oil recovery.

The Spindletop innovations

Technological innovations

Spindletop produced two technological innovations that finally allowed the well to be completed, which then became essential technology of all oil drilling rigs to follow. Drilling mud was the first technology addition, followed by a newly created back-pressure valve. Rotary drilling that was applied here by the Hamilton brothers was not a new innovation introduced at Spindletop, but it was the first significant application—a necessary, though not sufficient technology for success.

Patillo Higgins, who was the instigator of the project, having exhausted all local prospects for investors to support the effort, advertised and found a "believer" to support the drilling. Higgins had already sold all of his rather large lease acreage at Spindletop (called Big Hill) retaining a mere 35 acres of his leasehold. Captain Anthony Lucas, an engineer with salt mining experience in the Louisiana bayou area responded to the Higgins newspaper ad for partnering. He leased 600 acres at Bill Hill and Higgins joined him in the drilling effort by giving over his 35 acres for a 10% stake in the joint effort. They brought in a rotary drilling rig operated by the Hamilton brothers of Louisiana who had gained rotary drilling experience in the Corsicana play of north-central Texas, and they knew how to put extra-large casing in the well allowing water to be circulated to cool the drill bit. When the drilling hit the soft sandy formation, which had stopped Higgins' early attempts, the team tried an experiment with drilling mud—a thick liquid substance generated locally in a water pit. The mud was piped down the well bore and it sealed the well against cave-ins in the sandy formation.

But the mud failed to work when the drilling hit a layer of loose gravel and sand. The team then forced 8-in. pipe down the hole to hold the well against cave-ins in this strata. But then another problem arose. A natural gas pocket blew mud and sand back up the hole. The solution was the second innovation created on the fly by Lucas—a back-pressure valve. The well was completed a few days later when the unprecedented blow-out occurred, sending pipe, drill steam, mud, and oil into the sky. History was made and Spindletop found its way into the history books. But the two technological innovations—drilling mud and a back-pressure valve—were not the only innovations growing out of Spindletop (Goodwyn, 1996, p. 179).

Business model innovation

The business model begun at Spindletop integrated the talents and interests of wildcatters, drillers, investors, bankers, royalty-owners, landmen, roustabouts, scouts, roughnecks, land-owners, and professional engineers and geologists. The engineers and geologists came from both private practice and at university engineering and geology departments. It also cemented the important industry divide between independents and majors, an industry structure that many believe was critical to the progress of innovation by a special class of risk-taking entrepreneurial investors.

The important divide of industry roles between independents and majors drove the debates that followed which became a significant factor in the regulatory system that developed in the years following Spindletop. The recognition of the importance of Independents became a major policy driver at the Texas Railroad Commission discussed in some detail in what follows.

Institutional innovation

The main concern of public policy initiatives in the era leading up to Spindletop was anti-trust legislation at both Federal and State levels. The concerns about competition that resulted in the breakup of Standard Oil in the 1890s led the way. But the need for regulation of the industry was not in play until after the turn of the century, and although Spindletop did not itself generate a compelling need to regulate, the experience made it very obvious that unfettered markets built on basic property rights and the rule of capture, tied to the surface area of land, would led to massive "waste" of the oil and gas resource. The experience that would result in the depletion of the resource at Spindletop within 2 years, from wells drilled often on "postage-stamp" tracts of land, set the stage for the development of a regulatory system that ultimately resulted in well-spacing and market-demand prorationing. This regulatory system was developed in the 1930s following discovery of the East Texas Field, an event that made the chaos of the Spindletop experience look like child's play. (This is the second major innovation experience discussed in Chapter 4). Finally, and perhaps most importantly, Spindletop created an enhancement to the cultural attitudes of risk taking that was the beginning of industry's "independent" oil men. This institutional industry structure—Majors and Independents—was well on its way in the Spindletop era.

In summary, there are several aspects of the Spindletop story that instruct us about innovation—aspects the importance of which go way beyond the oil production that flowed from this well. Innovation came in several important ways. First, there was

Higgin's "idea" based more on his intuition than any objective data—the way ideas often emerge. Oil seeps coming to the surface were common in Pennsylvania and other older regions but were not proven to indicate large deposits of oil. Second, there were two technology contributions that allowed the well to be completed following several failed drilling attempts. Third, the business model at the time used the ability of an informal "venture capital" system for funding risky ventures. It would take several decades to develop an institutional system of venture capitalists that we know today. But from the Spindletop experience there was an important contribution to the business model for finding and producing oil. Fourth, and perhaps most important of all, Spindletop created an enhancement to the cultural attitudes of risk taking that was the beginning of industry's "independent" oil men. This Spindletop experience was instrumental in structuring the industry into two important factions—the Majors and the Independents. Finally, the most important innovation was institutional. Spindletop began an awareness that the uniqueness of oil production in a completely unfettered market would led to great inefficiencies and "wasting" of the resource that could only be corrected through innovation in political, legal, and regulatory domains, mostly limitations on the institution of property rights that finally came to pass after the East Texas experience.

While the technology and economic promise of major oil production that would transform the Texas economy did not emerge until 29 years after Spindletop, the completion of this one well set up the structure of market participants. The labels of participants that began here remain today, including labors known as "roustabouts" and "roughnecks" working around the equipment called "drilling rigs," businessmen known as "drilling contractors," "landmen" who negotiated and set up drilling and leasing arrangements, farmers and ranchers who participated by trading royalty payments for providing access and the right to drill on their land, and "scouts" who searched out and found individuals willing to lease.

In large measure Texas university petroleum engineering and geological sciences became disciplines associated with the set of industry players listed above. For example, the Bureau of Economic Geology at the University of Texas was formed in 1909 and became the Texas geologic survey which the industry relied on for guiding exploration (Bureau of Economic Geology, 2017 at http://www.beg.utexas.edu). The Santa Rita well drilled on University of Texas land in the Permian Basin in 1923 generated the formation of the Petroleum Engineering Department at UT Austin in 1930 (The University of Texas, 2017 at http://www.pge.utexas.edu). The first geology classes at Texas A&M University began in 1903 as a part of the Chemistry Department; today's Department of Geology and Geophysics was formed in 1994 and is housed in the Michel T. Halbouty Geosciences building complex, a facility named after this alumni and well-known Texas wildcatter (Texas A&M University, 2017 at http://geoweb.tamu.edu). The College of Engineering at Texas Tech University has existed since 1925 while the Department of Geology was established in 1926 (Texas Tech University in Lubbock, 2017 at http://www.geosciences.ttu.edu/ and http://www.depts.ttu.edu/coe). Perhaps the most important research work following Spindletop was the focus of geology expertise in identifying structures that were promising for oil deposits. Because of Spindletop, Texas universities now have a hard focus on the science of finding oil.

The early interplay of policy, markets, and technology

The Spindletop event was the primary contribution of a **culture** that drove a few risk-taking individuals acting alone at first, then combining their visions until success finally resulted in an outcome that exceeded their wildest imaginations. These individuals were high risk-takers willing to "bet the farm" on the prospect that they would hit oil. Their enthusiasm was contagious enough for them to obtain funding from outsiders after exhausting their own resources, and that of their neighbors around Beaumont.

The wildcatters at Spindletop faced a great incentive—the prospect of failure. It is often said that "necessity is the mother of invention." That was certainly the case at Spindletop. The drilling team first solved the well cave-in problem in soft-sand strata by the innovation of drilling mud that plastered the hole against cave-ins. But a few days later the drilling mud did not prevent the cave-in in a rock and sand strata, raising the near certainty that the well would fail. Guffey, faced with the prospect of failure and the certainty that his funding source would not support another well attempt, developed the back-pressure tool that allowed the well to go deeper to "pay-dirt" (Goodwyn, 1996, p. 17).

The **market** for crude oil in the wake of Spindletop had several complicating dynamics. First, there was the leasing of acreages which allowed drilling for oil. The completion of the first well brought investors from everywhere wanting to get a piece of the action. A woman sold her pig pasture for $35,000 and soon land that had sold for $10 per acre 2 years earlier now went for $900,000 per acre. Within months, 214 wells had been drilled on the Hill owned by a hundred different companies. Big money and those with political power followed. One Joseph Cullinan, a former employee of Standard Oil's pipeline business, left their employee and formed his own oil company in Pennsylvania. Cullinan made a visit to Corsicana, Texas in 1897 and then decided to settle there. When Spindletop broke in 1901 he rushed to Beaumont, where he had already built a nearby oil storage facility—he had a specialty in oil purchasing and marketing. Soon he gained control of valuable oil leases and created a syndicate led by the former Texas Governor James Hogg. Part of Cullinan's strategy was to keep Standard Oil from using their familiar strategy of control of production tied to downstream parts of the industry; pipelines, refining, and retail outlets. Cullinan then brought in New York and Chicago money sources to finance development of the leases. Cullinan's Texas Company using his storage facilities, sold oil for 65 cents that he had bought at 12-cents per barrel. Smart money wrote good contracts and James Guffey, a successful wildcatter from Pittsburgh with experience in Corsicana, organized the drilling of

Spindletop. Following the success of the well he negotiated a contract with the new firm, Shell Transport and Trading company in London. The contract called for Shell to take at least half of Guffey's production at a guaranteed price of 25 cents per barrel (Yergin, 1991, pp. 84–86).

The Spindletop experience resulted in a glut of oil in a limited consumer market causing the market price to plunge to $0.03 per barrel (Yergin, 1991, p. 89). It would be one of several events in years to come that the wide swings in price would wreak havoc in oil markets.

The public policy that made Spindletop possible was the rule of capture. Secondarily, the antitrust laws that were responsible for breaking up Standard Oil in the 1890s caused the majors to be more conservative in their deals than were the independents. This left a gap that a growing number of independents filled. Spindletop and the other major oil field discoveries that followed over the next 30 years resulted in such turmoil and wide price swings in the oil market that limits were imposed on the rule of capture lead by the Texas Railroad Commission and defended in the courts.

Impacts of the Spindletop innovations

Spindletop came about as the product of Higgin's "idea" based more on his intuition than any objective data—the way ideas often emerge—**and a culture** of risk-taking individuals, operating in a framework of the rule of capture **policy** supported by a **market** price that made the prospect of great wealth possible. But the prospect of failure lead to **technological innovation** that created the first great Texas oil success of the century. Spindletop also was responsible for creating the industry structure (**institutional innovation**) that would bring the talents of a dozen forms of labor and capital together. It also set the stage for a revision of the policy structure to bring a chaotic market to stability some 30 years later.

While it would not be correct to attribute all of the future oil and gas industry development to Spindletop the event certainly had a major influence on the future Texas economy. The economic impacts of oil market development on Texas employment and gross domestic product (GDP) during the century that followed has been tremendous. The industry, including the combined value of products in oil, natural gas, electricity, pipelines, refined products and chemicals registered 26% of the Texas GDP in 1981 and about the same share again in 2013. Exploration and development of oil resources most certainly would have continued without Spindletop but the event was like a magnate drawing investment and exploration activity to Texas in the immediate aftermath of the discovery. The invention of down-hole casing and drilling mud followed by the back-pressure valve made Spindletop happen. These tools and the industry support system of wildcatters, drillers, investors, bankers, royalty-owners, landmen, roustabouts, scouts, roughnecks, land-owners, and professional engineers and geologists, can, in significant part, be attributed to Spindletop.

CHAPTER 4

From chaos to order in the East Texas Field

Contents

The event

The culture of risk-taking wildcatters that produced Spindletop had the most dramatic impact in East Texas near Kilgore in what became the East Texas Field in 1930. The East Texas experience was developed following the enthusiasm created by discoveries in several areas of the State, especially from early successes in West Texas. The earlier discovery of oil at Corsicana, Texas in 1894 created a good deal of interest in Texas oil prospects, but the wells were small producers. Spindletop in 1901 was followed by major wildcatter efforts that produced other historic strikes. The main ones included a discovery well in West Texas in 1920, followed by a well named Sana Rita No. 1 at a place called Big Hill in Mitchell County in 1923. The following wells, about 70 in all, drilled on University of Texas land, opened up a major find in West Texas known as the Yates Field. The well, Yates No. 30A was the most productive well in the world as of 1929 producing 204,682 barrels per day (Goodwyn, 1996, p. 29).[1] These discoveries formed the beginning of the Permian Basin oil and gas contributions to the Texas energy story, the beginning of the Permian Basin events that are the focus in Chapter 6, West Texas and the Permian Basin Early Innovations.

The well that began the historic discovery of the giant East Texas Field was the Daisy Bradford No. 3, so named No. 3 because it followed the first two wells that

---

[1] The Yates development ultimately contributed major funding to an endowment fund called the Permanent University Fund (PUF) created by the Texas Constitution of 1876 (Texas Legislative Council, Research Division, 2015). The assets of PUF at first were land grants in West Texas. The discovery of oil in West Texas on University land on this land grant made the University of Texas one of the best endowed universities in the U.S. (Permanent University Fund Financial Statements, Independent Auditors' Report, 2015, p. 8.).

Innovation Dynamics and Policy in the Energy Sector
DOI: https://doi.org/10.1016/B978-0-12-823813-4.00004-X

failed. Daisy Bradford was the lease holder where the wells were spudded. The Daisy Bradford No.1 was lost to a stuck pipe and, the other, the Daisy Bradford No. 2 was lost to a twisted off drill pipe. The Daisy Bradford No. 3 was drilled to a depth of 3592 ft. in the Woodbine sand. The initial production was measured at 300 barrels per day. The exploration that followed the completion of this well quickly established the fact that the formation was not an isolated pocket of oil but the largest single reservoir ever discovered in the world. This giant field was later determined to contain 5.5 billion barrels of oil (Clark and Halbouty, 1972, p. 109). The giant East Texas oil field extended into parts of Smith, Upshur, Gregg, Cherokee, and Rusk counties, ultimately with 30,340 historic and active oil wells. Its size is roughly 45 miles north to south, and 5 miles wide. It became the largest oil field in the United States outside of Alaska at the time (Joinerville Texas and Daisy Bradford Discovery Well, 2017).

The East Texas Field became the economic base for Tyler, as well as Longview, Kilgore, Marshall, Henderson, Gladewater, Van, Jacksonville, and several other cities and towns in the area. But the economic boom was quickly to become a market out of control. The rule of capture and the active market for leasing of mineral rights tied to the surface area created nothing short of chaos. Lease prices went from $10 per acre to $1,000 per acre "overnight." The market for drilling rigs and supplies of all kinds skyrocketed as well. Disagreements and lawsuits multiplied as owners of production and royalty interests tried to protect their share of the profits.

The credit for the discovery of the East Texas Field was a 70-year-old wildcatter named C.M. "Dad" Joiner. Joiner believed oil was present in the Woodbine formation some 3600 ft. below the surface in Rusk County. As was the case for both Spindletop at Beaumont and Big Hill in West Texas, Joiner spent every dime he had along with all the money he could raise, at first in Overton and Henderson, then in Dallas and Houston. He sold smaller and smaller shares, used volunteer labor, and ran up credit balances all around before finally completing the well.

The story of Daisy Bradford No. 3 soon spread throughout the region and finally got the attention of scouts of the major oil companies. Before the majors got serious, however, new wells were completed by other independent oil men, nearby at first, then further away eventually covering five counties. It took some time for the knowledge of the extent of the field to develop. But as each new well was completed, all in the Woodbine sands, it became clear that the East Texas Field was in fact a very large pool of oil, larger by far than anything the oil industry had produced anywhere in the world (Goodwyn, 1996, p. 32–39). (The Fig. 4.1 map below shows the location). As related by Prindle, "drilling rigs were hauled in from everywhere possible, and wells were put down in pastures, near front porches, in flower beds, and over parking lots" (Prindle, 1981, p. 25).

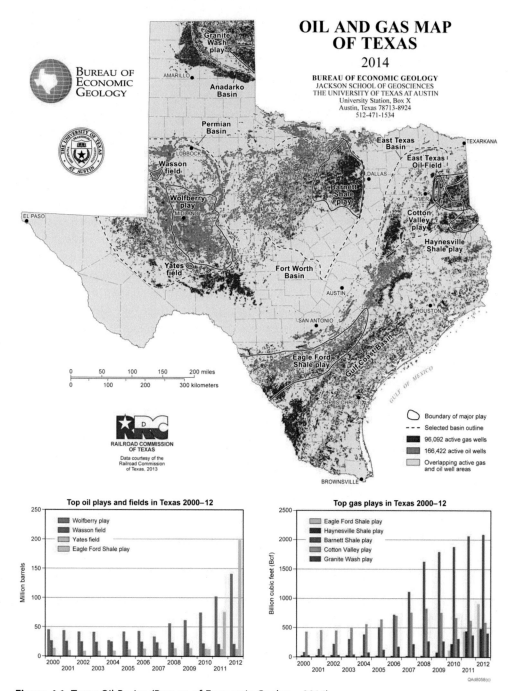

Figure 4.1 Texas Oil Basins (Bureau of Economic Geology, 2014).

Since crude oil and natural gas can flow within reservoirs from beneath one land property to another, it is the case that one landowner can pump a neighbor's oil and gas because it is impossible to distinguish one owner's oil and gas from the others. The basic rule of petroleum ownership and production then became the "rule of capture." In order to protect their mineral property, the neighbors must also pump or risk losing the asset (Prindle, 1981, p. 23).[2]

The reservoir in the Woodbine formation is what is known as a water-drive reservoir. In the East Texas Field water pushed the oil into the eastern stratigraphic trap at about 16,000 pounds per square inch (Prindle, 1981, p. 27). Oil pumped too rapidly from such a formation will result in water pushing through the oil leaving oil in place that cannot be recovered. Controlled pumping rates allow the water pressure underneath the oil to continue efficiently recovering the oil deposit. Knowledge of the eventual outcomes of unfettered production under the rule of capture soon spread throughout and the understanding was reinforced by the continual decline in actual reservoir pressure.

Reservoir engineers and geologists knew that the rapid rate of production from the field would drastically reduce reservoir pressure leaving large quantities of oil unrecoverable. Such rapid pumping raised an additional problem. Crude oil exists in formation under pressure from either water or natural gas, known respectively as a "water-drive" or a "gas-drive" reservoir. Although the oil flows into a depression formed by pumping or by lowered pressure from the well bore, the viscosity of oil is relatively high and the flow rate is relatively slow. If the pumping rate is too rapid, the water or gas breaks through the oil and moves to the well bore and the maximum oil recovery from the reservoir over time may be drastically reduced. That is, the natural drives may dissipate, leaving much of the oil trapped underground, possibly forever (Prindle, 1981, p. 6).

At the end of 1931, given the level of pumping and loss of down-hole reservoir pressure it was estimated that the reservoir would ultimately produce one-billion barrels of oil. Following the eventual implementation of "prorationing to market demand" which preserved down-hole reservoir pressure (discussed in detail later in the chapter) the estimate in 1942 was 6 billion barrels (Goodwyn, 1996, p. 54).

Before long the declining pressure issue became the focus of the Texas Railroad Commission (TRC) as supporters pushed for prorationing of production. But the pathway to eventual success toward prorationing was a long and sorted one, perhaps the most eventful interplay of courts, legislatures, regulation and internal industry-conflicting

---

[2] The rule of capture is a doctrine based in English common law. Originally, the idea in England was that if a game animal or bird from one person's estate migrated to another, the owner of the latter had a perfect right to kill the game. American judges, when first confronted with the oil ownership problem, reasoned that this problem was fundamentally the same as the English game problem (Yergin, 1991, p. 32).

Figure 4.2 East Texas Oil Field.

interests the country had ever faced.[3] The property rights legal framework of the rule of capture prevailed when the pressure to generate revenue to service debt became essential following 1930; the courts did not support attempts of government regulation to enforce production controls. Land owners with accompanying mineral rights tied to the surface area had every incentive to sell leases of very small acreages resulting in wells placed back-to-back across the countryside (see Fig. 4.2). Once the production rates were finally controlled in 1942 and the practice of reinjecting salt water into the reservoir implemented, reservoir pressure actually increased.

[3] The Railroad Commission had authority under conservation statues to discipline the industry to prevent waste but the agency was caught in a web of competing interests in the industry, the result was changing legislation and repeated constraints imposed by the courts (both federal and state courts). In addition, the Commission lacked the staff expertise to understand the unique behavior of this water-drive reservoir not experienced anywhere else. Furthermore, the attempts to design rules and implement them lacked hard data to support the enforcement of rules. The potential solution to the challenge was to "unitize" the oil field. If the reservoir was essentially under one unified production program under unitization, the optimum rate of production could be defined and implemented (Prindle, 1981, pp. 25–39).

The confluence of the rule of capture, the wildcatter's successful discoveries of major oil reservoirs, and new small refiners all pushed the supply of oil and oil products up rapidly. While the demand side was also growing due to the growing number of automobiles and the beginning of an airline industry, consumption and exports did not match the increase of a supply surge brought on by East Texas, especially when adding in the major contributions from Oklahoma and Louisiana. Standard Oil lost its control over production that had kept crude oil prices somewhat stable (Yergin, 1991, p. 95).

The East Texas Field induced innovations

The innovations that evolved out of East Texas during the early 1930s were among the most interesting in the oil industry history in Texas. The unprecedented rate of production, fed by the property ownership rule of capture, created a climate of chaos. Participants scrambled to "calm the waters" on the one hand, but grab their share of the bounty on the other. Producers, royalty owners, oil field service providers, government regulators, the US Congress, the Texas Legislature, as well as Texas and Federal courts, including the US Supreme Court, tried (depending on their role in the matter) to protect legal rights, change the statutory law, provide regulatory rules, generate revenue to service debt—or simply to grab their money and run. But each move by one group brought up a response from others, and various forms of innovation grew out of this dynamic.

The East Texas Field and TRC policy development was a difficult period as the process worked out a system of well-spacing and prorationing to market demand. But the already difficult process was made even more challenging by the "hot oil" conflict and the belief that the reason for production controls was to set the price of oil above competitive market levels. The Texas statutes that empowered the TRC to act were based on a somewhat ill-defined concept of "conservation" of the oil resource. The rules that developed actually did both things—increased the efficiency of total recovery from reservoirs, and set the oil price high enough to support small producers (independents) that otherwise would have been shut out of the market.

The pressure for policy change was immense. Production allowed under the rule of capture added to the illegal production, called "hot oil" flooded the market sending prices through the floor. "Hot oil" was sometimes simply production beyond what the TRC allowed, and at other times actually stolen oil. Oil men ignored legislation and regulations of the TRC—regulation they believed to be unconstitutional—which kept the feeding frenzy going until drastic measures were called for. Governor Ross Sterling declared martial law and sent in the National Guard to restore order in August of 1931. US Interior Secretary Harold Ickes sent in Federal agents to East Texas in the fall of 1933, and thus helped the TRC shut down hot oilers. Ickes acted under an Executive Order of President Roosevelt. He relied on the National Industrial Recovery Act, which made it illegal for pipelines to transport oil in interstate commerce if the oil was produced in violation of state statutes. This approach by the Federal government avoided the problem of the Feds overriding the state regulatory bodies and state legislatures by relying on the Interstate Commerce clause of the US Constitution, a power that is the clear purview of the federal government (Prindle, 1981, p. 37).

The primary institutional innovations from the East Texas event were built on two Railroad Commission rules that eventually disciplined the rule of capture. Well spacing (Rule 37) was the first and prorationing to market demand was the second. While the technologies of drilling and production were perfected during the 1930s following Daisy Bradford No. 3, by far the most important innovation flowing from the East Texas event was institutional. It took more than 10 years for the industry, the regulators, the legislatures and the courts to work through the conflicts and to perfect a new order. But in the end the regulatory system of prorationing to market demand was institutionalized, leveraging the well-spacing rule, and was effective in not only enhancing the ultimate recovery of oil from East Texas, but the practice was implemented statewide. Like a one-two boxing hit, prorationing was implemented in conjunction with the well-spacing rule (Rule 37).

The spacing rule had been in place since initiation in 1919, but was so poorly enforced that it was of little effect until East Texas (Goodwyn, 1996, p. 54). The well-spacing rule was hard to implement—property owners and their drillers simply ignored the order. But the prorationing policy was even more of a challenge.

Part of the disagreements among the industry, regulators, and the courts were fed by an indirect effect of prorationing and well spacing. The effect of, if not the intent of the TRC implementation of the rules, was the control of the price of oil above production cost—an outcome many believe was the primary, unwritten objective of prorationing. The influence was so great that the TRC prorationing system set the world oil price, because in the 1930s and 1940s Texas produced about one-half of the US supply and the United States produced the majority of world output—70% in 1925, and 63% in 1941 (Encyclopedia of the New American Nation, 2017).

As the TRC tried to implement prorationing, agents were sent into the field to turn off wells that were exceeding their allowable production. But creative producers, convinced of their right to produce, installed left-handed cut-off valves so that the agent attempting to turn off a well was actually opening it on full bore. Other producers hired truckers to haul during the night-time hours to move oil from the wells to market without detection. Still others installed buried pipelines for the same purpose (Goodwyn, 1996, p. 42).

The extraordinary production from East Texas also created another challenge—the massive production of salt water that required disposal. Salt water production also caused the reservoir pressure to dramatically decline threatening the long term recovery of oil from the reservoir. By perfecting the technique of **reinjecting the saltwater** for disposal into the reservoir, the reservoir pressure was recharged, and the saltwater disposal problem was solved (Goodwyn, 1996, p. 54). The practice continues today.

A third institutional innovation flowing out of East Texas was the creation of the Texas Independent Producers and Royalty Owners Association (TIPRO). The interindustry divide between the independents and the majors solidified during the decade following Daisy Bradford No. 3. TIPRO became the organization that has developed and kept the interests of independents, and their important wildcatters in the dialogue of public policy and has provided a platform for organized political activity ever since. In almost every debate the often differing views of the independents and majors are kept alive by this organization.

The practical outcomes of the TRC intentional practice and policies that protected the small guy—a balance between production limits that stabilized the world price of oil above competitive levels and special provisions that protected the independent producers, land owners, and royalty owners—is exemplified in the data that illustrate the importance of the policies that maintained a viable set of risk-taking independents. Of the 241 giant oil fields discovered in the United States containing a hundred million barrels or more, through 1963, 122 or 50.6 percent had been discovered by independents. Fifty percent of US reserves and yearly production from these fields came from independents (Prindle, 1981, p. 194–195).

A fourth institutional innovation came as industry representatives pushed for federal intervention. The hot oil problem became such a challenge that some proponents of federal controls were willing to create a federal "oil czar." The National Industrial Recovery Act of 1933 contained a section outlawing hot oil. But the US Supreme Court declared the provision unconstitutional. A follow-up statute, the Connally Hot Oil Act, was then passed making it a violation of interstate commerce to transport illegally produced oil. The outcome of this series of events was the creation of the Interstate Oil Compact Commission (IOCC). The IOCC an entity still active today, armed with the Connally Hot Oil Act, helped make state regulation possible.

Another innovation that might have happened, but in the end did not, was compulsory unitization. Such a policy was widely supported by some majors, and supported by economists and engineers. Such a policy would make a field a single unit, maximize the recovery of oil, and minimize the investment, vastly reducing the number of wells and pipelines needed to drain a reservoir, which was especially important at the point of implementing secondary recovery operations requiring the use of some existing wells for water or gas injection. The East Texas event effectively ended interests in unitization, and institutionalized regulatory well spacing and prorationing as the policy of Texas. Other producing states followed suit.[4]

---

[4] Compulsory unitization became a possible solution to the excess production and waste problem with support from the majors and large independent producers. The small producers who had a large stake in continuing under the raw openness of the rule of capture understandably opposed a policy of unitization, and the courts tended to support them. Conservation of the resource and an associated maintenance of reasonable market prices that would potentially flow from unitization were in conflict with the rule of capture. Technological innovation would not solve this problem; only innovation in the form of policy and political compromise with support from the courts would provide the way forward.

The interplay of policy, markets, and technology

The dynamic interplay of policy, markets, and technology brought on by the East Texas Field was something to behold. The legal production of oil allowed under the rule of capture expanded as producers exceeded allowable levels. As both legal and illegal production exploded in East Texas the oil price was driven to $0.03 per barrel and lease prices to $1000 per acre. Such rapid production created the threat of wasting vast quantities of oil left in the reservoir.

A full decade passed before the Texas Legislature and the US Congress, the courts (both Texas and the United States), a maturing of the private industry (the firm participation of both Majors and Independents), and the TRC regulators stabilized the oil market. Once the process evolved from 1930 to 1942, the market was stable, the TRC was firmly in charge of implementing the well-spacing and prorationing policies and the courts acceptance of the constrained rule of capture that limited the domain of property rights.

In summary, the East Texas event created incrementally improved technologies of drilling, pipelining, refining, and the reservoir engineering and geological expertise put in play at Spindletop. The major technological innovation added by East Texas was the method of **salt water disposal by reinjection into the reservoir** that not only solved the disposal problem, but enhanced the reservoir pressures allowing more efficient recovery of oil. But **the most important East Texas innovation was institutional and policy**. The rule of capture was overlain with a new legal and regulatory system of production controls supported by well-spacing and reservoir-level production limits, followed by statewide prorationing to market demand. New and more powerful institutions changed the playing field and brought stability in production and prices that prevailed until the Arab Oil Embargo of 1973. The power of the TRC was solidified, supported by, rather than replaced by, federal authority. The early interest in an organization that became TIPRO was generated by the East Texas event. Built on interest growing out of East Texas, TIPRO was finally organized in March 1946. The final incentive for forming the organization was a push-back response to an attempt by the majors to set up a world organization called the Anglo-American treaty to control and allocate production around the world (Goodwyn, 1996, p. 61).

Finally, another institutional innovation flowing from the East Texas experience helped unify policy agreement in the United States. The IOCC was formed in 1935 and became a permanent institution aiding stability in the interindustry balance between majors and independents, and the balance between state and federal regulatory jurisdiction. In the meantime a separate approach to oil development and production in the other world markets, especially markets in the Middle East, was in progress. The innovations in the United States and the Middle East in the decades following East Texas are the subject of the following chapter.

Impacts of the East Texas innovations

The East Texas events created market disruptions so pervasive that it became obvious that a competitive market for oil based on property rights defined by the rule of capture could no longer serve the interests of the industry or of consumers. The chaos drove the price of oil to $0.03 per barrel and created a conflict between federal and state authority over the hot oil situation among other things. The result was the market ordering by the TRC in the form of **well spacing and ratable take rules** that brought order to the market and stabilized the price at a sustainable level. The system eventually became the practice statewide, and other states learned from the Texas experience and adopted similar market rules. The resulting market order continued for 40 years until the Arab Embargo in 1973.

CHAPTER 5

Concessions abroad and the disciplined rule of capture

Contents

The event

The struggle to define the market structure and control of access to the oil and gas resources in Texas following Spindletop and East Texas did not occur in a vacuum. The struggle that ultimately resulted in a definition of the role of government agencies, settlement of court battles, and revisions of legislative statutes was led by events in Texas but spread throughout the United States. But a different dynamic took place in other parts of the world during this time frame as the dominating share of world reserves shifted from Texas and the United States to the Middle East, Africa, and Russia. Not only were these foreign sources more plentiful, but they were also available at a lower price. The US production in 1925 was 70% of world production, about 63% in 1941 and just over 50% in 1950 (Painter, 2002, p. 2). By 1973 the US share of world production declined to 12% while that of Organization of Petroleum Exporting Countries (OPEC) rose to 51%. This shift in production dominance took place in the period that included two world wars, the Great Depression, and rapid growth in oil demand through a revolution in manufacturing sector power sources, and land, water, and air transportation based on oil products.

The experience of World War II greatly set the stage for US energy policy in the decades that followed. The role that the United States played in the war was heavily dependent on the United States' ability to supply oil for the war effort supporting the United States and her European allies. By the end of the War the United States supplied 80% of the Allied oil requirements (Painter, 2002, p. 5). The UK and France took great care to also secure oil supplies from North Africa and the Middle East. The German failure to secure enough oil and oil substitutes to support their efforts led to their failed invasion of Russia, a disaster that set up the conditions for Hitler's defeat. The importance of oil in the war effort was thus a defining influence on postwar policy that encouraged the importation

Innovation Dynamics and Policy in the Energy Sector
DOI: https://doi.org/10.1016/B978-0-12-823813-4.00005-1

of oil to the United States so as to keep adequate domestic supplies available for national security reasons. War and the threat of war has thus become a key element of geopolitical influences on energy sector policy at home and abroad (Painter, 2002, pp. 4−6). The US government supported a joint British−US company agreement to form a multinational consortium, the Iraq Petroleum Company. The agreement included a provision disallowing the members from developing oil in the Old Ottoman Empire, detailed by a "red line" on a map. But Standard Oil of CA (SOCAL)—then the Texas Company not members of the Iraq Petroleum Company—signed concessions with Bahrain (1930) and Saudi Arabia (1933). World War II events resulted in the Anglo-American Petroleum Agreement in 1944—and the Anglo-American treaty.[1] The agreements' purpose was the same as the regulatory policies of the Texas Railroad Commission (TRC)—to balance supply and demand, manage surplus, and bring order and stability to a market having perpetual over supply (Yergin, 1991, p. 402).

The combination of war-driven energy policy and the geologic location of oil supplies set up the conditions for rising oil imports to the United States in the years following World War II. But these conditions also drove a deep division within the industry in the United States. The major oil companies of the United States emerged from World War II with a commanding presence and control of oil supplies in the Middle East and Africa, as well as Mexico and Venezuela, all supported by the federal government foreign policy that helped secure concessions by these foreign governments. The independents in the US industry, however, grew increasingly opposed to rising imports that they believed diminished their role in US oil production. As a result the TRC struggled to get a handle on the Texas oil market, and the inherent conflicts among the Independents and the Majors. The solution was to develop allocation rules through well spacing and per well-production limits that protected the small producer. The overall setting of production to match market demand kept prices stable, and somewhat above what an unfettered market would have produced, thus also protecting the small producer participation in the market. But production limits also encouraged imports from the Major's activities in the Middle East, Africa, Mexico, and Venezuela. So, two basic systems of property rights and production incentives coexisted in the world ultimately setting up the conditions for the Arab Oil Embargo of 1973. This is the topic of Chapter 7, Panhandle Field and Natural Gas Flaring.

---

[1] The agreement provided for an eight-member International Petroleum Commission. The Commission would regularly estimate world oil demand and then allocate suggested production quotas to countries based on available reserves, sound engineering, economic factors, and the interests of producing and consuming countries—all with a view toward satisfying expanding demand. It expected that governments would then follow the recommendations. The Commission would report to the two founding governments on how to promote the development of the world petroleum industry (Yergin, 1991, p. 402).

Private property ownership of oil and gas resources linked to the land surface area ownership structures of US law has been and remains a novelty in the world. This fundamental part of US capitalistic democracy has tested and guided the capabilities of federal and state governments to deal with the development of the oil and gas resources for the last 150 years. This set of conditions is pretty much unique to the United States. Such conditions have created the interesting and dynamic development of the legal and regulatory policy centered in the activities of the TRC as well as in its counterparts in other producing states and the federal government. On the other hand, most of the access to explore, drill, and produce these natural resources abroad is, and has always been, controlled by governments who own the resource.[2] These foreign governments either form state-owned production companies (e.g., Mexico) or negotiated concessions (e.g., Saudi Arabia) with major oil companies, or consortiums of companies, to explore and produce oil and gas in exchange for a share of the revenues or from taxes and other payments.

Concessions are contracts giving private companies or consortiums the right to do business in the country, usually for a given period of time and within a defined territory. The agreements include provisions guaranteeing the concessionaire exclusive rights. The terms are negotiated and typically include agreements for payments of taxes, royalties, or a percentage of the revenues from the sale of oil and gas products which become the property of the oil company. Concessions became the common practice in the Middle East early on because the countries were poor and lacked both the financial ability and expertise to develop the oil reserves. International oil companies provided both financing and expertise.

As the discoveries of oil during the early 1900s revealed the enormity of the oil and gas resources in the Middle East the multinational companies moved in with the help of US foreign policy and naval support to protect ocean-based transport. The dominant companies operating in the Middle East under these concessions were known as the Seven Sisters. Five of the companies were American (Chevron, Exxon, Gulf, Mobil, and Texaco), one was British (British Petroleum), and the other Anglo-Dutch (Royal Dutch/Shell; Oil Concessions in Five Arab States 1911–1953, 1989).

Government-owned oil companies remain the rule in Mexico, Russia, and Argentina. There is a mixed arrangement in Canada where the dominant production area is controlled by Alberta under agreements with multinational companies. Alberta is the location of a large oil sands project that is the primary source of oil production in the country.

The concession-based approach to the exploration and production of oil and gas has encouraged collaborative organizations that have had, and continue to exert considerable

---

[2] There are also considerable oil and gas resources in the United States owned by and located on state and federal lands (both on-shore and off-shore) where access to explore and produce oil and gas is managed by competitive leasing agreements between the several governments and private companies.

influence on world oil markets and consequently on the geopolitical balance in the world. The most prominent of such arrangements is the OPEC. Classified as a cartel that cooperates to reduce competition the organization's consultations are protected by the doctrine of sovereign immunity under international law. OPEC is an intergovernmental organization that currently includes Algeria, Angola, Ecuador, Equatorial Guinea, Gabon, Iran, Iraq, Kuwait, Libya, Nigeria, Qatar, Saudi Arabia, United Arab Emirates, and Venezuela (US Energy Information Administration, retrieved May 28, 2017).

The struggle for control of oil ownership and production during the period when the TRC was getting control of the Texas market was also in a constant state of flux. Technology, markets and political structures were ever changing creating both opportunity and major trade conflicts. The two world wars increased the pressure for control of oil supplies as a national security matter.

Thus, because of the rule of capture for access to oil in the major producing regions of the US crude oil and natural gas producers during the first half of the 20th century faced a set of problems not faced by producers in most other countries. The problems stem from the combination of geologic and physical characteristics of crude oil and natural gas and the foundation of US capitalism—private property rights—discussed at length in the prior chapter.

Innovation US stile and the follow up by OPEC

The 10-year long battle that raged in Texas following the discovery of the East Texas field in many ways set the stage for the shift of market power to OPEC in 1973. Using methods and policy following that of the TRC, the United States developed and implemented policies that controlled production levels. The result was world oil price stability from the mid 1930s until 1973. But the exploration and production technologies of the era did not maintain a reserve base adequate to serve US long-term needs. The price level was high enough to encourage development abroad. The result was rising imports of oil from abroad so that by 1977 the US imports were almost equal to domestic production, and actually exceeded domestic production in 1993 as the US production continued to decline while demand and imports continued to rise. OPEC was formed in 1961 and by 1973 had commanded 54% of world oil production giving them the power through production limits to set, or at least have a major influence on world prices. Indeed this event changed the geopolitical balance in the world.

The United States through the TRC approach had achieved conservation of the resource and price stability but lost control of the world market price to OPEC. It should be noted, however, that OPEC has never had the level of control on production that the TRC obtained, and therefore the world price has varied drastically during three periods since the Embargo in 1973. OPEC often reaches joint agreement to limit production in order to boast prices, but the organization does not have regulatory power to enforce the agreements. There are many examples of individual member countries violating the production agreements since 1973. Saudi Arabia alone has had the power acting alone to influence world prices, but does not typically exercise such since it would simply reduce their market share. Instead she uses other indirect means to try to get cheaters to live up to the agreements, but with limited success.

The United States, led by the TRC, succeeded in creating order and price stability in the oil industry during the 1930s, 1940s, 1950s, and 1960s. During this period, however, OPEC countries, Russia, and the UK realized that they had adequate reserves to overtake the US in annual production. Russia, the UK, and Venezuela relied heavily on state-owned oil companies to discover, develop and produce oil. The OPEC countries, being poor, relied on concessions with multinational oil companies.

The effect of the rule of capture on the development of natural gas markets has the same basic problem as oil concerning the tendency to over produce. Often, though certainly not always in significant quantities, gas is present with oil deposits and must be produced in order to produce oil. Since it is not ordinarily economical to store and transport gas by truck or rail, producers vented and flared the gas at the well head during the early years, prior to development of a natural gas pipeline network. But the geology of the resource presents a different set of problems from that of crude oil. The natural gas industry, which began as a mostly unwanted byproduct of oil production, eventually developed into a profitable enterprise as major pipelines were developed to move the product to market as either a fuel for commercial, industrial, and residential use, or a feedstock for petrochemicals. This topic is developed in Chapter 6, West Texas and the Permian Basin Early Innovations.

The interplay of policy, markets, and technology

Innovation often has many sources, at times growing out of events that are quite unpredictable. The period of 1930−73 is an extreme case of interaction of policy, markets, and technology. The policy influence on a global scale evolved in large part due to the two world wars and the need to have control of adequate oil supplies to support the war efforts. An ongoing influence from the oil/war-efforts dependency has made oil supply control a national security matter ever since. As the United States developed oil reserves that dominated world markets with adequate supplies to support the allies in World War II, the TRC gained control of Texas oil production through well spacing and prorationing to meet market demand. But the war effort and US concerns for maintaining domestic production capacity led to increased foreign imports adding to the ability of concessionaires to develop enormous supplies in the Middle East and Africa. By 1973 the power to influence world markets shifted from the TRC to OPEC.

The dynamics of the innovation process, however, proved once again that one idea usually brings up its opposite. The Arab Embargo of October 1973 set up the conditions for development of alternatives to primary recovery technology for oil, and for energy and conservation alternatives to oil. This seminal event that set policy, markets, and technology on a new path for modern energy requirements has taken many forms and continues today. The era of alternative energy technologies and struggles to deregulate markets is the innovation story of the rest of the book. As is often the case, innovation grows out of a necessity to address unacceptable current conditions. Necessity is the "mother of invention." So in a strange way, but not uncommon in a new round of innovation, the Arab Embargo of 1973 set the stage for most all of the energy sector innovation that followed. The institutional innovation in Texas and in Washington that followed the Embargo is the subject of Chapter 7, Panhandle Field and Natural Gas Flaring.

Impacts of the institutional innovations

The combined effects of oil market stability with accompanying market prices in the United States held below the long-term replacement cost, paired with reserve additions of enormous proportions at low cost in the Middle East and Africa, set up conditions for the Arab Embargo of 1973. The US support of Israel in the Arab-Israeli War of early 1973 added to the Arab decision to impose the Embargo. The shift in oil reserves to the Middle East, in the presence of rapidly growing US demand for oil, sent imports to the United States to historically high levels with concentrations from the Middle East. In short, the oil concessions for American companies with Middle East countries, especially Saudi Arabia, joined with stable but low US oil prices resulted in rapidly rising oil imports to the United States from politically unstable OPEC countries. Ultimately this era set up a mandate in the United States to rethink government's role in energy markets.

CHAPTER 6

West Texas and the Permian Basin early innovations

Contents

The event

The Permian Basin in West Texas and Eastern New Mexico has a long and phenomenal history that has continued for most of a century, and during the last decade has brought a surprising influence over world energy markets. While the East Texas field experience was the event that brought Texas onto the world stage as a major source of crude oil and brought important institutional innovation, the Permian continues to produce innovation that has upended the geopolitical balance in the world. The first set of innovations that are the focus of this chapter—secondary and tertiary recovery of crude oil—developed by oil industry leaders with support from university research and private sector demonstration projects following the era of recovery from primary production. The focus on the newest Permian innovation—hydraulic fracturing (fracking)—is the focus of Chapter 9, Electric Industry Deregulation and Competitive Markets.

The enhanced oil production in the Permian is built upon the important technologies developed in a major way by work of the University of Texas Bureau of Economic Geology (BEG) research and a following industry education program in partnership with the Texas Independent Producers and Royalty Owners Association. The effort grew out of an ongoing research program in "unconventional" oil prospects beginning in the late 1980s and continuing into the 1990s. BEG's program of economic geology focused on a strategy of oil and gas recovery by use of enhanced recovery techniques used to unlock previously unrecoverable oil. The focus was on recovery from existing plays rather than from wildcats. This emphasis was continually voiced by Dr. Bill Fisher, Director of the Bureau of Economic Geology on many occasions at industry conferences. Eventually the Bureau and Texas Independent Producers and Royalty Owners Association hosted a series of educational seminars all

Innovation Dynamics and Policy in the Energy Sector
DOI: https://doi.org/10.1016/B978-0-12-823813-4.00006-3

across the state. In a follow-up the US Department of Energy (DOE) engaged with funding and along with the bureau helped inform the rest of the nation, an action that did a great deal to advance enhanced oil recovery (EOR) and then fracking, concentrated heavily in the Permian (Bureau of Economic Geology, 2017).

The Permian early experience did not have the same kind of impact as did the East Texas field, which was the largest proven oil deposit in the world in the 1930s. East Texas, because of its size and characteristics, set the stage for the well-spacing and production control rules that stabilized the oil market for the next four decades—a key institutional innovation. More than any singular event, this prolific field and the wildcatters of Texas were responsible for the industry structure and primary market performance that prevailed until the Arab Oil Embargo of 1973. While the policies of well-spacing and control of production to meet market demand can be criticized for creating market inefficiencies they provided price stability and kept the little guy in the industry.

But East Texas was by no means the whole of the Texas oil experience during the first half of the 20th century. During this period oil was discovered throughout the state from the Panhandle Field in the Panhandle region of the High Plains to the Yates Field in West Texas; Snyder, Ranger, and Electra fields in North Texas; Mexia, Melrose, and Goose Creek fields in East Texas; Piedras Pintas and Kelsey fields in South Texas; and several others scattered throughout the state (see Fig. 6.1 map).

The most prolific among the Texas oil discoveries has been the combined set of fields that form the Permian Basin spanning 38 counties in West Texas, the extent and productivity of which was not fully understood for decades of drilling, scientific study, and experimentation. The Permian discovery was different in many ways from the experience of the East Texas field, but similar in that it resulted from the extraordinary determination of a risk-taking wildcatter.

The beginning in West Texas was a discovery well in Mitchell County in 1920—a modest 50 barrel per day well. But the discovery that opened the path to the future of this complex geologic system that has today become the largest oil field in the world is known as San Rita No. 1. The well drilled in 1923 by wildcatter Mike Benedum in Reagan County opened up the oil field named Big Lake. While the well was only a modest 100 barrel per day producer, the follow-up agreement with investors to drill out the play resulted in a series of eleven wells that proved out the field. One well came in at 5000 barrels per day followed by number eleven at 8000 barrels per day. Big Lake became to turning point for West Texas which at the time had the reputation of being the "graveyard of the wildcatter." This developed field, among other great outcomes, funded a major university endowment that later became the Permanent University Fund (PUF). This fund has in large measure made the University of Texas and Texas A&M great research universities. Big Lake was on University of Texas land (Goodwyn, 1996, pp. 24–29).

The Big Lake success led a rancher, Ira Yates, to explore an area some 100 miles to the west across the Pecos River. An agreement to drill four wells, that turned

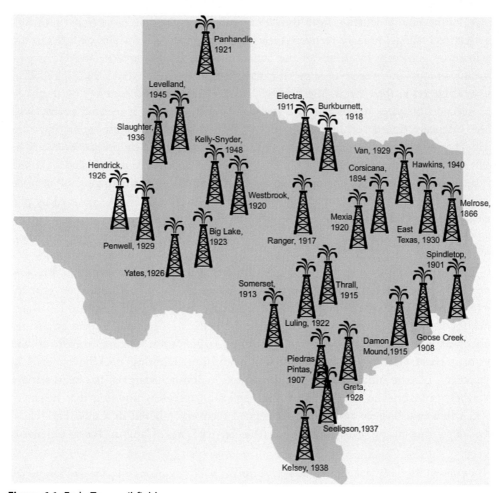

Figure 6.1 Early Texas oil fields.

out to be failures, finally was expanded to allow drilling a fifth one on the Yates ranch. The result was Yates No. 1 that came in at 4000 barrels per day at a depth of only 1000 ft. It was followed by a well drilled to 1500 ft. that came in at 71,000 barrels per day. Finally, Yates No. 30A came in at 204,682 barrels per day, the largest well in the world as of 1929 (Goodwyn, 1996, p.29).

The years following Yates No. 30A defined the Permian Basin as a sedimentary basin 250 miles wide by 300 miles long, from just south of Lubbock to south of Midland-Odessa, east to Snider, and west into Eastern New Mexico. The Permian is not one geologic formation but multiple layers that define eight basins, including the Midland Basin (the largest), the Delaware (the second largest), and the Marfa Basin (the smallest)—containing more than 7000 fields (Vertrees, accessed July 19, 2017).

All of the early production from the Texas fields across the state came from "primary" recovery technology. Primary recovery means pumping from wells where oil flows to the wellbore by gravity or by means of the natural pressure created in water-driven or gas-driven reservoirs. But it soon became clear that primary recovery would not exploit all of the available oil in these fields. "Secondary" recovery enhances oil recovery by means of the injection of water or gas into the reservoir, usually via use of selected nearby well bores originally drilled for primary recovery, now abandoned to use as an injection well. The resulting pressure drives some of the oil not recovered by primary recovery to wells surrounding the injection well. "Tertiary" or "Enhanced" recovery includes the use of thermal, gas injection, or chemical injection to free up more of the original oil in place (OOIP).[1] The DOE estimates that generally, only 10% of the OOIP is recovered by primary recovery. Use of secondary recovery brings the recovery to 20%−40% and tertiary recovery brings the total recovery to 30%−60% of the OOIP (US Department of Energy, Office of Fossil Energy, 2017).

During the 1930s, 1940s, and 1950s traditional methods of discovering and recovering oil from throughout the Permian Basin continued with production by primary recovery means. As wells began to decline secondary recovery methods, through water injection into the reservoirs, followed during the next two decades. But the origin and perfection of secondary recovery technology and application was not unique to the Permian Basin. The major Permian Basin innovation (ignoring fracking covered in Chapter 9, Electric Industry Deregulation and Competitive Markets) has been tertiary recovery, primarily the development and application of CO_2 flooding that began in 1972 when two projects using CO_2 enhanced recovery ushered in a new era. These two experimental projects, known as the Scurry Area Canyon Reef Operators

---

[1] The primary, secondary, and tertiary recovery shares of OOIP vary among oil reservoirs. One of the largest oil fields in the United States, the Wesson Field in the Permian Basin is described by Oxy Petroleum Co. in a filing with EPA of a MRV plan for CO_2 sequestration in the Wesson. In a typical sedimentary formation, like the San Andres reservoir in the Denver Unit in the Wasson Field, primary production produces only a portion of the OOIP. The percentage of oil recovered during "primary production" varies; in the Denver Unit, primary production recovered approximately 17% of the OOIP, and approximately 83% of the OOIP remained in the pore spaces in the reservoir. Water injection may be applied as a secondary production method. This approach typically yields a sizeable additional volume of oil. In the Denver Unit, water injection led to the production of another 33% of the OOIP, leaving approximately 50% still in the pore space in the reservoir. The oil remaining after water injection is the target for "tertiary recovery" through miscible CO_2 flooding. Typically, CO_2 flooding in the Permian Basin is used as a tertiary production method and it entails compressing CO_2 and injecting it into oil fields to mobilize trapped oil remaining after water flooding. Miscible CO_2 flooding can produce another 20% of the OOIP, leaving the fraction of oil remaining in the pore space in the reservoir at approximately 30%, which is the target of fracking covered in Chapter 9, Electric Industry Deregulation and Competitive Markets. Source: OXY Petroleum Company, Technical Review of Subpart RR MRV Plan for the Denver Unit, April 9, 2015).

Committee (SACROC) converted a waste stream of byproduct CO_2, from natural gas wells in the southern portion of the Permian Basin. The CO_2 was sent by pipeline to Scurry County where it was injected into the reservoir (Tenaska Trailblazer Partners, LLC, 2011, p. 5). The Permian experience of this technology application then spread to projects in Kansas, Mississippi, Wyoming, Oklahoma, Colorado, Utah, Montana, Alaska, and Pennsylvania in the United States as well as abroad.

The SACROC project success in the Permian convinced the major oil companies that CO_2 EOR was economically viable. As a result, pipelines for delivering CO_2 from out-of-state, underground, natural CO_2 sources at Sheep Mountain and McElmo Dome in Colorado, and Bravo Dome in New Mexico were developed bringing the gas to several points in the Permian Basin (see Fig. 6.2).

The CO_2 pipelines were developed during 1983 through 1985 to deliver CO_2 to the Permian Basin's CO_2 EOR projects, owned mostly by the Kinder Morgan CO_2 company operated by Bravo Pipeline Company. Other joint owners included Occidental Permian, XTO Energy, and Kinder Morgan CO_2 also operated some of

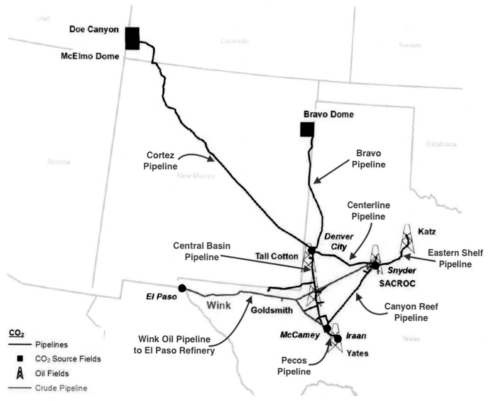

Figure 6.2 CO_2 supply and pipelines to Permian Basin. *Tenaska Trailblazer Partners, LLC, 2011. Bridging the commercial gap for carbon capture and storage. Global CCS Institute, p. 32.*

the pipelines (Kinder Morgan CO_2 website at https://www.kindermorgan.com/pages/business/co2).

Success is contagious. The CO_2 EOR industry was born and has thrived ever since the SACROC project. By 2011 the Permian Basin CO_2 EOR industry had grown to more than 60 producing projects purchasing more than 1.7 billion cubic feet of new CO_2 per day and producing more than 65 million barrels of oil per year (Tenaska Trailblazer Partners, LLC, 2011, p. 25).

But the context for this large investment in CO_2 production capacity and pipeline network is important. Federal and state policies following the Iranian Crisis in 1979 helped industry start seriously down the CO_2 recovery technology path. There were federal regulated oil price advantages relative to oil from new ordinary wells, federal tax credits for EOR investment and Texas state reduction of severance taxes along with franchise tax credits for tertiary recovery oil (Texas allows an 80% severance tax reduction for CO_2 recovery; National Enhanced Oil Recovery Initiative, 2012, pp. 21−22; Advanced Resources International, 2006). These influences are discussed in the following section.

The contribution of the Permian Basin in ushering in CO_2 EOR has been enormous. While primary and secondary recovery technologies and related institutional structures [mainly the Texas Railroad Commission (TRC) rules] were similar to that elsewhere in Texas and the United States, the progress of enhanced recovery through CO_2 injection is the unique contribution of the Permian Basin players. And importantly, the beginning of CO_2 EOR was primarily by the major oil companies who could finance the projects with longer term prospects of pay-out that included the building of pipelines from Colorado and northern New Mexico.

The example of the Denver Unit experience that led the CO_2 EOR technological innovation is a unique example of a development that could only have been done by the majors or the larger independent producers. The SACROC project success led to enough confidence in long-term outcomes to induce investment by the majors, perhaps only matched in oil development history by the decisions that brought in the Prudhoe Bay Oil Field in Alaska. Although the Prudhoe Bay Oil Field was discovered and confirmed in 1968, production did not begin until June 20, 1977 when the Alaska Pipeline was completed. The pipeline was built between 1974 and 1977 after the 1973 Arab Embargo produced a sharp rise in oil prices in the United States (Banet, 1991).

The expectation of high future oil prices made exploration of the Prudhoe Bay Oil Field economically feasible. Similarly, the oil price expectations following the price spike that accompanied the Iranian Crisis in 1979 made the CO_2 EOR in the Permian Basin economic. Oil and gas price increases, coupled with federal and state tax and price preference incentives justified investment in the CO_2 infrastructure in 1983−85. Fig. 6.3 shows the result which is Permian Basin dominance of world CO_2 EOR production.

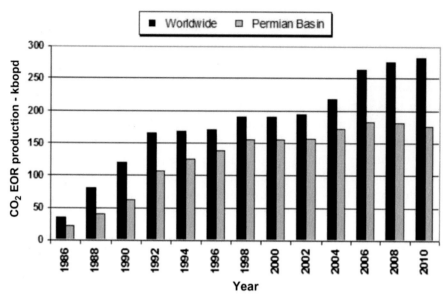

Figure 6.3 Worldwide and Permian Basin CO_2 EOR production. *Tenaska Trailblazer Partners, LLC, 2011. Bridging the commercial gap for carbon capture and storage. Global CCS Institute, p. 24.*

The Permian Basin is home to more than 50% of the world's CO_2 projects and about two-thirds of the oil production from this gas flooding technology. While the early decades-long development has focused on CO_2 flooding using natural CO_2 sources piped-in from Colorado and New Mexico, there has been a growing interest in expanding the sources by capturing CO_2 from industrial and power generation industries. This technology promises to solve two problems—reducing CO_2 emissions to the atmosphere (which contribute to global warming) and providing a source of CO_2 for EOR. The industrial sources include cement plants, chemical plants, refineries, paper mills, and manufacturing facilities.

By far the largest CO_2 EOR project in the Permian Basin is the Occidental Oil and Gas Corp's project at Denver City, Texas (known as Denver Unit CO_2 Recovery Plant) in the Wesson Field. The project began in 1983 and is still operating today. It operates under the approval of US EPA's Underground Injection Control rules and TRC regulations. Occidental Petroleum is also developing a $1.1 billion natural gas processing plant in West Texas that will capture about 265 billion cubic feet (13.5 million metric tons) of CO_2 per year for use in CO_2 EOR operations in the area (Carbon Capture and Sequestration Technologies, 2016).

The Denver Unit project produced 120 million barrels of crude oil by 2008. There are approximately 2200 production wells and 600 injection wells in the project.

Some of the CO_2 that is injected is sequestered in the reservoir. Forecasts indicate that 603 billion cubic feet of CO_2 will be produced from 2014 through 2021. Oxy estimates the total CO_2 storage capacity to be about 14,700 billion standard cubic feet (bscf) or 775 million metric tons. Oxy forecasts that over the lifetime of EOR operations, stored CO_2 will fill approximately 25% of the storage capacity. Oxy has injected 4.035 Bscf of CO_2 into the Denver Unit through the end of 2013. Of that amount, 1593 Bscf was produced and 2442 stored. The project Monitoring, Reporting and Verification (MRV) plan was approved by EPA in December 2015 with a beginning date of 2016 (OXY Petroleum Company, 2015).

The National Energy Technology Laboratory (NETL) constructed the following Fig. 6.4 to explain (by example) the production cycle from primary recovery through secondary and then CO_2 flooding. The NETL example is from the Denver Unit that began production in 1938 and has continued for a period of 60 years. The primary production stage lasted from 1938 through 1965 and secondary production by water flooding through about 1983 when the CO_2 EOR began.

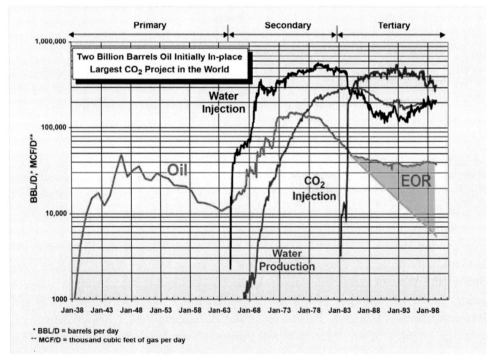

Figure 6.4 Primary, secondary and CO_2 EOR production sequence of oil. *US Department of Energy, National Energy Technology Laboratory, 2010. Carbon dioxide enhanced oil recovery untapped domestic energy supply and long term carbon storage solution. Available from: <https://energy.gov/fe/science-innovation/oil-gas-research/enhanced-oil-recovery> (accessed 31.07.17.).*

The graph below shows oil production from 1938 through 1998 for primary, secondary (water flood), and tertiary (CO_2 EOR) technologies. The production is for the Denver Unit of the Wasson Field in West Texas. Incremental oil production due to EOR is represented by the green area under the curve at right (Fig. 6.4).

The Denver Unit flood reservoir had 2 billion barrels of OOIP and residual oil saturation after water flooding of 40%. The Unit produced about 31,500 barrels per day in 2008, of which 26,850 was incremental oil from the CO_2 flooding (U.S. Department of Energy, National Energy Technology Laboratory, 2010, pp. 8–14). The remaining oil in place following primary, secondary, and current generation CO_2 EOR is an obvious target for further recovery through improvement in the efficiency of CO_2 flooding. DOE has included a number of R&D efforts to advance the CO_2 technology by competitively selecting seven Next Generation CO_2 EOR research projects. Four of the projects include techniques for mobility control of the injected CO_2 and novel foams and gels to prevent highly mobile CO_2 from channeling through high-permeability areas, which often leaves unswept, unproductive areas of the reservoir.

All of the early CO_2 EOR projects involved CO_2 sources from naturally occurring reservoirs. Newer projects (from various locations around the globe) depend on CO_2 from power plants and industrial applications including natural gas processing, fertilizer, ethanol, and hydrogen plants. An example is a North Dakota gasification project that delivers CO_2 to the Weyburn oil field in Saskatchewan, Canada via a 204-mile pipeline. Two Carbon Capture and Storage (CCS) projects developed through the late planning stages in the Permian Basin would have provided CO_2 as a major addition to CO_2 supplies from power plants, but both have been canceled. These CCS projects include clean coal electric power stations with carbon capture and sequestration components described briefly here but explained more thoroughly in Chapter 12, Capture and Global Warming: The Technology and Regulation Debate.

The first of the two canceled projects was to have been a $3 billion 600 MW, super critical pulverized coal power plant known as Tenaska Trailblazer, capturing 90% of the CO_2 was to be located in Sweetwater, Texas (Fig. 6.5). The project, developed to a final planning stage, was the joint effort of Tenaska, Arch Coal, and Fluor Corporation. The CO_2 demand for EOR was to have been a major revenue stream to support the project that would have produced electricity sold it into the Electric Reliability Council of Texas (ERCOT) electric market (Tenaska Trailblazer Partners, LLC, 2011, p. 5).

The second project intended to expand CO_2 sources (among other objectives) was known as Texas Clean Energy Project (TCEP). It was a joint venture of Summit Power Group Inc, Siemens, Fluor, Linde, R.W. Beck, Blue Source, and Texas Bureau of Economic Geology located in Penwell in Ector County, Texas. The $3.98 billion, 400 MW clean coal project would have been a major source of CO_2 for EOR in the Permian Basin. The project, if completed, would have relied on an initial DOE

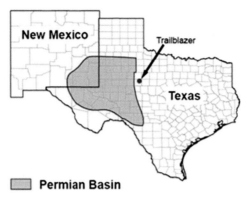

Figure 6.5 Location of the Permian Basin and the Trailblazer project. *Tenaska Trailblazer Partners, LLC, 2011. Bridging the commercial gap for carbon capture and storage. Global CCS Institute, p. 5.*

grant of $450 million. The project planning was initiated in October 2009 and canceled upon DOE defunding in May 2016 (Carbon Capture and Sequestration Technologies, 2016).

The project had a sizable DOE commitment. On December 4, 2009 Secretary Stephen Chu announced that TCEP would receive a $350 million award, which substantially reduced TCEP's effective costs. This award was the largest made under the Department of Energy's Clean Coal Power Initiative, enacted and funded by Congress. The US DOE made an additional $100 million award to TCEP in August 2010.

A Record of Decision and environmental impact statement for TCEP was issued by the DOE on September 29, 2011 in accordance with National Environmental Policy Act, which requires an assessment of the environmental impacts of any federally subsidized project. TCEP received its air quality permit from the Texas Commission on Environmental Quality on December 28, 2010—a major milestone that is the key state governmental approval that is required to move a project forward to be privately financed and built. TCEP was scheduled to achieve financial closing in April 2015 and commence construction in the summer of 2015. Commercial operation was scheduled for 2018. The project would have begun sequestering carbon during start-up and testing in 2018.

A third Texas CCS project providing CO_2 for EOR (not in the Permian Basin but encouraged by the two prior Permian projects) is Petra Nova CCS, a 240 MW slip stream from a 610 MW coal-based power plant that is now operational. The technology is a postcombustion carbon-capture facility. The Petra Nova project is a 50/50 partnership between NRG Energy and JX Nippon Oil and Gas Exploration Corp in Fort Bend County, Texas. Technology added to an existing power plant is designed to capture carbon before it is emitted into the atmosphere. The $1 billion project removes 90% of the CO_2 emissions from the coal-based electric power plant and provides CO_2 for EOR operations at mature oil fields in the Gulf Coast region. At a capture rate of 1.4 Mtpa, this is the world's largest postcombustion capture facility at a power plant (the Boundary Dam postcombustion CO_2 capture facility in Canada has a capacity of 1.0 Mtpa). The captured CO_2 is used for EOR purposes at an oil field not far from the Houston area. (Carbon Capture and Sequestration Technologies, 2016).

A fourth project, Air Products and Chemicals, also in the Gulf Coast area, is designed to capture carbon from a hydrogen production facility in Port Arthur. The project is a joint venture of Air Products and Chemicals, Denbury Onshore LLC, University of Texas Bureau of Economic Geology and Valero Energy Corporation in Port Arthur, Texas. It began capturing CO_2 in December, 2012. By June, 2016, Air Products it had successfully captured more than three million metric tons of CO_2 at Port Arthur after 3½ years of operation. The project is made up of two Steam Methane Reformers located within the Valero Refinery in Port Arthur, Texas.

All of the last three projects received federal support from the American Recovery and Reinvestment Act (ARRA) administered by the DOE. Much of the background for these projects grew out of an early effort in Texas to win a DOE sponsored project called FutureGen originally developed in 2003. As originally envisioned, the project (later called FutureGen 1.0) would have been the world's first coal-fired power plant to integrate carbon capture and sequestration, producing electric power, capturing the CO_2 and storing it permanently in underground saline formations. Two Texas sites were among the four finalists in the national DOE contest to be structured as public-private partnerships. Due to cost and other factors the project was canceled during the Bush Administration in 2008, but brought back in a modified form (FutureGen 2.0) by the Obama Administration in 2010. FutureGen 2.0 was made a part of the ARRA of 2009 (ARRA, P.L. 111-5) by appropriating almost $1 billion for the project.

There were two Texas proposals for FutureGen 2.0, one at Jewett, Texas and the other at a site west of Odessa, Texas. TCEP would have been located at one of the former FutureGen 1.0 finalist sites—the 600-acre Penwell site—situated 15 miles west of Odessa, Texas. TCEP was one of several public-private partnerships funded by ARRA that were intended to build and operate an integrated CCS projects in the United States. The modified idea under the ARRA was to fund several projects of smaller scale than the original, large-scale FutureGen 1.0. The successful projects were to have been comprehensive projects combining (1) the capture and separating of CO_2 from other gases, (2) compressing and transporting CO_2 to the sequestration site, and (3) injecting CO_2 into geologic formations for permanent storage (Folger, 2014).

The Permian Basin innovations

Prior to the development of hydraulic fracturing (fracking) the most important innovation contribution of the Permian Basin experience has been the development and implementation of CO_2 EOR technologies. The focus, lead in a major way by the

BEG in the late 1980s and throughout the 1990s, was enhanced recovery techniques from previously unrecoverable oil, which the Bureau called "unconventional oil." The development of this reorientation joined with fracking technology has expanded the economically available supplies of oil throughout the state and nation but heavily in the Permian Basin. This innovation, along with fracking, has elevated the Permian Basin to the largest oil play in the world. Hydraulic fracturing which has turned the energy world upside down is the subject of Chapter 9, Electric Industry Deregulation and Competitive Markets.

The innovation of CO_2 EOR technologies was led by the majors in the industry who made the investment in pipelines and an EOR infrastructure that became the SACROC project. This is in contrast to the discovery activity of the 1920s and 1930s which were led by the wildcatters—the "little guys" of the industry.

But the experience of the development of the Permian Basin energy-based industries goes far beyond the contribution of oil and gas supplies from primary and secondary recovery and the follow-on development of CO_2 EOR. The Permian Basin has also made a major contribution to the development of the petrochemical industry and to an understanding of the challenges of CCS-related technologies.

Innovation in downstream petrochemical industries became yet another contribution of the Permian. Hundreds of millions of dollars have been spent on petrochemical refineries and supporting construction work in the region, which has been rated the largest inland petrochemical complex in the United States. The Texas Petrochemical Industry that developed into the world's leading manufacturer of petrochemicals became a strategic supplier of synthetic rubber and synthetic chemicals for explosives during World War II. Following the War the state's share in the American petrochemical industry increased dramatically and much of the early development that is now concentrated in the Gulf Coast was in the Permian Basin, principally in Odessa and Big Springs.

During the 1960s Texas played an increasingly diversified role in all phases of the petrochemical industry. The region developed manufacturing capacity for basic petrochemicals including ammonia, ethylene, propylene, polypropylene, butadiene, benzene, isoprene, carbon black, and xylenes, which are the building blocks for innumerable chemical products including plastic, rubber, and synthetic fiber industries that are the basic ingredients of consumer products—clothing, construction materials for homes and offices, household appliances and electronic equipment, food and beverage packaging, and many others (Handbook of Texas Online, 2017a; see also Handbook of Texas Online, 2017b).

Major company participants in the early petrochemical industry development in the Permian included Cosden Petroleum Corporation, El Paso Products, Dart Industries, W.R. Grace, General Tire, Shell, Sid Richardson, and Cabot Oil and Gas Corporation (carbon black; Pace Company Consultants and Engineers, Inc., 1974, pp. 6−8).

Two factors drove the development of the petrochemical industry in the Permian. First, the production of oil and natural gas brought with it a need for refineries and ample supplies of crude oil, natural gas and natural gas liquids (NGLs) which needed a market. NGL is the feedstock for olefins (including ethylene and propylene) and aromatics (including benzene, toluene, and xylene isomers). Oil refineries produce olefins and aromatics by fluid catalytic cracking of petroleum fractions. Chemical plants produce olefins by steam cracking of NGLs in the form of ethane and propane. Aromatics are produced by catalytic reforming of naphtha. Olefins and aromatics are the building blocks for a wide range of materials such as solvents, detergents, and adhesives. Olefins are the basis for polymers and oligomers used in plastics, resins, fibers, elastomers, lubricants, and gels. In short, the production of crude oil and natural gas also brought NGLs, and it made economic sense for part of that production to be processed into downstream products in the area.

Second, the World War II effort needed massive supplies of these downstream petroleum products, especially rubber and explosives. Thus, the Permian Basin became a strategic supplier of synthetic rubber and synthetic chemicals for explosives during World War II. In short, the downstream product demand and federal government policy (the war effort) drove the development of a petrochemical industry in the Permian Basin.

A third innovation that is a major contribution of the Permian Basin players is development of CCS technologies. Although not yet economic the current advanced state of CCS innovation is in large part an outgrowth of the CO_2 EOR development in the Permian Basin. The geologic conditions and industry/TRC rules that gave birth to CO_2 EOR in the Permian Basin also had a great influence on the related CCS technological innovations that followed and are still in progress.

The demand for CO_2 for EOR when joined with efforts to capture CO_2 before it is released into the atmosphere drove CCS innovation potentially supporting two primary objectives—a source of CO_2 for EOR and the sequestration of CO_2 in order to remove it from the emissions from coal and natural gas power plants, cement manufacturing, and other industrial operations. This innovation effort that is heavily driven by environmental objectives is creating a host of technologies that promise to produce clean coal and crude oil that are being processed into more environmentally friendly products.

As the worldwide concern over CO_2 emissions into the atmosphere (that contributes to global warming) has grown, the most promising technological solution is with major support of the federal government is CCS. CCS will potentially serve as a source of CO_2 for EOR projects but also reduce future global warming by reducing emissions into the atmosphere. A promising focus is capture and sequester of CO_2 emissions from electric power and industrial sources into underground formations. This set of innovation contributions is the topic of Chapter 12, Capture and Global Warming: The Technology and Regulation Debate.

In summary the major innovations flowing from the Permian Basin experience included CO_2 EOR recovery technology, a solid beginning of the petrochemical industry and the early learning from research, development and demonstration projects that may eventually create a viable industry that removes and stores carbon from coal use in the electric power and industrial sectors. That is, the Permian may become the source of the development of the technologies to create a viable industry from the removal and storage (CCS) or use of CO_2 to make market products out of the captured carbon.

A unique institutional innovation that grew out the Permian Basin experience is the PUF. This university permanent fund was created and continues to be funded from royalty payments paid by producers of oil and gas on the University of Texas lands in the Permian Basin. The PUF has helped make great universities of the University of Texas and Texas A&M University.

The interplay of policy, markets, and technology

Policy changes combined with oil market price improvements created incentives that were instrumental in driving the CO_2 EOR result we see today in the Permian Basin. A tertiary oil price category was part of the federal oil price deregulation program in the late 1970s and early 1980s. Price regulations on oil were put in place originally as part of economy-wide price controls in 1971 during the Nixon Administration but restructured following the Arab Oil Embargo in October 1973. Separate categories of oil production in the years following 1973 were defined as oil production from "old" wells and "new" wells. Oil from new wells and enhanced recovery were deregulated to encourage new production. By December 1973 old oil was priced at $5.25 per barrel ($28.38 in year 2017 purchasing power dollars) while new oil was $10.35 ($55.95 in 2017 dollars). Deregulated oil from new wells followed world oil prices. The Congress, following President Carter's recommendation, removed oil price controls (phased out over a period of several years for some categories) in 1979 but enacted a "windfall profits" tax. Under the new tax, tertiary oil production was exempt. Tertiary oil production was later granted a federal investment tax credit (a 10% tax credit for EOR expenditures) that made investment in projects after 1990 less risky. Texas also reduced the severance tax rate for tertiary recovery oil (Sherlock, 2011).

The run up of oil prices following the Embargo, followed by a second major price increase during the Iranian crisis of 1979−80, plus the exemption of enhanced recovery from the Windfall Profits Tax—all on the heels of the successful SACROC project in 1972—gave the CO_2 recovery technology the incentives needed to launch EOR projects in the Permian (see Fig. 6.6). A summary of these policy developments is contained in Chapter 7, Panhandle Field and Natural Gas Flaring.

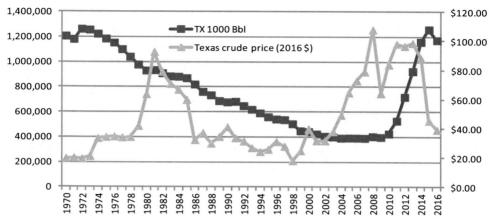

Figure 6.6 Texas oil production and price: 1970—2016. Price is First Purchase Price in 2016 Dollars (inflation adjusted by CPI all Urban consumers). *Energy Information Administration, Texas First Purchase Price 1978—2016; EIA U.S. First Purchase Price 1970—1977; CPI from U.S. Department of Labor, Bureau of Labor Statistics.*

The same factors when negative (current market price declines and removal of policy incentives) invariably have a negative effect on investments of both large corporations and collaborative groups to develop and implement new technology. Prices and public policy also directly determine government commitments to fund large, long-term projects. The drastic drop in oil prices starting in the mid-1980s stalled further expansion of CO_2 EOR in the Permian Basin. But oil and gas price increases beginning at about year 2000 greatly increased the investments in CO_2 projects based on the technologies developed following SACROC. The result was to significantly dampen the rate of oil and gas production declines, that when combined with hydraulic fracturing success (discussed in Chapter 9, Electric Industry Deregulation and Competitive Markets), reversed the downward trends that began in 1972. Oil production in Texas and the US increased for the first time in 40 years, and by 2016 reached levels of production that matched the prior peak of 1972.

The Permian's two major contributions to oil industry innovations discussed in this chapter (CO_2 EOR and CCS) amount to a study in the reinforcing incentives of policy, markets, and technology that have to be maintained over a long period of time to succeed. There is no sure means of sustaining these complex factors long enough to fully test a set of innovations. The unfolding of policy, markets, and technology that brought CO_2 EOR to a successful outcome developed over a period of about 20 years. First, the government response to the Arab Embargo of 1973 became a powerful driver by setting a regulated incentive market price for EOR. The major oil companies took advantage of a low-cost demonstration (the SACROC project) to test the market viability of CO_2 EOR. Large supplies of CO_2 in Colorado became recognized as a large enough source to justify pipeline construction. The rest is history.

The CCS experience is not assured of a successful outcome because the complementary support from markets, policy and technology that exists for CO_2 EOR does not seem likely at this point in time. The technology part is available, but the policy needed to help drive the innovation is not supportive. Policies and technology development in the natural gas industry (low prices driven down by fracking) make coal-based power generation uneconomic, and therefore the justification for capturing and storing CO_2.

Tenaska Inc. announced in June 2013 that it would cancel the development of two coal-fired power plants and focus on generation from natural gas and renewable sources. The projects included the Taylorville Energy Center in Christian County, Illinois, and the Trailblazer Energy Center in Nolan County, Texas discussed above. The explanation given was that both policy and market prices had made the projects no longer viable. Tenaska cited the decreasing costs for generating electricity from renewable sources, the low cost of natural gas, and the lack of a government climate and energy policy needed to make sure clean coal projects are economically viable.

The federal government environmental policy related to global warming and project funding of CCS found a ready interest in the Permian during the Obama Administration. The demand for CO_2 for EOR and easy access to the ERCOT electricity grid for a new clean coal power plant initially drove the company consortium for the Trailblazer project. But the prospects for economically viability disappeared, in part by the dramatic decline of natural gas prices and the associated price of wholesale electric prices. Tenaska said market and political conditions had forced them to stop work on the Taylorville Energy Center in Christian County, Illinois, and the Trailblazer Energy Center in Nolan County, Texas.

The lack of federal funding and interest in the capture and sequestration of CO_2 from existing and new power plants and industrial plants drove several failed attempts to deploy a CCS project in the Permian, although two projects are going forward along the Gulf of Mexico in Texas. The commercial sale of CO_2 for EOR in the Permian may only be economically viable for a source of waste CO_2 that is relatively pure and/or high pressure. Coal-based power plants usually do not meet this pure or high pressure CO_2 product source advantage, thus requiring large investments in equipment to compress the CO_2 prior to transport. It seems abundantly clear that simply capturing and storing the CO_2 in a tight underground formation will not be economic, unless, of course a future policy imposes a high carbon tax or a restrictive cap and trade program.

The Permian Basin experience with CO_2 EOR and CCS is a textbook example of the **complex interaction of markets, technology, and government policy** that make it difficult to prescribe a process that is certain to drive innovation, even though we know that "innovate we must" or the United States will fail to remain the world's leader of innovation. Our evolution both in the energy sector and elsewhere has brought us to the point that innovation is the main driver that determines economic growth.

Impacts of the Permian Basin innovations

The most important innovations from the Permian Basin development over the period from 1923 until the advent of fracking in 2005 (discussed in the following Chapter 10: Hydraulic Fracturing: The Permian Basin Challenges OPEC Leadership) is the perfection of **secondary oil recovery** by water flooding of the reservoir, and **tertiary recovery** with the injection of CO_2 Enhanced Oil Recovery (**EOR**). Not only has this technology greatly expanded the oil and gas recovery in the Permian but the development has also spawned a CO_2 industry that is developing beneficial uses for CO_2, but also development of technology for removing CO_2 from emission stream and storing it underground thus reducing the CO_2 contribution to global warming.

CHAPTER 7

Panhandle Field and natural gas flaring

Contents

The dynamics of the natural gas market development is arguably the most complicated of energy markets because of the requirement to integrate several resource, technology, and institutional factors, and to work out the market and political interests between oil interests on the one hand, and natural gas interests on the other. Each of these markets has been driven by different downstream markets, developed at multiple geographical locations, and on different time paths. These fundamentals have driven the Texas legislature and the US congress to reflect these varying interests, and to craft legislation to set up the regulatory and legal parameters of each resource.

The natural gas market lagged the oil market development in Texas and elsewhere during the first half of the 20th century because the options for transportation and storage were more limiting and downstream final product markets were not developed. Crude oil can be stored in tanks and shipped to refineries by truck, rail cars, and pipelines. Then the refined products can be stored and shipped by the same means. Also the downstream crude oil products, especially gasoline and fuel oil had early market demand (automobiles and residential/commercial heating, respectively) that drove oil market development while the uses for raw natural gas were more dependent on the development of petrochemical plants that, today derive many industrial and consumer products from the raw natural gas stream (paints, insecticides, plastics, etc.) the demand for which came later in the century. The immediate products derived from raw natural gas produced at the well head required some processing on-site by separating liquids from the gas stream.

Innovation Dynamics and Policy in the Energy Sector
DOI: https://doi.org/10.1016/B978-0-12-823813-4.00007-5

The primary immediate demand for the separated liquids from the raw natural gas stream includes natural gasoline (condensate) and carbon black (the later produced by controlled burning to produce a product for the rubber and printing industries). So, gasoline and carbon black plants often developed near the oil and gas reservoir locations. In the early stages of development there was an immediate industrial demand for the remaining dry natural gas after the liquids were stripped, but transportation to industrial locations required construction of natural gas pipelines. This large-scale demand for dry natural gas was in large industrial locations like Houston and Chicago. Thus, the development of the natural gas market depended mightily on the construction of expensive, often long distance intrastate and interstate pipelines. Interstate pipeline development was complicated by multijurisdiction regulatory systems.

The immensely important natural gas market of today advanced during the 1930s in large measure because of two Texas innovations—the discovery and development of the Panhandle gas Field, and the control of natural gas flaring both in the Panhandle and throughout the state by the Texas Railroad Commission (TRC). The Panhandle Field provided access to a large enough resource to justify expensive pipelines, and the flaring control made gas available for a large and distant market for heating and boiler fuel for industry rather than wasting the resource in pursuit of crude oil production. But these innovations came about incrementally over the two decades of the 1930s and 1940s.

The largest natural gas discovery in the United States

A clear understanding of the importance of the Panhandle Field requires knowing the larger, dynamic process of which this huge resource was a part. The first fundamental is that natural gas can only be efficiently transported to markets from a remote field like the Panhandle via pipeline. Second, the natural gas can only be efficiently stored in underground cavities like salt domes. So, in order for natural gas to serve a load that is variable due to seasonal and daily demand shifts, storage became an integral part of a pipeline operation to deliver the right flow of energy to the user. Third, the high demand for natural gas is not usually near the source (the gas field) requiring a pipeline to deliver the gas to a user.

The next complication of the natural gas systems is that sales of the product across state lines (boundaries) comes under the purview of the Federal Power Commission responsible under the commerce clause of the US constitution for regulating interstate natural gas commerce. Therefore, the natural gas market is regulated under two governmental systems—the various state governments and the federal government. The two governmental systems often differ in the extent, and rule of regulation. That is certainly true for the State of Texas and the federal government.

The Panhandle Field is an area of some 200,000 surface acres stretching 125 miles in length and approximately 25 miles wide within eight High Plains counties—Hartley, Potter, Moore, Hutchinson, Carson, Gray, Wheeler, and Collingsworth (Prindle, 1981, p. 61). The discovery was an indirect outcome of a water-resource survey by a geologist named Charles Gould begun in 1904 followed by completion of a gas well called Masterson No. 1 by Amarillo Oil Company in 1918 (Olien, 2016). The field is primarily a natural gas deposit but the early discoveries included crude oil.

The first oil was found in the Panhandle Field in Wheeler County in 1910. Early oil production was supported by railroad tank car shipments to refineries before pipelines were built in 1926. Natural gas was of little value in the beginning because there existed no pipelines for moving the gas to users—the primary interest was in recovering crude oil. There were no natural gas pipelines at the time for exporting gas from the area but a pipeline was developed to move natural gas into Amarillo for local consumption in 1923. Casinghead gas from the oil wells was processed to obtain "natural" gasoline (condensate) at the first local gasoline plant constructed in 1925 and a carbon black plant was first set up in 1926 to make this a second useful product from the raw natural gas steam (Smith, 2018).

At the time of discovery of the Panhandle Field it was the largest natural gas reserve in the nation estimated to contain 15−25 trillion cubic feet (TCF) of natural gas. That is a quantity two to three times the aggregate, annual Texas natural gas production at the peak production period of 1972. It turned out that the Panhandle Field contained a huge deposit of natural gas but it was also a concentrated source of helium. The field quickly became the primary helium source in the United States. The federal government acquired 50,000 acres in the area and developed an extraction plant with storage near Amarillo in 1929. Together the natural gas and helium became the strongest driver for development. Although many legal and regulatory impediments had to be overcome by 1994 the field became the largest (by volume) gas field in the United States with annual production of 165 billion cubic feet of gas from 4499 wells. Over the 20-year period ending in 1993, the field yielded a cumulative total of 8.1 TCF of natural gas and 1.4 billion barrels of oil (Handbook of Texas Online, 2018).

The beginning TRC natural gas policy development was to classify the Panhandle Field deposit as one reservoir, which it did in 1940. The production of oil and gas was from several geologic horizons where the primary recovery of oil was supported by gas pressure in the reservoir. Following the initial discovery in 1918 and the drilling of several wells over the next 4 years, the fight began over how, when, and where the market for natural gas would be developed while the primary interest of oil production was accommodated.

The ongoing regulatory and legislative battle that developed because of the Panhandle Field in the 1930s happened at the same time period that the TRC was struggling to order the oil market in East Texas by well spacing and prorationing rules. The oil interests in the Panhandle opted to focus on being able to deal with the natural gas problem by stripping the liquids from the gas stream and flaring the dry gas—burning the gas at the well head. In 1933 the legislature, with oil producer interest support, passed a bill allowing Panhandle operators the strip and flare gas. The authority to use this practice, which was wasting about a billion cubic feet per day, was modified by follow up legislation in 1935 that allowed the practice only on associated gas. The revision recognized that there was an important difference between stripping and flaring of casinghead (associated) gas in support of oil production and stripping and flaring of unassociated gas. The later practice would be a great waste of dry natural gas that amounted to about 70% of the natural gas in the Panhandle Field (Prindle, 1981, p. 56). To produce this gas only to recover condensate gasoline and make carbon black, and then flare rest would amount to a great waste—a practice that the TRC was obligated to avoid under the conservation statutes the agency had fought hard to implement. (The condensate gasoline was mostly used to blend with refined gasoline at refineries. The carbon black was used in the manufacture of automobile tires, ink for printing, paints, and other products). So, the oil interest support singled out the casinghead (associated) gas challenge and the legislature passed a bill (H.B. 266) in 1935 forbidding the practice for unassociated gas. But the outcome was to continue allowing the practice for associated gas (Prindle, 1981, p. 61).

The stripping and flaring of associated gas in the Panhandle set up a president that grew in importance as new large oil fields were discovered and produced. Such new fields included Conroe in 1931, Tom O'Connor in 1934, Wasson in 1936, Levelland in 1938, and Hawkins in 1940. These fields all produced large quantities of associated gas. The flaring light up the night skies for years before the practice was finally ended.

While the legislature, the courts, and the TRC dealt with the flaring practice throughout the 1930s, the Panhandle gas field continued its development. The development of the Panhandle natural gas resource began seriously with the development of seven major interstate pipelines early in the decade of the 1930s. A 926 miles long pipeline from the Texas Panhandle to Chicago was completed in 1931. A second pipeline 940 miles long from the Panhandle to Indianapolis was completed in late 1931 (Patch, 1931).

The 1930s and 1940s saw the TRC gain prominence in the Texas and US energy sector by finally creating order out of chaos in oil markets in East Texas. There were two key policies of the TRC (discussed in the previous chapter) that gave the TRC a solid footing as the leading oil market regulatory body of the United States, and created stability in the industry. Following passage of enabling statutory authority from the Texas Legislature, and reviews by the Texas and US Supreme Courts the TRC developed and enforced well spacing and prorationing to market demand for crude oil. But the major unaddressed problem of natural gas flaring developed in the 1930s and continued through the 1940s. Ironically, this period with a focus on development of a natural gas market, and the eventual shut-down of flaring followed with the challenge of shortages and excess supply in a maturing natural gas market in the 1970s and early 1980s. But the roots of the natural gas challenge began much earlier in the Panhandle Field.

The TRC issued an order to proration based on market demand in the Panhandle Field on October 10, 1929, and on December 13, 1930 issued an order limiting production to 25% of the potential production. On August 12, 1931 the Texas legislature amended an 1899 statute (which had limited gas use to lighting, fuel, or power uses) to allow use for any practical use the TRC determined to be in the public interest. The Act defined "physical waste" and forbid the TRC from limiting production to market demand. On December 30, 1932 the TRC issued an order for the West Panhandle Field based on market demand and fixing the daily production allowable to match daily demand. This was the beginning of an ongoing process of legislation, TRC rule making, legal challenges, and US and Texas Supreme Court decisions that did not end until May 1, 1935 when the Legislature enacted a comprehensive gas regulation statute. The act provided detailed provisions for apportioning natural gas market demand throughout the state. It prohibited production that causes underground waste, prevents releasing gas into the atmosphere before removing its gasoline content and preventing the use of sweet gas for manufacturing of carbon black, while allowing sour gas to be used for that purpose (History of the Railroad Commission 1866–1939: Chronological Listing of Key Events in the History of the Railroad Commission of Texas, 1866–1939).

In the end the natural gas policy process, which was driven heavily by the conditions in the Panhandle Field, resulted in rules promulgated field by field, prorationing to Texas-wide market demand, distinguishing between sweet and sour gas, setting well spacing, and limiting use among different uses (especially restricting use of sweet gas for carbon black manufacture), apportioning fairly among producers, distinguish between sweet and sour gas, setting production allowable and the uses of each, and allowing for special exemptions—all derived from the TRC conservation doctrine. Then followed the Texas Supreme Court decision upholding TRC authority to prevent flaring of gas in 1947. (Note: sour gas has high concentrations of hydrogen sulfide and carbon dioxide; sweet gas has little of such contaminants.)

The developed TRC rules fundamentally define oil fields separate from gas fields and thus the limits of well spacing (density allowable typically are 40 acres for oil wells and 160 acres for gas wells). These TRC guidelines differ somewhat through "diveations" to the base rule but the idea is that oil and gas are efficiently "drained" from the reservoir from greatly different well concentrations. The natural gas flows move easily to well bores over a long distance while the crude oil flows are much more confined.

Interstate pipelines

The Texas deposits of natural gas—the combination of associated and nonassociated gas—provided much more production than the Texas market demand by industry and commercial/residential users could absorb. The rest of the production relied on out-of-state users. So, the interstate pipelines, joined with underground storage facilities, had to be an integral part of the Texas gas market. A major share of the Texas interstate gas sales came ultimately from the Panhandle Field. The gas producers in the Panhandle of Texas found the best markets to be in the Midwest, especially the Chicago and Indianapolis industrial complex. As the Texas market matured, interstate on-shore sales by 1976 made up 45% of the market with 55% staying in Texas for Texas users. Total Texas gas production increased by 10-fold from 0.4 TCF in 1929 to 4.4 TCF in 1950 and then to 19.3 trillion by 1980 (Energy Information Administration, December 1981, pp. 38–40, An Analysis of the Natural Gas Policy Act and Several Alternatives; Part I: The Current State of the Natural Gas Market).

The complexity of natural gas regulation

The Panhandle Field, along with its next door neighbor, the Hugoton Field that extends from just north of the Panhandle Field near Amarillo into Oklahoma and Kansas, had a great impact on how the complexity of the gas market developed. These fields became a primary source of natural gas for the interstate market, a development that set up the dynamics for a major political battle that developed during the decade following the 1973 Oil Embargo.

The complexity of the natural gas market begins in 1938 when the federal government began regulation of the transportation and sale for resale of natural gas in the interstate market. Interstate sales are covered by the commerce clause of the US constitution and the 1938 Natural Gas Act (NGA) declared such and assigned the regulatory authority to the Federal Power Commission (FPC—now the Federal Energy Regulatory Commission). The NGA granted the FPC authority over all gas moving in interstate commerce, sales jurisdiction over wholesale gas sales in interstate commerce, and general jurisdiction over gas companies engaged in interstate transportation or sales. Intrastate sales and transportation, direct sales to industrial customers, and production and gathering were exempt—thus a bifurcated market destined for the chaos that came later in the 1970s. A law case known as the Phillips case and a subsequent ruling, by the US Supreme Court in 1954, extended the FPC jurisdiction to the regulation of well head prices of interstate natural gas and specified that gas reserves first engaged in sales to an interstate pipeline were dedicated to interstate commerce for the life of the field (Texas Energy and Natural Resources Advisory Council, 1982).

The well head price controls, which held prices below the market prices being traded in the unregulated intrastate market, caused shortages (deliverability deficits) in the interstate market beginning in the early 1970s as oil and gas prices began a rapid increase following the Arab Embargo. Although the price increases in the natural gas market were not directly affected by the Embargo, energy products are to some extent substitutes in consumption (e.g., natural gas for fuel oil), so the Embargo had an indirect effect on gas prices, and more so on intrastate than interstate gas. So the market allocation among the two markets left the interstate market short. Producers naturally choose to make sales of newly produced gas to higher priced alternatives in the intrastate markets. The result was deliverability shortfalls in the interstate market of the United States that rose rapidly from 0.3 TCF in 1971 to 3.2 TCF in 1976. The 3.2 TCF amounted to about 11% of the interstate gas demand. (Governor's Energy Advisory Council, 1977a,b,c, p. 7).

The federal government focus on energy policy during the decade following the Arab Embargo of 1973 covered the full spectrum of energy issues. The policy changes of the Nixon, Ford, Carter, and Reagan eras reviewed in the following chapter dealt with every imaginable aspect of energy and its relationship to national security. The rising shortages of natural gas, our most preferred alternative to imported oil, became a major issue and as a result the Carter National Energy Plan of 1978 included two statutes that dealt with natural gas. There was the Powerplant and Industrial Fuel Use Act of 1978 (FUA) that focused on mandatory reduction of natural gas use as a boiler fuel in electric utility and industrial sectors. The second act was the Natural Gas Policy Act of 1978 (NGPA). The focus of NGPA was to set price ceilings on some 22 different categories of natural gas production in both interstate and intrastate markets, where the various categories were defined by the vintage of the wells (so-called "old" gas and "new" gas), whether the gas was in the interstate or intrastate market, was on-shore or off-shore, defined as deep gas or hard to recover gas, etc. The Act then scheduled a phase out of price controls over a several year period of time (usually 1985) for most of the 23 categories. This era of national energy policy that included the regulation of natural gas markets served to continue the debates, court rulings, statutory revisions of the legislature and TRC rule changes finally allowed this important market to develop in a more orderly fashion. But the dynamics of the market still over- shadowed by the discovery and development process of the past and historically low prices lead to the continual decline of Texas natural gas production through the 1990s. But the world changed beginning in 2005 with the serious beginning of fracking technology coupled with historically high oil and gas prices.

While the energy crisis resulted in federal policy that reached into the domain of state regulation by setting price controls for a large set of defined gas production categories and limited natural gas use by attempting to phase out the use of gas for boiler fuel, Texas retained the right to regulate the natural gas market under the state's conservation statutes. The long-standing authority of the state to set well spacing and allocate, and limit production to meet market demand was maintained and respected by the courts. A major part of the remaining state authority provided the flaring policy that greatly increased the availability of gas for beneficial use. The TRC developed and enforced rules for preventing the flaring of gas at the well head.

Natural gas flaring

Crude oil was the economic product and natural gas was a waste byproduct during the early stage of petroleum industry development. Crude oil, and its stream of refined products, all being liquids can be stored easily and transported by various means—trucks, pipelines, barges, and ocean-going tankers. The products of crude oil refining including gasoline, heating oil, jet fuel, kerosene, greases, and other products are very versatile in the range of uses they serve and there are several storage and transportation options available to move them from the refinery to wholesale distribution centers, and finally, retail outlets to consumers.

Natural gas, on the other hand, is more difficult to store and transport and is less versatile in its uses. Following the development of a major delivery system of natural gas pipelines to the nation's factories, electric utilities, and residential/commercial buildings, natural gas eventually became a premium fuel. But this was not so in the beginning. Since a pipeline system did not exist, storage was very expensive and no useful market existed; natural gas was indeed a "waste" product of crude oil production. As such, the substance was either vented to the atmosphere or burned (flared) at the well site by way of open ended pipes extending into the air away from the well. This was the least costly way of disposing of an unwanted byproduct.

Gas flaring therefore became the answer to two problems for crude oil producers. First, gas from "associated" oil wells is an inescapable consequence of producing crude oil since the gas is dissolved in the oil. As the oil reaches the surface of the well and the natural pressure is dissipated, natural gas is released and is known as "casinghead" gas. Pipeline systems and markets for the gas did not exist, so the least costly solution for the producer was to flare the gas (Stockton et al., 1952, pp. 231–236). At the peak of gas flaring the entire country side was light at night in many of the large oil fields of Texas.

Second, prior to development of major pipeline systems, a producer would not purposely drill for natural gas, but sometimes he found it anyway. Although there was no market for the gas, such wells typically produced a fine grade gasoline as a consequence of the expansion of the gas at the wellhead, for which there was a market. Therefore, companies set up plants in the gas field to "strip" and market the gasoline from the gas flow and then flare the remaining gas. This process captured a useful product from about 10% of the gas; the remaining 90% was wasted (Cheek, 1938, p. 279).

During the late 1940s, a series of TRC actions led to the eventual end of gas flaring in 1949.[1] It became clear in the reporting of gas flaring by producers that the extent of the waste was being grossly underestimated. A full scale study in the late 1940s accurately documented the size of the waste problem and confirmed the under-reporting of the companies. Beginning in 1947, the commission began shutting down oil wells until cycling plants or pipelines could be installed. In the following months, the Commission shut down 17 Texas fields for flaring. The orders were appealed to the courts but TRC was upheld and the gas flaring essentially stopped in 1949 (Railroad Commission vs. Flour Bluff Oil Co., 1949, error ref'd., p. 508). The gas flaring issue has found its way into the current debate again, however, as the Permian Basin pipeline shortage puts the flaring issue in the spotlight. In existing production areas, flaring may be necessary because existing pipelines may have no more capacity. Commission staff issue flare permits for 45 days at a time for a maximum limit of 180 days (Collins, 2018).

Since the TRC was charged with implementing the state's conservation law, the agency struggled during the 1930s and 1940s to deal with the problem of gas flaring. The problem was fundamentally different from the oil problems, and during most of these years the commission was still struggling to maintain its authority to regulate and to deal with oil problems. The gas problem was different in that it was not a matter of adopting policy that settled differences among the various functions of the industry. All producers, majors, and independents alike wanted to flare gas because it fit their short-term economic interest in oil recovery. Since the commissioners needed industry support for reelection, it was difficult for them to stop gas flaring for conservation purposes. In the meantime, TRC engineers were working to build the case for eventual elimination of flaring (Prindle, 1981, p. 58).

The practice of gas flaring ran unchecked until it was finally stopped by the Railroad Commission in 1949. One can only guess at the magnitude of natural gas wasted through this process prior to enforcement of the new rules. Data is not available for the early years of flaring until 1952. Vented and flared natural gas was 0.354 TCF in 1952 declining to 0.059 TCF in 1972. For comparison, at the peak of Texas production in 1972, the state's annual production rate was 9.6 TCF (US Department of Interior, Bureau of Mines, 1974). The volume of wasted gas was enormous before the practice was eventually reduced and stopped for most oil production.

Gas flaring problems were of two types and their history developed separately. Early conservation laws—the 1899 law and later revisions—prohibited the flaring of gas if it

---

[1] Bill Murry served as Chairman of the Texas Railroad Commission during this period and is widely credited with bringing natural gas flaring under control. Murry was a petroleum engineer by training and was finally instrumental in solving the flaring problem by getting a few companies to experiment with reinjecting gas (after removing liquids at the surface) back into the oil reservoir to maintain reservoir pressure and thus increase oil recovery, and to learn that the process was an economic proposition. He chaired a Gas Conservation Engineering Committee at the TRC in 1945 which estimated that 57% of Texas' total gas output was being burned.

was classified as unassociated gas.[2] In 1925, the legislature passed a law allowing gas to be flared if it was classified as associated gas from an oil well. This dual law became impossible to implement, however. One problem was that gas changes from a gas to a liquid state, and vice versa, as temperature and pressure changes occur. Therefore, it was difficult to determine what should be classified as an oil well versus a gas well. The first success in preventing associated gas flaring by TRC was possible because of the ambiguity of such definitions and the changes in liquid/gaseous state of natural gas.

In a series of court orders, known as the Clymore case, the commission was allowed to execute an order to reclassify hundreds of wells as gas wells rather than oil wells, as originally classified, and thus stopping the flaring of gas (Clymore Production Co. vs. Thompson, 1935). These wells produced a clear oil liquid at the wellhead, which if counted as crude oil for purposes of classifying the well, caused the ratio of gas to oil to be less than the 100,000 to one rule, thus classifying the wells as oil wells which allowed the gas to be flared. Commission studies, however, allowed that if the oil liquid was pressurized and heated to reservoir conditions, it became a gas. The wells, it was argued, should therefore be classified as gas wells instead of oil wells and flaring would have to be stopped. The courts upheld the commission order and the operators were faced with shutting down the wells or finding another way of disposing of the gas.

The solution finally implemented was to cycle the gas. By stripping the liquids from the gas at the surface and returning the stripped (dry) gas to the reservoir, it would help maintain reservoir pressure and thus produce more crude oil and oil liquids, and so on. Thus, the commission succeeded in stopping flaring in some cases by reclassifying the wells as gas wells and encouraging industry to set up cycling plants for recycling the gas.

In another important episode, producers in the Panhandle convinced the legislature in 1933 to pass a law specifically allowing the flaring of gas in the Panhandle gas field. This field originally contained 15−25 TCF of gas, and under this law, over 1 billion cubic feet per day was being flared. The public outcry and interest of the pipeline companies in the gas for commercial use caused the legislature to reverse itself in 1934, making the law stronger than before and giving TRC the authority to prorate gas (Acts, 44th Leg., regular session, 1935, Chap. 120).

Natural gas proration and ratable take

The TRC unsuccessfully struggled with mechanisms to solve problems associated with excess supply (or excess deliverability as it was called in the gas industry) of natural gas during the late 1970s and throughout the 1980s (Murry, 1983). The effects of

---

[2] Unassociated gas is natural gas from a well that is not an oil well. Texas statutes defined an oil well as one having a gas to oil ratio of less than 100,000 cubic feet of gas per one barrel of oil produced. All other wells were classified as oil wells. Gas from oil wells is called associated gas.

government intervention in oil and gas markets, and the unusually high world oil prices following the Arab Embargo of 1973 and the Iranian crisis of 1979—80, left the gas market in the intrastate market in a condition of excess supply. The surplus may have been as large as 1.5 TCF of gas (8% of the US market) at the peak.

Gas shortages during the cold winters of 1975 and 1977 resulted in the belief by many policy makers that the United States was woefully short of natural gas resources and that government had to take the initiative to reduce demand and encourage supply—but the emphasis was on reducing demand. The use of gas for "low priority" boiler fuel in industrial and electric utility uses was to be phased out. The TRC first adopted Order 600 mandating such reductions in gas use in 1975. The federal government followed with a more aggressive schedule of phase out under provisions of the Powerplant and Industrial Fuel Use Act (H.R. 5146), 95th Congress (1977—78). The Act called for the complete phase out of natural gas use in electric utility boilers by 1990 and a slightly less onerous schedule for industrial boiler fuel. This was the principal demand-side government policy initiative.

The supply side policy was set up under an incentive price structure in the NGPA of 1978. The NGPA set up some 23 categories of gas, each with its own future price schedule. High cost gas was given a special incentive price that escalated beyond $5 per mcf by the early 1980s. New gas (gas from newly discovered reservoirs and new wells) under certain conditions was to be deregulated in the late 1980s.

These government interventions on both the supply and demand sides of the market, along with the influence of end user energy prices, created the major surplus of gas on the Texas market. Drilling effort expanded in response to incentive prices, and electric utilities built coal and nuclear generators rather than gas units to meet electric demand growth.

It was in this climate of market adjustment that the TRC undertook to use their regulatory powers to bring order to the gas market in Texas. The problem of the gas market was much different from that of the oil markets of the 1930s and 1940s, however, and such attempts were ultimately to fail. Some industry interests characterized the problem of the 1980s gas market as depressed prices, while others were concerned about equity of market access among producers.

The TRC could not legally restrict production to raise prices. The emphasis, as was the history of oil proration, had to be pursued in terms of preventing economic waste of the resource, a case much harder to make for gas than for oil. Furthermore, given the competition for gas markets from interstate and Canadian sources, in a growing climate of federal deregulation, the TRC could not practically raise prices by prorationing to market demand. The gas market continued with depressed prices for well over a decade (see Fig. 7.1).

The attempts of TRC to order the natural gas market failed because the market was always outside the commission's control due to the dual regulatory system of state

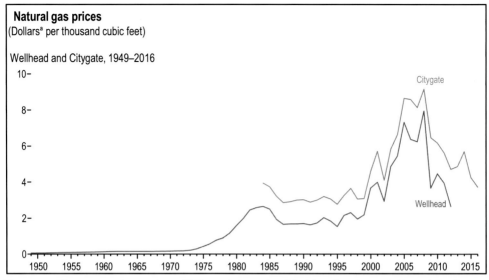

Figure 7.1 Texas natural gas prices. *US Energy Information Administration, Domestic Natural Gas Prices at the wellhead and city gate (Dollars per mcf) Available at https://www.eia.gov/dnav/pet/pet_pri_dfp1_k_a.htm.*

and federal governments. The Phillips case of the 1950s established the federal government's power to regulate the price of interstate gas at the wellhead and to require the lifetime dedication of reserves to interstate pipelines (Phillips Petroleum Co. v. State of Wisconsin, 1954). This price regulation amounted to a system of price ceilings for interstate gas, which became more and more restrictive, strengthened the intrastate market where prices were unregulated. The federal government acted unintentionally to strengthen supply and price conditions in the intrastate market. The bankruptcy of the federal regulatory system eventually caused the US Congress to deregulate the industry, however, removing any remaining power the TRC had to order the market.

Industrial marketing programs and deregulation

The deregulation of the natural gas markets in the United States occurred primarily due to state-level approval of a series of actions beginning with "industrial marketing" programs, much of the leadership coming from Kenneth Lay of Enron during the 1990s. The excess deliverability of natural gas created by the market response to federal (and in some cases, state) wellhead price ceiling removals created an incentive to undercut the prevailing price of pipeline commercial/industrial prices built into the pipeline regulated price for commercial/industrial consumers. The industrial marketing programs expanded the use of surplus gas to lower prices designed to meet the competition from oil and coal.

The Texas market became the national poster child for the industrial marketing programs and Kenneth Lay was the high-profile leader of the idea that greatly changed the market for natural gas. But the natural gas market is closely integrated with the electric markets because natural gas is more environmentally friendly for power generation than the alternatives like coal. The interplay of the natural gas and electric markets became ever so obvious in one of the most notorious scandals in modern history—the California energy crisis that came to a head in year 2000 electric blackouts that were unprecedented in the history of the electric industry. He also lead a similar effort in California which resulted in chaos in that natural gas market that in turn impacted the deregulated electric market. The California chaos resulted in federal prosecution laying the blame for fraud at the doorstep of Enron and the market design in the state that allowed the abuses.

The innovations in the Texas natural gas markets

The Texas contribution to the development of the US natural gas market is primarily the result of two major episodes of the Texas energy experience. The first is the contribution of supplies of natural gas from the Panhandle Field that has proven a valuable resource for the country for 90 years, but more importantly, the lessons learned, and innovations that have grown out of the Panhandle experience have amounted to another innovation. What East Texas was to ordering the oil market, the Panhandle was to the ordering of the gas market. While the well spacing and prorationing to market demand did not provide price stability as the similar rules did for crude oil, the gas market rules provided a means to share the market benefits among industry participants, and most importantly, to prevent the waste of gas through venting and flaring. The level of detail attention to the distinction between associated and unassociated gas, and between sour and sweet gas, and attempts at statewide prorationing to market demand, provided a stability outcome while preventing waste that was a major contribution of the experience of the Panhandle Field.

The second, related contribution has been the success of policy and technology that closed down the wasteful practice of natural gas flaring. The TRC's success in finally bringing the wasteful flaring of natural gas under control was ta key innovation in the Texas natural gas market in the last half of the 20th century. The innovation was an institutional, technical and market based achievement—institutional because the Texas regulatory body oversaw a solution that remains institutionalized even today and technical and market based because it demonstrated that natural gas could be economically reinjected into the reservoir for oil recovery.

Attempts of the TRC to order the natural gas market as was accomplished in the oil market, however, proved to be an unattainable goal. The split federal/state regulatory approach to regulation, which for a key period imposed wellhead price controls and demand-side boiler fuel phase out, in both the interstate and intrastate markets, respectfully made the TRC attempts through prorationing to market demand fail, at least if the measure of success is assumed to be stable market prices. The eventual deregulation of wellhead prices and removal of the off-gas rules on the demand-side of the market served to eventually stabilize the natural gas market. The TRC could not achieve market stability acting alone as it had done for crude oil.

The interplay of policy, markets, and technology

The Embargo was the major energy issue in the 1970s but it certainly was not the only one. There were natural gas curtailments beginning in the winter of 1969–70 on the Eastern Seaboard, and in Texas during the winter of 1972–73. The Texas gas market then developed a surplus of natural gas as the joint effects of rising natural gas prices and regulated use of gas for boiler fuel use.

The natural gas shortages were, for the most part, quite independent of the events that allowed the Arabs to shift oil market power from Texas to Saudi Arabia. The federal government had maintained a heavy hand in the natural gas industry since the 1930s when the major interstate pipelines were developed. Antitrust concerns about the operation of the large natural gas pipeline companies eventually led to wellhead price controls dictated by a Supreme Court decision in 1954 (Phillips Petroleum Co. v. State of Wisconsin, 1954). Gas reserves, once under contract to interstate pipelines, were considered dedicated to interstate commerce for the life of the gas field. The producers and pipelines that served only the Texas market were not subject to such restrictions, being guided instead by the contractual agreement between producer and pipeline and pipeline and consumer. The result was the creation of two markets, one for out-of-state (interstate) and one for in-state gas (intrastate).

In the face of a strong demand growth in all energy markets, but especially natural gas, the federal wellhead price controls on interstate gas greatly restricted the industry's willingness to add new reserves by drilling new wells. Nationwide reserves relative to production were in decline. But in general, the Texas natural gas market had ample reserves to supply current and future consumption needs, while other states dependent on interstate pipelines and interstate gas saw their consumers face major shortages in the winters of 1972–77. It was estimated that 15% of the interstate market was curtailed during the peak shortage year of 1976. Curtailments amounted to 5% in 1973, the year of the Embargo (Holloway, 1977).

Although the Texas market was spared significant statewide shortages, a major south Texas firm, Lo-Vaca, a Coastal States pipeline subsidiary, made contract commitments to supply natural gas to most of central and south Texas without adding sufficient reserves to its holdings. Lo-Vaca committed to sell gas at an attractive, fixed long term contract price which could not cover the costs of new gas reserves when well head prices began to rise. Rapidly growing markets in the face of inadequate reserves brought shortages (curtailments) to major communities in central and south Texas during the winter of 1973—74. In short, Lo-Vaca sold more gas than it could deliver, hoping to add reserves as needed. This event was only remotely related to the federal ceiling prices on interstate natural gas. The Lo-Vaca problem persisted for several years before TRC got the matter under control. The crisis created by this incident was in progress at the time of the Arab Embargo (Coastal Corporation, 2017).

Impacts of the natural gas market and institutional innovations

Institutional, policy and technological innovation that developed the modern natural gas market happened importantly in Texas. These developments included the development of complex rules first developed and implemented in the Panhandle Field and the technology and institutional innovation that **stopped the flaring of natural gas** created the major impact on the natural gas market development and it happened in Texas. The results developed an orderly process that is the Panhandle Field's ongoing energy contribution that not only enhanced oil recovery by reinjecting the gas into the reservoir it provided supplies for commercial use that has been a relatively clean fuel for both industrial boiler fuel and electric power plants. Natural gas is currently the best available "bridge" to other technologies to slow the emissions of CO_2 that are contributing to global warming.

CHAPTER 8

Upheaval in the energy markets: the Arab Oil Embargo and the Iranian Crisis

Contents

What happened

On October 20, 1973 the Arabian nations of the Middle East imposed an embargo on the shipment of crude oil to the United States and several other countries. Because the United States supported Israel in the 1973 Arab-Israeli War, Arab states of the Organization of the Petroleum Exporting Countries (OPEC) instituted an oil embargo against the United States. Initially the Arab exporting countries reduced production by 5%, which was cut by an additional 5% each month, and banned oil exports to the United States and the Netherlands. Subsequently, the list of embargoed countries included Portugal, South Africa, and Rhodesia. By December 1973 the Arabs had cut 5 million barrels per day from the world market, or about 10% of the total free-world consumption levels (Yergin, 1991, pp. 613–614).

In February 1979 a renewed energy crisis came out of turmoil in Iran. The Shah of Iran, Mohammad Reza Pahlavi known as Mohammad Reza Shah, a long time ally

Innovation Dynamics and Policy in the Energy Sector
DOI: https://doi.org/10.1016/B978-0-12-823813-4.00008-7

of the United States and Britain, was overthrown by the Iranian Revolution on February 11, 1979. Ayatollah Khomeini who opposed the shah returned triumphantly to Iran. He had organized the revolution in exile from Paris. He established the Islamic Republic in Iran (Energy Policy, 1981, p. 22).

These two events—the Arab Embargo and the Iranian Crisis—set up a new geopolitical balance that transformed US involvements in the Middle East, and surprisingly set the United States on a new path of energy innovation and defined a new interdependency between energy and national security policy. The pathway to new sources of energy and national security policy did not happen right away as such fundamental changes take time. We are still playing out the pathway of innovation. Texas has been a big part of these transitions.

Because of the military importance of oil and the lack of adequate supplies in Europe there was a major effort at the end of World War I to tie up oil resources in the Middle East. The French and British struggled over influence and access to known and potential Middle East oil supplies that grew out of the "spoils" of victory. The primary focus of these struggles was the oil reserves in Iraq.

The US involvement in the Middle East oil supplies during the period following World War I that intensified after World War II was built around the evolution of the Aramco Oil Company. Aramco was originally a joint venture between Texaco and Standard Oil Co. of California (Socal), two American oil companies to work with Saudi Arabia to develop oil supplies.[1] The arrangement was known as a concession.[1] King Ibn Saud of Saudi Arabia was irrevocably committed to maintaining the concession for access to the Saudi oil by American companies as begun with Aramco.

Texaco grew out the Spindletop oil boom in 1901, and Socal was one of the seven major entities that became the successors to the breakup of the John D. Rockefeller's Standard Oil Company in the late 1890s. Aramco's vision of developing Middle East Oil (primarily Saudi Arabian oil) meant they had to develop all of the downstream infrastructure in Europe (their target market) and they did not have the financial resources to develop the full supply chain. The evolution of options for expansion included joining with the European companies that were already heavily committed to the Middle East, but such a joint company would have run against the rivalries that had developed among American, British, and French companies doing business in the Middle East. Eventually, much to the chagrin of Socal owners, Aramco brought in Standard Oil of New Jersey and Standard Oil Company of New York (Socony). The series of developments over the following 55 years kept the relationship between Saudi Arbia and the United States intact until the Embargo in 1973. Leading up to the Embargo the Saudi government acquired a 25% stake in Aramco. The Saudi share

---

[1] An oil concession is an agreement—a grant by a country to permit a company to explore for and produce oil.

was increased to 60% in 1974, and then 100% in 1980. Aramco is now the most valuable company in the world valued at about $2 trillion (Aramco History, 2018).[2]

As a result of the World War I experience Britain recognized that oil was the greatest resource deficiency Britain faced for maintaining an adequate national security defense. US involvement in the Middle East was secondary at the time. By the time of World War II that all changed. National security and oil became inseparable for the United States as well. As the post—World War II era unfolded the discoveries of oil reserves in the Middle East, Venezuela, Kuwait, Iraq, Iran, and other OPEC countries gave OPEC a large enough market share to greatly influence the world oil price. The embargo was targeted at nations perceived as supporting Israel during the Yom Kippur War of 1973.

While the Yom Kippur War was a contributing factor to the 1973 Embargo, OPEC's market power to set prices by production controls gave them the power they had been waiting for; hence, the Embargo. The Arab countries had carried out a prior embargo following the Israeli Six-Day War of 1967 but it failed to have much impact because the United States, Venezuela, Iran and Indonesia immediately ramped up production to make up for the Arab shortfall of 1.6 mmbd (Yergin, 1991, p. 557). But the Embargo of 1973 was different—the United States no longer had spare capacity to fill the void created by the 5 mmbd embargo shortfall (Yergin, 1991, p. 614).

The US spare capacity had prevailed for more than two decades with the "prorationing" policies of the Texas Railroad Commission (TRC) that controlled production below the maximum capacity. This policy kept a spare reserve available for emergency purposes (estimated at 4 mmbd during the period). The reserves served their security purpose much like the World War II effort and lesser crises of the Six-Day War and the Suez event. While the TRC policy kept prices stable and high enough to support development and short-term production, eventually it deterred investment and new discoveries—the spare capacity had disappeared by 1972.

As the Embargo of 1973 unfolded the major oil companies who were producing crude oil in the Arab countries acted to allocate the shortfalls among countries in order to avoid unequal economic disruptions in the world economies; they needed to avoid criticism for national favoritism. By agreement among the oil companies the shortages of crude oil were allocated proportionately among oil importing countries throughout the free-world economies of Western Europe, Japan, and the United States, based on each country's current consumption of oil. In the end the loss to Japan was 17%, to the United States 18%, and to Western Europe 16%. Still, the embargo caused shock

---

[2] Aramco has recently begun the process of putting the company up as a publically owned entity in order to raise capital for other infrastructure purposes. As of January 2018 Aramco had not yet decided which of the world's stock exchanges as the preferred venue. The value of Aramco may be as much as $2 trillion (see Wall Street Journal, 2018).

waves throughout the developed world economy as nations scrambled to reduce consumption and manage skyrocketing prices of gasoline and other products. By December 1973, oil traded on the international markets was short by 14.4%. World oil prices reached $17.14 per barrel in December 1973, up from $2.18 for Arabian oil in 1971 (Yergin, 1991, pp. 620–625).

The Arabs targeted the United States with the "oil weapon" with the intent of changing US–Israeli policy in the Middle East. The Egyptian conflict with Israel, which developed into the Egyptian and Syrian War with Israel, was reinforced by the use of oil as an economic and political instrument. The use of oil for these purposes was made possible by an agreement between Anwar Sadat of Egypt and King Faisal of Saudi Arabia (Yergin, 1991, p. 593). The US excess oil-producing capacity which had been available since World War II to supply oil to our allies and maintain stable oil prices was no longer available. Texas, Louisiana, and Oklahoma were producing at 100% of oil production capacity and could no longer put additional supply into the world market. The long talked-about "oil weapon" of Middle East politics became a reality. The agreement between Saudi Arabia and Egypt was to use oil as a means of achieving a change in the balance of power in the Middle East. Sadat, who became the leader of a humiliated Egypt following the Six-Day War of 1967, searched for and found a way to break the stalemate of his country with Israel—a stalemate held together by US policy of support for Israel. Sadat achieved his purpose through the Egyptian and Syrian War with Israel and the oil embargo. He got the attention of the United States, which eventually resulted in a more balanced US role in the region.

King Faisal had previously rejected the use of oil as a weapon of politics and economics in the Middle East because Saudi Arabia's future and that of the United States were economically and strategically linked. He also knew that any cut-off of Persian Gulf oil supplies would be met with increased production from the United States (especially Texas) and was thus a useless gesture. The Saudis had actually tried such a tactic during the 1967 Six-Day war which failed for exactly that reason—the US excess production capacity joined with additional production from Iran, Indonesia and Venezuela filled the shortfall (Yergin, 1991, p. 594).

Suddenly, however, the rapid growth of world demand for oil running at 7.5% per year (Yergin, 1991, p. 614) and the inability of the United States to increase production to maintain stable prices in a crisis had changed. At last Saudi Arabia had replaced Texas as the marginal producer of world oil supplies. King Faisal recognized that this market change of the loss of excess US production capacity, long anticipated in the 1985 time frame, had suddenly been pushed back to 1973. In support of the alliance of Arab states and the Palestinians, and to foster his own position in Middle East politics, King Faisal changed his attitude about the "oil weapon" in the Spring of 1973 (Yergin, 1991, p. 595). A series of actions and public statements progressed from this point until the embargo was enacted on October 20, 1973. This event surprised the political leadership in Washington and sent Texas scrambling for a new energy policy (Clements, 1999).

The Arabs had cut off exports in a tight world oil market which sent prices soaring. Texas average oil prices doubled from $3.48 in 1972 to $6.90 in 1974 (Stevens and Cummings, 1977) even though price controls were in effect.[3] Prices in the world market reached $17.04 per barrel at the peak of the embargo in December, 1973 while the posted price[4] in the Middle East was $5.40. Bids at auctions in the Middle East were reported as high as $22 per barrel.

OPEC met in Tehran in late December 1973 and set the "marker" price of Arabian light crude (the premium grade) at $11.65 per barrel. The sequence of posted prices in less than a 4-year period had been raised from $1.80 in 1970 to $2.18 in 1971, $2.90 in mid-1973, $5.12 in October 1973, and $11.65 following the December 1973 meeting of OPEC (Yergin, 1991, pp. 620−625).

The economic effects of the embargo were staggering. To get a good idea of how important such price increases were, consider that the free-world oil production was running at 50 million barrels per day; at $2 per barrel world oil production would be worth $36 billion per year; at $17 per barrel at 45 million per day, oil production would have been worth $279 billion per year, if prices everywhere were allowed to move to market levels. Since the OECD countries Gross Domestic Product (GDP) was $3.3 trillion at the time, such oil price increases would amount to 7.4% of OECD GDP. (Since oil prices were controlled in most of the world, the actual impacts occurred as a result of economic consequences of shortages rather than price increases.) Economists often described the event as having the same effect as placing a huge tax on importing countries to the benefit of the Arabian exporting countries.

The Arab embargo caused major economic disruptions throughout the developed world economy. Secretary of State Henry Kissinger said that the embargo cost the United States 500,000 jobs, more than $10 billion in national production and rampant inflation (Congressional Quarterly, Inc., Energy Policy, 1979, p. 3-A). The federal government's response to the shortages caused by the embargo was intended to cause the pain to be shared equally across the country and among user classes. Gasoline shortages fell more heavily on the ordinary motorist, however, because the government

---

[3] Nationwide price controls were imposed by the Nixon Administration under the Economic Stabilization Program of 1971. The Economic Stabilization Act (P. L. 93−28) passed the Congress in August 1971; future amendments allowed the President to continue price controls on crude oil following a lifting of price controls on the economy in April 1974. Source: Congressional Quarterly, Inc. **Energy Policy** (Washington DC, April 1979, p. 12−A).

[4] The posted price is a term that evolved from the early years of the oil industry where producers "posted" their offering price periodically, a price that represented the closest thing available to a measure of the market price. Companies still use the posted price but the current availability of daily spot prices and the futures market prices for oil arguably provide a more accurate measure of market prices.

set up an allocation program that provided an equal percentage reduction in oil products to regions and classes of users, while exempting certain classes. The impact was forced to fall more heavily on ordinary motorist because all kinds of emergency users were supplied all of their needed supplies. Federal rules imposed a 55 mph speed limit but contrary to expectations actual gasoline consumption in 1974 fell by only 1% or 100,000 barrels per day.[5]

The world economy was shaken by oil shortages for a second time in 5 years as Middle East conflicts unfolded in 1979. The Iranian revolution shut down most of the Iranian oil fields taking 4 million barrels per day of supply off the world market. Iranian oil production was a major contributor to world oil supplies, amounting to 8.8% of total world and 11.3% of the noncommunist world oil production in 1978. As a share of the total export market for crude oil, this crisis was smaller than that of the embargo in 1973, but the effect on oil prices was considerably greater.

The period following the Arab Embargo, which ended in March of 1974 was filled with regular OPEC meetings and disagreements over the price of oil. The embargo caused a recession throughout the world economy with a resulting decrease in the demand for OPEC oil. OPEC production declined from 30.7 mbd in 1974 to 27.2 mbd in 1975 (Energy Information Administration, 1986). Inflation rates were also high, pushed up by oil prices, which reduced the purchasing power of OPEC member revenues. Reduced revenues severely strained the fabric of OPEC as key members began cheating on their production quotas. Saudi Arabia, being the dominant producer of the group, managed to keep OPEC from falling apart by adjusting its production downward by 1.4 million barrels a day. The posted price of oil was set at $11.65 during the December 1973 meeting of OPEC. The average price of OPEC oil dropped slightly to $11.02 during 1975. The July 1977 meeting of OPEC resulted in an agreement to forego a scheduled 5% price increase. Demand for OPEC oil slackened again in 1977—78, in part because of new production from Alaska and the North Sea which added about 1 million barrels per day to the world supply. OPEC implemented an 18-month freeze on prices (Energy Policy, 1981, p. 21).

The shah of Iran, Reza Pahlavi, had left the country in exile (eventually living in Mexico) and set up the government of Premier Shahpur Bakhtiar in 1978. But by February 1979, the Ayatollah Khomeini returned triumphantly to Iran from exile and

---

[5] Several states, including Texas tried to set up low speed limits to reduce gasoline consumption. Texas Governor Dolph Briscoe proposed a 55 mph limit that was imposed by the Texas Highway Commission in late November 1973, but the Attorney General found the Commission exceeded their authority—it belonged to the Legislature to set limits. President Nixon soon after set a 55 mph limit using the power to cut off Federal highway funding to get states to comply—which they did. The limits were loosely enforced, however, and in 1982 the federal limit was raised for some locations to 65 mph; the law was finally repealed in 1995.

the Bakhtiar government was overthrown. Khomeini had organized the revolution from Paris. He established the Islamic Republic in Iran. The shut down of Iranian oil production, in the face of recovered economic growth and low inventories throughout in most of the developed world economies, sent prices soaring. The world economy was finally recovering from the 1973 to 1974 embargo and economic growth spurred oil demand upward from 50 mbd in 1978 to 62 mbd in 1979. OPEC moved the oil price to $14.54 in March of 1979, but for the first time allowed members to add surcharges to the official price (Energy Policy, 1981, p. 22).

Spot market prices of oil began spiraling upward to $25 and $30 per barrel as countries scrambled to store up their inventories. Americans were lining up at gasoline stations for the second time in 5 years. The shortages caused gasoline lines first in California and by May 1979 the shortages had spread to the East coast (Energy Policy, 1981, p. 22). OPEC first limited their official price to $23.50 per barrel in June of 1979. But rising interests in increased revenues and strong demand prompted OPEC to raise prices further. At the OPEC meeting in Algiers in June of 1980 the range of allowable prices was increased, with the base price at $36 a barrel and a ceiling price on premium oil at $41 a barrel (Energy Policy, 1981, p. 23). Spot prices ranged from $40 to $50 a barrel; the Iranians boasted of charging $50 prices to Japanese importers (Yergin, 1991, p. 704).

Hostilities in the Middle East were heightened by Iran's taking of US hostages in November 1979 and the Soviet invasion of Afghanistan in December 1979. The taking of US hostages and the Russian invasion threw the oil market into a state of panic. Buyers paid outlandish prices because a build-up of inventories was a form of insurance. It appeared that the oil exporters had really taken the political and economic leadership away from the West. It seemed to some that the United States and the West "were truly in decline, on the defensive, and it appeared unable to do anything to protect their interests, whether economic or political" (Yergin, 1991, p. 701).

President Carter's response to the hostage taking was to impose an embargo on Iranian oil imports to the United States and to freeze Iranian assets. Iran countered by imposing a prohibition on the export of Iranian oil to any American firm. The US embargo on Iranian oil was of no consequence to Iran since any oil they managed to produce could be immediately sold in the frantic world market; the freezing of assets was a minor imposition (Yergin, 1991, p. 702).

President Carter's failed attempt to rescue the hostages in April of 1980 added to the impression that the oil exporting countries were in control. The presence of Russian forces in Afghanistan, the apparent fulfillment of Russia's long awaited opportunity to extend its influence into the Middle East, added to the uncertainty in the oil market. This event was a major failure of the Carter Administration and further reduced any remaining respect he may have retained among Texas politicians.

While the Camp David Accord was a major political accomplishment that helped secure peace in the Middle East among Israel and her enemies, Carter's role in weakening the power of the Shah of Iran no doubt helped heighten his downfall, and usher in the crisis that caught the United States up short. The economics of the oil market soon began to overpower the politics, however, and the greed of oil exporting countries proved to be a sobering experience for members of OPEC. As Yamani of Saudi Arabia had unsuccessfully argued at the OPEC meetings in Caracas, Venezuela in December 1979 the rising prices reduced demand, threw the world economies into recession and forced oil prices down again (Yergin, 1991, p. 704). Yamani told the OPEC ministers, "Prices go up, and demand goes down, it's simple, it's ABC." This second run-up in world oil prices initiated a renewed set of market dynamics, the end results of which greatly reduced OPEC's share of the world market, and the price collapse of 1986 to as low as $10 per barrel in the United States. OPEC's share of the world oil market fell from 55.9% in 1973 to 30% in 1985 (Energy Information Administration, 1986, p. 237).

The federal response

The Embargo forced the United States to face a new reality. We no longer had spare oil capacity to provide national and international security. So, US policy makers had to figure out how to respond to this new reality. The proposals for reorganizations of Federal government agencies by Nixon, Ford, and Carter all included elements of (1) policy development, (2) conservation and emergency preparedness, (3) research and development (R&D), and (4) regulatory control. The proposals included creating new authority for policy and transfer of existing programs to a new central organization. No existing agency had authority for policy development. Existing R&D was woefully underfunded given the magnitude of the problem and the existing regulatory programs resided, without good coordination, in several agencies. The Department of Interior had several energy divisions dealing with oil and natural gas and coal resources. The Atomic Energy Commission (AEC) dealt with civilian uses of nuclear energy; oil imports were handled by the Mandatory Oil Import Program set up by President Eisenhower in 1959; and the regulation of natural gas and electricity was managed at the federal level by the Federal Power Commission, and by state regulatory agencies at the state and local levels.

A long series of ideas for government reorganization emerged from several sources as The energy problems became a reality in the early 1970s. President Nixon called for reorganizing the government's energy policy, regulatory, and research functions when he addressed the nation on November 7, 1973, following the Embargo by the Arabs on October 20. The president, in a televised speech to the nation, outlined

"Project Independence." The proposal included a number of legislative recommendations in the context of his general call on the American people:

> Let us set as our national goal, in the spirit of Apollo, with the determination of the Manhattan Project, that by the end of this decade we will have developed the potential to meet our own energy needs without depending on any foreign energy sources.

On specific issues, President Nixon said:

> We must have legislation now which will authorize construction of the Alaska pipeline...to encourage production ... of natural gas, ... legal ability to set reasonable standards for the surface mining of coal. And we must have organizational structures to meet and administer our energy programs ... I have stressed repeatedly the necessity of increasing our energy research and development efforts.
>
> **Nixon (1973c)**

The calls for organizational changes to carry out policies of increased domestic production, reduction in energy consumption and switching off of oil in order to reduce dependence on imported oil were not, however, a long-standing concern of the Nixon Administration. The attitude of the Federal Government during the last years of the Johnson Administration and the early years of the Nixon Administration are best characterized by a belief that "industry is responsible." In remarks before the Energy Committee of the Organization for Economic Cooperation and Development, J. Cordell Moore, Assistant Secretary of the Interior, stated that energy policy of the US Government meant:

> [T]hat industry is responsible for production, distribution, marketing and pricing, except in markets where fair prices cannot be guaranteed by competition, such as, for example, gas and electricity in interstate commerce. The Federal Government attempts to establish a climate favorable for the growth of the energy industries. It tries to stimulate initiative, to help advance technology, and to encourage and maintain competition. It monitors the overall energy situation to be sure that the national security and the broad interests of the public are protected; it applies constraints to the operations of the private sector where the public interest so require, and it makes liberal use of the instrument of persuasion at times to influence the course followed by the private sector But the Federal Government does not control production, it does not direct the efforts of industry, and it does not involve itself in company affairs. Even in the regulatory field its posture is mainly reactive rather than positive. Information on costs, reserves, processes and plans is generally closely guarded by the companies and they are not required to divulge it ... I stress our lack of authoritative knowledge concerning these matters because it is a basic part of our policy. (Energy Policy Papers, Committee Print, Senate Committee on Interior and Insular Affairs, 1974, pp. 350—351)

Basically, the attitude of the government prior to the embargo, was to monitor the situation for national security purposes, cope with emergencies if needed, and insure the consumer's interest through a favorable climate for industry growth, subject only to price regulation in industries where lack of competition would lead to unfair prices. Not surprisingly, the government was organized to reflect these attitudes. Reliable imports of oil and a strong domestic energy industry were complementary objectives.

Security was the responsibility of the Departments of Defense and State, primarily oil import and naval oil reserves policy.

The embargo of 1973–74 was such a major disruption for the United States, and for much of the rest of the world, that the government laissez faire attitude obviously needed to change. Suddenly, national security, economic stability, and the international flow of crude oil supplies became different parts of the same problem. Governments were not organized to manage the problem; therefore, most of the early political attention to the problem involved rearranging the government.

The Nixon, Ford, and Carter Administrations crafted dozens of organizational ideas and spent much effort trying to get legislative initiatives passed to implement these ideas, but often failed. Since many energy functions resided inside existing agencies and needed to be transferred to a new, more centralized agency, vested interests rose up to prevent such consolidation.

The proposals for reorganizations by Nixon, Ford, and Carter all included elements of (1) policy development, (2) conservation and emergency preparedness, (3) R&D, and (4) regulatory control. The proposals included creating new authority for policy and transfer of existing programs to a new central organization. No existing agency had authority for policy development; existing R&D was woefully underfunded given the magnitude of the problem, and the existing regulatory programs resided, without good coordination, in several agencies. The Department of Interior had several energy divisions dealing with oil and natural gas and coal resources. The AEC dealt with civilian uses of nuclear energy; oil imports were handled by the Mandatory Oil Import Program set up by President Eisenhower in 1959; and the regulation of natural gas and electricity was managed by the Federal Power Commission.

Recommendations for centralizing energy activities began with President Nixon. The Ash Council (Advisory Council on Executive Organization) reported to Nixon in late 1970 and recommended that the President press for a new Department of Natural Resources. This new department would handle coordination of the nation's energy policy and research programs paying attention to alternative energy sources, conservation, and energy/environmental conflicts, as well as traditional sources of energy. This recommendation grew out of the Energy Policy Staff of the Office of Science and Technology under the direction of S. David Freeman (Whitaker, 1976, pp. 61–68).

Soon after the Ash Council recommendation, Domestic Council Director, John D. Ehrlichman, created an energy subcommittee to make recommendations concerning the government's responsibility in alleviating a threatened, acute shortage of clean fuels for the winter of 1970–71 and to assess the long-term energy outlook. The subcommittee was headed by Paul W. McCracken, Chairman of the Council of Economic Advisors. It was primarily this group, with input from several sources, including a special study by the National Petroleum Council

(National Petroleum Council's Committee U.S. Energy Outlook, 1972), and a national energy study directed by John J. McKetta of the University of Texas at Austin (Advisory Committee on Energy, U.S. Energy, 1971) that formed the substance of President Nixon's first energy message to the Congress in June 1971. Nixon's message included a strong call for centralization of energy functions within a new Department of Natural Resources (Marchi, 1981a,b, pp. 409–410).

President Nixon's first energy message was largely ignored, however, as economic issues related to inflation and the recession took top priority. The idea of a natural resources department drew strong opposition from both industry and government sources. Nixon modified his proposal for a new department in his April 18, 1973 energy message to the Congress in which he promised draft legislation to establish a Department of Energy (DOE) and Natural Resources, as well as ordering several new energy-related offices to be established in the Interior Department. The message also called for a new Energy Research and Development Administration (ERDA) incorporating the research functions of the AEC and, along with it, developing a new energy R&D effort of $10 billion for the next 5 years. Civilian nuclear regulation would be handled by a new Nuclear Regulatory Commission (NRC) (Nixon, 1973b).

Much disagreement arose among the committees of the Congress concerning jurisdiction over the new agencies and loss of responsibility for functions of AEC and Interior programs to be changed in the definition of the new DOE and Natural Resources. The conflict became heated and irresolvable and the President was forced to take what he could get. ERDA was formed and the regulatory functions of the AEC were separated and assigned to the newly created NRC. Although discussion of a Department of Natural Resources or Energy and Natural Resources continued from time to time, it was never a serious possibility and the idea was eventually modified in the formation of the DOE in the early months of the Carter Administration.

In the months following the embargo, Nixon set up a policy office by Executive Order. The office was known as the Federal Energy Office (FEO), which later became the Federal Energy Administration (FEA). FEO was responsible for putting together and administering allocation plans for crude oil and for all petroleum products as required by the Emergency Petroleum Allocation Act of 1973 (P. L. 93–159). The Office was also responsible for putting together a plan for energy independence by 1980. The act also contained authority for gasoline rationing. FEO was charged in the executive order with advising the President "with respect to the establishment and integration of domestic and foreign policies relating to the production, conservation, use, distribution, and allocation of energy and with respect to all other energy matters" (Nixon, 1973a, pp. 990–991).

FEA was consumed with administering the oil price and allocation control mechanism and with producing a plan for energy independence called for in the President's October 17 "Project Independence" speech to the nation. The FEA became the instrument to side step the reorganization problem and to create a major policy initiative to mobilize the functions of government. This initiative became known as the Project Independence Blueprint. The Blueprint defined the policy, regulatory, and R&D efforts that would be needed to bring about *US energy independence by 1980*. A large computer model of the nation's energy and economic system was put together to be able to project the consequences of energy independence; the modeling system was known as the Project Independence Evaluation System (PIES) (Federal Energy Administration, 1974, pp. 195–204). PIES not only was a modeling exercise that produced analytical results useful in the policy debate, it also formed the making of a bureaucracy that perpetuated itself and the use of the modeling system for several future national energy outlook reports and special analyses that continues today. The State of Texas became concerned about the uses and abuses of PIES in the formulation of President Carter's National Energy Plan (NEP) in 1977 (Holloway, 1980). This Texas challenge is discussed in detail later in this chapter.

Gerald Parsky, Executive Assistant to FEO Administrator Simon, distributed a memo in late August of 1974 which summarized the tie between major energy policy decisions and the need for governmental organizations to deal with energy policy. He said that Project Independence must embody (1) decontrol of domestic crude oil prices, coupled with a windfall profit tax; (2) deregulation of natural gas; and (3) imposition of a gasoline tax. Parsky added that FEA should cease to exist as an independent agency and its functions should revert to various offices in the Interior Department. With the completion of Project Independence, there would be no continuing justification for the existence of FEA (Marchi, 1981a,b, p. 466). But Parsky's proposal did not prevail.

The Strategic Petroleum Reserve was established by the Energy Policy and Conservation Act of 1975 during Ford's administration. The legislation declared it to be US policy to establish a reserve of 1.0 billion barrels of petroleum. FEA was given the task of establishing the reserve. Eventually six, near-surface salt domes in the Gulf Coast areas of Texas and Louisiana were developed for the purpose and filled with approximately 600,000 barrels of oil, the largest stockpile of energy in the world.

The Ford Administration did propose legislation to create other institutional entities such as the Energy Independence Authority. The Authority was to be set up as a special corporation to provide Federal subsidies to companies for energy development projects. The Authority was to have $25 billion in equity and $75 billion in government-backed borrowing authority. Liberals in the Congress objected to the subsidy to big business and conservatives objected to the plan as a further government interference in the marketplace (Energy Policy, 1981, p. 150). The President, in his 1975 State of the Union message, had previously set a goal of achieving a capacity of 1 million barrels a day in synthetic fuels by 1985 (Ford, 1975a,b, blc. 1, p. 42). The Energy Independence Authority legislation failed to get the support of the Congress.

Early in the Ford Administration, the internal conflicts among agencies and offices with some stake and role in energy policy led Ford to announce the creation of an Energy Resources Board headed by Roger Morton, Secretary of the Interior. The Energy Reorganization Act of 1974, signed by Ford on October 11, provided for an Energy Resources Council. The Council was charged with developing "a single national energy policy and program" (Ford, 1975a,b, p. 231). The Council, however, turned out to have little influence and, in the end, FEA became the chief energy policy agency of the government, evolving eventually into the DOE under President Carter.

Foreign policy also became an energy issue in the early days of the Ford Administration. Henry Kissinger was apparently enthused over the domestic objectives of Project Independence and believed that consuming nations should ready agreements in order to counteract OPEC policy and to provide for stability in the event of another embargo (Marchi, 1981a,b, p. 527). Strong disagreements developed between Kissinger and Simon, Secretary of the Treasury, and Milton Russell of the Council of Economic Advisors, but Kissinger prevailed. The International Energy Agency was established in 1975 and agreements reached such that member nations would, upon the calculation of a significant shortage, implement predetermined conservation measures and agree to share the shortages by an agreed upon formula (Marchi, 1981a,b, pp. 530−533).

A major and long-standing national security policy initiative was implemented during the Ford Administration. The Strategic Petroleum Reserve was established by the Energy Policy and Conservation Act of 1975. The legislation declared it to be US policy to establish a reserve of 1.0 billion barrels of petroleum. FEA was given the task of establishing the reserve. Eventually six, near-surface salt domes in the Gulf Coast areas of Texas and Louisiana were developed for the purpose and filled with approximately 600,000 barrels of oil, the largest stockpile of energy in the world.

Finally, the Carter Administration succeeded in getting the Congress to bring together most of the energy policy, research, and regulatory functions of the Government into a new cabinet department—the DOE. The new organization combined the functions of the FEA, the ERDA and the Federal Power Commission. Certain programs from Departments of Defense, Interior, Commerce, Housing and Urban Development (HUD), and the Interstate Commerce Commission were also consolidated into DOE. The President signed P. L. 95−91 on August 4, 1977; the new Department came into existence October 1, 1977 with an employment level of almost 20,000 people, most of who were transferred from other agencies, supported by a 1978 budget of $10.6 billion (Energy Policy, 1981, p. 178).

The structure of DOE was clumsy and awkward, however, and produced many frustrations and caused further proposals for reorganization and abolishment in the years ahead. DOE was structured to provide energy policy, regulation, research, and information gathering functions, but certainly not all of such functions. As the record of the Reagan Administration will show, Interior, State, Defense, Council of Economic Advisors and Treasury Departments would continue to play a role in energy policy, in fact, a dominant role in some cases.

DOE incorporated the Federal Power Commission responsibilities for regulation of sales of natural gas (rate setting authority) and electricity (wholesale interstate purchases) under the direction of a five-member commission appointed by the President with Senate confirmation. The new sub-agency was named the Federal Energy Regulatory Commission (FERC). As a practical matter, however, the FERC was not set up to be responsible to the Secretary of DOE except for budget and administrative purposes. A second regulatory function, oil pricing and allocation, was assigned to the Economic Regulatory Administration under the direction of an administrator appointed by the President and confirmed by the Senate. The NRC, responsible for the regulation of civilian nuclear power use and waste disposal, was left a separate energy regulatory agency, however.

P. L. 95−91 also set up the Energy Information Administration directed by an Administrator, appointed by the President and confirmed by the Senate, with authority to require reporting of energy information from private firms, with requirements to make the information available to other branches of DOE and the public, and with protection of the administrator from having to obtain approval from departmental supervisors in the collection, analysis, and publication of data and information (Energy Policy, 1981, p. 181).

This act transferred from Interior the power to set economic terms of leases for energy on public lands but reserved to the Secretary of the Interior the sole authority over the issuance of leases and enforcement of their regulation. Also transferred from Interior were the various power administration organizations (Southeastern, Southwestern, Alaska, Bonneville and Bureau of Reclamation power marketing) and the authority to gather data on energy resources, and technology research on production from solid fuel minerals and coal. Authority over standards for new buildings was transferred from HUD; regulation of oil pipelines from the Interstate Commerce Commission; authority over three naval oil reserves and three oil shale reserves from Defense, and van pooling and car pooling from Transportation. The act created positions of a deputy secretary, a general counsel, an under-secretary and eight assistant secretaries; all Presidential appointments required Senate confirmation (Energy Policy, 1981, p. 181).

Along with the creation of DOE in 1977 and the establishment of the NRC as an independent agency in 1976, Congress abolished the long-standing Joint Atomic Energy Committee in 1977. The Atomic Energy Act of 1946 (P. L. 79−585) created both the AEC and the Joint Atomic Energy Committee to foster development of the new atomic energy technology. Congress transferred military nuclear concerns to the Committee on Armed Services, general regulation of the nuclear industry to the Committee on Interior and Insular Affairs, R&D questions to the Committee on Science and Technology, and military nuclear matters to the Senate Armed Services Committee, thus ending 31 years of the joint committee's responsibilities, and with it, the promotional activities of the Congress.

There was also a major focus on R&D for promising new energy technology from coal, shale, and geothermal. The Synthetic Fuels Corporation (SFC) (known as the Energy Security Corporation as proposed by Carter) was another version of the Ford Administration's idea of a major federal program of subsidies to develop synfuel projects. Although the mechanism was slightly different, the concept and objectives were the same.

The Congress and President Carter set up the SFC to create substitutes for crude oil and natural gas, an effort funded by taxpayer dollars. On June 30, 1980, the SFC was formed and given authorization to use $20 billion through 1983; by 1984 up to $68 billion in additional funds could be authorized by Congress. Through loans, loan guarantees, price guarantees, and purchase agreements, the SFC encouraged the private sector to produce a target production of the equivalent of at least 0.5 mbd of crude oil by 1987 and 2.0 mbd by 1992. The SFC was directed by a seven-member board appointed by the President and confirmed by the Senate (Energy Policy, 1981).

James R. Schlesinger, in early 1977, was responsible for drafting President Carter's NEP and for designing the structure of the new DOE. Curiously, Schlesinger served in the Nixon budget office, becoming Assistant Director of the Office of Management and Budget (OMB) in 1970. He served as Chairman of the AEC from 1971 to 1973. During February 1973 through July 1973, he served as Director of the Central Intelligence Agency and then Nixon appointed him Secretary of Defense in the summer of 1973. Schlesinger is a Harvard trained economist who apparently did not get along well with Nixon in the earlier days of his service in OMB (Cochrane, 1981, p. 552). But Schlesinger and Carter had lengthy discussions prior to Carter's inauguration and identified energy as an issue of sufficient urgency and character as to form the basis for a major unifying challenge to the American people.[6] James Schlesinger and Jimmy Carter developed strong, personal bonds.

Schlesinger put together a relatively small group of fifteen professionals—economists, public administrators, and lawyers—all from university or public sector backgrounds, to prepare the Carter energy plan. The group worked almost in isolation both from other government departments and officials and the private sector. They produced a complicated NEP which the President unveiled to the American people

---

[6] During Schlesinger's tenure as Secretary of Defense, William P. Clements served as Deputy Secretary and it was the incompatibility of this relationship that partly prompted a strong response to the Carter energy plan by Clements when he became Governor of Texas in 1979.

and the Congress on April 20, 1977. It appears that the choice of preparing the plan in a "back room" had its weaknesses. The House split the complicated package of legislation, known as the National Energy Act of 1978, into five bills,[7] abolished most of the tax related measures, and stalled over disagreements on natural gas. The longest legislative battle in history raged for 18 months over the natural gas issue before the Congress and the Administration finally agreed on greatly modified versions of the five bills.

One last effort was made by the Carter Administration to put an organizational structure in place to deal with the energy crisis. Since the five bills of the National Energy Act took 18 months of heated debate to pass out of the Congress, 1979 was no time to deal with more energy issues. But then the Iranian crisis erupted, refueling the debate over government managed fuel allocation and gasoline rationing. Another energy emergency that brought the country's focus back to energy was the Three Mile Island incident on March 28, 1979. The partial melt-down of the Unit 2 reactor raised a national alarm over nuclear power. In 1980, the DOE's NEP II produced three major pieces of energy legislation: the SFC, the Windfall Profit Tax, and the Energy Mobilization Board proposals.

The SFC (known as the Energy Security Corporation as proposed by Carter) was another version of the Ford Administration's idea of a major federal program of subsidies to develop synfuel projects. Although the mechanism was slightly different, the concept and objectives were the same. On June 30, 1980, the SFC was formed and given authorization to use $20 billion through 1983; by 1984 up to $68 billion in additional funds could be authorized by Congress. Through loans, loan guarantees, price guarantees, and purchase agreements, the SFC encouraged the private sector to produce a target production of the equivalent of at least 0.5 mbd of crude oil by 1987 and 2.0 mbd by 1992. The SFC was directed by a seven-member board appointed by the President and confirmed by the Senate (Energy Policy, 1981, p. 227).

The creation of SFC in 1980 was a symbol of the mood of the Congress and the country to finally deal with the nation's vulnerability to unstable sources of imported crude oil. Chances are that the SFC would never have been created except for the Iranian crisis and related gasoline shortages that, for the second time in the 1970s, spread across the United States. It happened in the spring and summer of 1979. For a few months in early 1979, the United States imported in excess of 50% of domestic oil use and the prospect was that the percentage would continue to grow unless major policy changes were made.

---

[7] The five bills included: (1) Public Utility Regulatory Policies Act (PURPA) (Pub.L. 95−617), (2) Energy Tax Act (Pub.L. 95−618), (3) National Energy Conservation Policy Act (NECPA) (Pub.L. 95−619), (4) Power Plant and Industrial Fuel Use Act (Pub.L. 95−620), and (5) Natural Gas Policy Act (Pub.L. 95−621) (Energy Policy, 1981, p. 178).

Nuclear waste disposal became an important issue during the Carter Administration, a policy and technology issue indirectly related to the Embargo and the development of the NEP. The natural gas shortage problem and rising prices of natural gas was the focus of the Natural Gas Policy Act, one of five Acts included in the National Energy Plan. Directly related was the Powerplant and Fuel Use Act, a second Act in the NEP that was designed to (among other things) phase-out natural gas use for boiler fuel by 1990. The continued development of nuclear power and coal promised to be a major replacement for natural gas electric power generation. A major challenge for the long-term development of nuclear power, however, is the permanent storage of nuclear waste from spent fuel rods. The Federal government from the beginning of nuclear technologies (at first for military use) took on the responsibility for waste disposal and therefore has for decades investigated alternative technologies and locations for a national high-level nuclear waste storage solution. It is the long-term requirement to make the nuclear power industry viable. The salt- beds that underlie the High Plains of Texas were one such option. The other major consideration was Yucca Mountain in Utah. The Texas response to the High Plains option is discussed later in the chapter.

The natural gas market developments leading up to, and beyond the Carter Administration's NEP had other policy impacts. Since the focus of the natural gas debate centered on the demand side (the Administration believed the supply-side had nowhere to go but down), it focused on reducing the use of natural gas for boiler fuel. That meant industrial use for manufacturing and petrochemicals and electric power generation. This federal policy impact fell heavily on Texas as both a major producer and user of natural gas.

Energy innovation growing out of the 1970s energy crises continues even four decades later to influence the geopolitical balance in the world economy. These innovations continue to impact overall economic growth. Such innovation also determines (in part) the distribution of economic activity and related environmental impacts among the nations of the world. The energy markets and energy innovation also are often the underlying triggers for military conflict.

One such conflict was the Iranian crisis of 1979. President Carter's failed attempt to free 52 US embassy hostages in April 1980 was followed by the Iran—Iraq War. In September 1980, the Iraqi military invaded Iran. These events were in large part responsible for sending the world price of oil to $40/bbl (over $100/bbl in today's dollars).

As the events of the Iranian crisis unfolded and a consensus grew that the government's policy of price controls and allocation were more of a cause of the energy problem than a solution, the Reagan Administration took a very different posture (from Carter) toward energy policy. Eventually price controls on oil and gas were removed, mandated oil allocation among refineries was stopped, the gasoline allocation program was dismantled, the use of natural gas for boiler fuel use was left to the market place, and the SFC was abandoned. And, the search for, and completion of a permanent high-level nuclear storage facility still remains unfinished business. But the DOE and all of its functions remain and is the primary source of reliable data on the energy sector of the nation and states. DOE has an ongoing policy recommendation role, oversees the set of National Energy Labs and operates an ongoing R&D program focused more on early stage research and less on large-scale projects that attempt to carry through to demonstration and commercialization. DOE did, however, carry out the oversight and implementation of the federal $130 million Stimulus program called the Recovery Act—Smart Grid Demonstrations during 2009—14 that was all about demonstration and commercialization.

The Texas response

Changes in the nature and seriousness of energy problems also produced policy changes and a series of institutional changes in Austin. The Governor's Energy Advisory Council (GEAC, 1973), a special commission established by Executive Order by Governor Briscoe, was created in the fall of 1973 to address the energy crisis for Texas generated by the Arab Embargo. The Council was chaired by Lt. Governor Bill Hobby with board member participation from most statewide elected and appointed agency heads plus a number of private sector leaders. Under the leadership of Abe Dukler of the University of Houston, serving as the Executive Director, GEAC carried out over 40 studies produced by teams of government employees, university researchers and private company experts all helping to identify policy recommendations.

One of several GEAC recommendations at the end of 1975 was to follow the Council's work by creating a statutory agency for Texas energy policy development. The 65th legislature created a statutory form of the council by the same name, the Governor's Energy Advisory Council (GEAC, 1975). GEAC, lead by Executive Director Alvin Askew had a 2-year statutory life beginning in September 1975 was followed by the Texas Energy Advisory Council (TEAC, 1977) in 1977 and Texas Energy and Natural Resources Advisory Council (TENRAC, 1979) in 1979 both lead by Milton Holloway. Holloway managed the Texas Energy Development Fund, the first ever energy R&D effort set up outside of university programs in Texas. It was funded by a $5 million appropriation by the Texas Legislature.

As in Washington, there was a recognition that the need for policy changes existed even before the Embargo of 1973—74 (General Land Office of Texas, 1976, p. 67). During late 1973 and throughout 1974, the gasoline shortage created long lines of automobiles at service stations, interrupted commercial transportation, and disrupted agricultural practices. The federal government passed an emergency act to address the shortages allowing the President to allocate fuel supplies for emergency and necessary purposes (Emergency Petroleum Allocation Act of 1973). This act required the President to set up a comprehensive allocation program for oil and oil products, and to do so within 30 days. The program made supplies available to the government for allocation to hardship and priority users by requiring suppliers to set aside a given percentage of the monthly inventories for such purposes. State governors could also carry out a portion of the responsibilities of allocation (taking the responsibility was optional) by managing up to 5% of monthly supplies held by suppliers serving their respective states. This delegation of allocation authority applied to gasoline, propane, and diesel fuels.

Governor Briscoe accepted the state program for Texas and set up a fuel allocation program within his office. Alvin Askew, an attorney employed at the TRC, was appointed to head up this new Fuel Allocation Office (FAO). At the peak of the program in 1974, about 45 employees were hired or transferred from other agencies to manage the large monthly flow of fuel supply allocation requests from public and private sector consumers.

Since the Governor had direct responsibility for this politically sensitive and explosive energy function, it was quite natural that the FAO was set up directly under his control. GEAC was a commission charged with studying and recommending policy; it was not in a position to operate a daily fuel allocation program. No other agency in the state bureaucratic system was well suited for the task either. Therefore, by 1974, Texas had two new energy organizations in addition to the long-standing TRC; GEAC for policy development and FAO for emergency fuel allocation.

GEAC completed its work in January 1975, reported recommendations to Governor Briscoe, and then disbanded. FAO allocated fuel through the severe shortage months of late 1973 and early 1974. As such programs typically develop, however, the monthly allocating continued long after the gasoline lines had disappeared in 1974; the staff size was considerably reduced by early 1975, but the program continued throughout the 64th Legislature in January through May of 1975.

The commission form of GEAC recommended the creation of GEAC as a statutory body. The idea was that energy policy and the energy industry was so important to the State that the Embargo and the Federal response were so important that it justified setting up a high-level body made up of most of the statewide elected officials, the top appointed officials, and selected industry leaders to forge a State energy policy. Importantly, that included protecting the State from the wrong-headed efforts of Washington. During the summer of 1975, Governor Briscoe appointed Alvin Askew to be the Executive Director of the new statutory GEAC. Askew set about in the fall of 1975 to hire a staff and to mobilize GEAC to address *national energy policy* issues that were rapidly developing in Washington.

The statutory form of GEAC, under Askew's direction, was provided a bi-annual budget of $1.0 million but given only a 2-year statutory life. The state leadership was uncertain whether there was a need for an ongoing energy policy function, or at least whether the council was the proper forum for its development. Therefore, a 2-year life for GEAC was set to guarantee that the question of need would be addressed by the next regular session of the Legislature. The structure of GEAC contained three groups: (1) the council,[8] (2) a citizens advisory committee appointed by the Governor, and (3) the staff directed by the Executive Director appointed by the Governor and confirmed by the Senate. The council was the decision-making board, the advisory committee provided advice on policy matters and the staff conducted research and managed a public awareness program. The council was mandated to hold quarterly public meetings (Holloway et al., 1979, p. 47).

In addition to setting up two operating divisions to provide policy analysis and a public awareness program, Askew began immediately to lay the ground work for an interstate lobbying effort to better represent the interests of the oil and gas production states in Washington. Askew spent most of the first 9 months of the new agency's life trying to reach agreement on this lobbing effort, which would have been funded by annual contributions from each member state and was to be located in Washington D.C. The states included Texas, Oklahoma, Louisiana, New Mexico, and Arkansas. The new GEAC had two functioning divisions, the Forecasting and Policy Analysis Division and the Public Awareness Program Division. In addition, the Administration group took on the task of overseeing development of a statewide Texas Energy Policy statement for use in both Texas and Federal venues to influence energy policy. This comprehensive energy policy development resulted in a 22 page document covering all the major energy issues of the day put together by consensus of public and private sector Texas leaders, made up of over 100 entities working with draft documents and finalized in an intensive 2-day meeting in Austin (GEAC, 1977a,b,c).

The lobbying idea was fully developed and Askew apparently received near informal agreement from individual council members prior to the quarterly meeting of the GEAC council in early 1976. Objections developed prior to the meeting, however, and the idea was unanimously voted down at the public meeting. Following this defeat Askew concentrated on the staff policy work and attempts to coordinate a Texas position informally with the other states in order to better represent a unified position in Washington. The Associate Director was given responsibility for the day to day operation and management of the staff.

---

[8] The Council was made up of the top elected officials of the state—the Governor, Lt. Governor, Speaker of the House, Chairman of the Railroad Commission, Commissioner of the General Land Office, Commissioner of Agriculture, the Comptroller, Attorney General, a member of the Senate and a member of the House of Representatives; the Council was chaired by the Governor with the Lt. Governor serving as Vice Chairman.

About one-half of the council's staff budget was allocated to the public awareness function and a highly visible energy conservation and information program was developed. By the summer of 1976, the public awareness program was in full operation but various members of the council began questioning the wisdom of this kind of visual exposure through alternate media programs. At the September 1976 council meeting the group voted to discontinue the public awareness program and ordered Askew to phase-out the program and staff devoted to it.

One motivation for the establishment of the public awareness program was the expected receipt of substantial federal funds for conservation programs. During the months that followed, an office was set up in Governor Briscoe's office to manage the Federal conservation programs. Askew resigned his position as Executive Director of GEAC and moved to the Governor's office to head up the new office and to continue his role as advisor and representative of Governor Briscoe on energy issues in Washington. Askew often relied on policy analyses and research at the University of Texas. This informal arrangement supported his effort to represent Governor Briscoe in Washington.

Briscoe was not comfortable with the Council forum and, in the later stages of the 2-year life of GEAC, asked Lt. Governor Hobby to chair the quarterly Council meetings. Thus, the first statutory forum of the Council came to its 2-year end with a rocky beginning plagued by an aborted lobbying organization effort, a partially dismantled staff, and a disinterested Chairman, the Governor. The policy division continued to function under these conditions, however, and produced several major policy studies and further developed an energy data base and forecasting capability. Serious efforts were being made in Washington to deregulate natural gas. A Texas natural gas shortage controversy (the La-Voca company shortage) had led to a series of Legislative committee hearings on the topic. The Forecasting and Policy Analysis Division, under Milton Holloway's direction, provided testimony and analysis papers for both Washington and Austin venues. Both legislative bodies held extensive hearings on the gas shortage crisis. By the end of the first biennium, GEAC became known primarily for its work in policy analyses, especially on the natural gas issue. By the end of the summer of 1977, a major debate developed over the Carter Administration's use of a large computer model for justification of the Carter NEP. This episode is discussed in detail in later in the chapter.

The 65th Legislature considered the passing of a new statute to continue the life of GEAC. Under the conditions described above, the only significant support was for the policy analysis function, and then only because of Lt. Governor Hobby's interest. The Embargo had long sent been lifted and the gasoline allocation system dismantled—markets settled down for a time. But the leadership knew the situation had not brought back the conditions of the past—the wolf was still at the door. Therefore, a new statute was passed for a second 2-year term creating the TEAC with the same structure and policy development responsibilities as GEAC but making the Lt. Governor Chairman of the council. GEAC was mandated to report to the Governor and the Legislature. The Lt. Governor was made Chairman of the council made up of the Speaker as Vice-Chair, and other members that included the Attorney General, a Commissioner of the Railroad Commission, a commissioner of the Public Utility Commission (PUC), the Commissioner of the General Land Office, the Commissioner of Agriculture, the Comptroller of Public Accounts, the Chair of the Natural Resources Committee of the Senate, the Chair of the Energy Resources Committee of the House, and the Chair and Vice-chair of a citizens advisory committee. The legislature also created a Texas energy research fund recommended by GEAC in 1975 and assigned responsibility for its administration to TEAC. Milton Holloway was appointed Executive Director by Governor Briscoe with Senate confirmation in 1977.

Upon the recommendation of the GEAC, the creation of a Texas energy research fund also passed the 65th Legislature and responsibility for its administration was assigned to TEAC. Holloway was appointed Executive Director with new responsibility for managing an energy research fund and a reduced budget and staff size less than one-half that of GEAC.

TEAC updated the Texas Energy Policy statement with the development of a consensus of government and industry published in February 1979 by reconvening a large group of public and private sector representatives, many of whom served the same function during the creation of the Texas Energy Policy statement of 1977. The updated statement resolved a number of issues that remaining unresolved during the 1977 agreement and addressed new issues in the fast changing energy sector markets and as a response to the policy directives issued from Washington D.C. One such issue was an agreement that electric utility pricing policy should consider marginal cost pricing as the driving consideration for utility rates.

The question of utility pricing was finally resolved in 1995 and following with the restructuring of the Texas electric industry discussed in detail in the following chapter. The TEAC policy statement also supported a major emphasis on enhanced oil recovery technology, a focus that would turn into the largest energy sector innovation of the century, namely the "fracking" effort that has turned the energy world upside down. This technology, discussed in detail in a following chapter has reordered the geopolitical balance in the world economy. The technology development came heavily from Texas and the Permian Basin of west Texas has become the leader in current oil production helping to return Texas and the US oil industry to the prior position of world leader in oil production.

The TEAC policy document, formed by consensus of public and private sector leaders included positions on major national issues, especially the NEP. The NEP as finally written into legislation and passed as the by the Congress was signed by President Carter included five individual statutes. The most continuous was the Natural Gas Policy Act which was debated for 18 months before final passage.

In the meantime, the office established in Governor Briscoe's office began receiving money and managing the federally funded conservation program, but Askew was frustrated by an inability to staff his office because of Briscoe's campaign commitment to reduce the size of the state bureaucracy. Most of the funding for establishing a conservation program was contracted out to a consulting firm established to provide these planning services. Askew resigned during the last year of Briscoe's term.

Under Hobby's chairmanship, TEAC operated as a well coordinated, but low-keyed, policy formulation forum. A second and revised major energy policy paper was adopted, consensus was formulated on major national issues (especially the NEP which included the Natural Gas Policy Act) and the state's first energy R&D program was established.

TEAC began operation in September 1977 in the midst of the Congressional debate of the NEP which had been announced by President Carter in April. The US House passed a single bill encompassing the comprehensive package of ideas in the NEP on August 4. The Senate considered five different bills ranging from conservation to taxation and decontrol of natural gas prices.

Word of the NEP had been leaked several weeks before it was unveiled by Carter in a nationally televised talk to the nation. Holloway had been preparing an economic impact analysis of the NEP for several weeks, anticipating that Texas would need to respond. The New York Times obtained a copy the Carter's plan the weekend prior to the President's speech and published a summary of all the major components. Holloway immediately put his staff to work on an analysis and completed a draft on the morning of Carter's scheduled talk to the nation. Bill Hobby's staff reviewed the study with him and he scheduled a news conference *an hour before Carter's speech*. Holloway received a call to bring copies for the press and to meet Hobby in the Senate Chamber. The news conference got major headlines and the focus was on the onerous impacts the major tax provisions of the NEP would have on Texas, the largest energy consuming state in the nation. It was a tax plan, not an energy plan. Crude oil price decontrol was not addressed at all, and natural gas decontrol was set up to phase in over a very long time with a very complex system of continued price controls on 22 different categories of natural gas. There was also a very restrictive mandate for large industrial and electric utility natural gas users to convert from natural gas to coal or nuclear power. Most of Texas industrial complex and all of the utilities were fueled at the time with natural gas. The impacts of the NEP, as introduced to the Congress would have been devastating to Texas—the message of the NEP appeared to be focused on Texas; keep price controls on oil and gas, mandate conversion from the use of natural gas so it would be available to residential and commercial users in the rest of the nation, and severely tax the continued use of oil and gas while the transition was in progress. The reaction from Texas was unanimously negative.

Another relatively unique policy issue developed during the debate over the NEP. It concerned the use of the energy and economic modeling system used to support the conclusions of the NEP. Analysts in Texas did not believe the results of the PIES which projected modest economic consequences of the NEP, or the related impacts on future imports of foreign crude oil. The analysis completely ignored regional and state-level impacts of the NEP. The NEP placed a heavy emphasis on new taxes to be imposed on energy consumption, continued price ceilings on domestic crude oil, and continued price ceilings on all natural gas production except for special high cost gas from tight sands or deep formations. Separate legislation imposing a windfall profits tax (WPT) on domestic crude oil would accompany a phased decontrol of crude oil prices.

Holloway tried to obtain the data and modeling system from the DOE but his inquiries were repeatedly ignored. Much effort was required to simply learn the exact origin of the analysis, and requests by Texas analysts at TEAC and the University of Texas went unanswered. The Texas political leadership was understandably upset that a data base and modeling system used to justify an analysis that had such profound implications for the largest energy producing and consuming state in the nation was unavailable for our examination.

Along with the very able help of the Texas State—Federal Office in Washington D.C. and a distinguished national advisory group, Holloway put together a prominent team of professionals to critique the PIES model and published a book on the topic. The modeling system was finally obtained with the help of the Freedom of Information Act and a personal visit by Lt. Governor Hobby and Holloway, with Secretary Schlesinger in Washington D.C.

TEAC set up a national-level critique of the DOE modeling system and challenged the conclusions it supported. Through this project Texas had a major impact on the l process of data collection and policy analysis that continues today. The entire data base and modeling system is periodically archived at the National Bureau of Standards and made available to the public (Holloway, 1980).[9] An overall result of the project was to improve the management and archiving of the modeling work so as to make it more readily available to interested parties in future energy debates. Through this work, the TEAC achieved an openness at DOE that was previously not there.

The Texas view of the NEP was totally negative. The NEP did not remove price controls on oil and gas, as needed to get increased drilling in the United States, but

---

[9] The records management process is now under the National Archives and Records Administration (NARA) shares responsibility with federal agencies to ensure that each agency creates and preserves records containing adequate and proper documentation of the organization, functions, policies, decisions, procedures, and essential transactions of the agency. The Archivist of the United States, as head of NARA, provides direction, guidance, assistance, and oversight through issuance of records management regulations and standards applicable to all federal agencies.

relied instead on heavy taxation of energy use to reduce consumption, and therefore, reliance on foreign oil. While economists often propose the use of taxes to encourage or discourage consumption, the typical view of politicians is that taxes are for raising revenues to fund needed government services, not as a means of punishments or rewards. This was one of the Texas views of the NEP. The other Texas view of the NEP was that it did very little to encourage US production. The circumstances called for an all-out production effort if we were to have any hope of decreasing our dependence on foreign oil.

TEAC and TENRAC funded a set of 120 energy R&D projects during 1978–82 through the funding from the Texas Energy Development Fund. The projects built on the work carried out under GEAC by focusing on lignite, biomass, solar, wind, geothermal, enhanced recovery of oil and gas, energy conservation and energy policy. The projects built an information baseline for resource assessments, and new technology prospects.

John Hill (former Attorney General) defeated Briscoe in his bid for the Democratic nomination for Governor in 1978. But Hill was defeated in November by Bill Clements, the first Republican Governor of Texas in 100 years. Clements owned the largest privately owned drilling company in the world, and had served as Under Secretary of Defense during the Nixon Administration under Secretary James Schlesinger.

The early stages of the Iranian crises were unfolding in early 1979 when William P. Clements took the Oath of Office in Austin. Clements examined the status of the Texas energy policy effort and immediately decided a much larger effort was called for. Holloway was Executive Director of the energy policy office at the time, then known as the TEAC, with responsibility to advise the Governor and the Legislature on energy policy. The Clements Administration decided to combine the policy office with the Governor's energy conservation office (Governor's Office of Energy Resources) and a nonfunctioning Natural Resources Council. The organization also incorporated a new version of the FAO which had been quickly revitalized by the Governor to carry out the federally mandated fuel allocation program. The Federal program was designed to dispense with long gasoline lines. A statute to accomplish this combination passed the Legislature and Clements appointed Ed Vetter of Dallas to direct the office. Vetter was one of the Clements "dollar-a-year" men. The energy policy office was known as the TENRAC.

Under Clements' initiative in the 66th Legislature, the Texas energy policy office took on a new life and with vastly expanded funding and function. The agency had responsibility for (1) policy development, (2) energy R&D, (3) energy conservation, (4) fuel allocation and (5) coastal zone management planning. The structure of TENRAC was essentially the same as GEAC and TEAC except for the addition of a "natural resources" policy function and the management of the FAO. The FAO had been housed in the Governor's office during the embargo period. After a short interim, Holloway was appointed Executive Director of the new energy council.

For 4 years TENRAC functioned as a "Board of Directors" for Governor Clements, who was accustomed to operating as the "Chairman of the Board" in the corporate world. He used the forum to organize a consensus on national energy issues among the set of Democrats that made up the Council. The politics of reelection, however, began to impinge on the operation of the Council as one Council member, Mark White, who was Attorney General at the time, decided to run against Clements in the 1982 elections.

Clements quickly reached the conclusion that Texas had not been effective in energy policy in Washington in the years following the embargo. He aimed to change that, and took several actions to accomplish the goal. TENRAC's organizational structure was the same as that of GEAC and TEAC, that is, the organization's chief decision-making entity was the Council, a statutorily defined set of officers of state government plus representation from the private sector. A staff function under the direction of the Executive Director supported and reported to the Council. The Council was required by law to hold quarterly public meetings to conduct its business. The structure also involved a large number of industry and interest group representatives that served on advisory committees, with membership appointed by the Governor, Lt. Governor and Speaker.

Clements' first action intended to increase Texas influence in Washington was to appoint leaders in the oil and gas industry to the four citizen positions on the Council—Louis Beechrel, CEO of Texas Oil and Gas, Perry Bass from the Ft. Worth Bass family, Hank Harkins, President and CEO of Harkins and Company of Alice, Texas and Michael Halbouty, an independent oil man from Houston.[10] He also shared the appointment of people to various advisory committees with Hobby and Clayton by appointing oil and gas, utility, university, environmental, and service group representatives.

The second Clements action was to immediately begin holding informal energy meetings at the Governor's mansion while the new organization was being set up by act of the Legislature. He held a series of meetings on the topics facing Texas in early 1979—federal oil price controls, managing the gasoline allocation program, and the prospect of a WPT on oil producers being promoted by the Carter Administration, to name a few. The meetings included

Lt. Govern Hobby, Speaker Clayton, Comptroller Bob Bullock, Chairman of the Railroad Commission, Mack Wallace, Ed Vetter and Milton Holloway (then Executive

---

[10] Governor Clements was widely credited with appointing highly capable business people to serve on numerous boards and commissions throughout state government. Many of them served without pay as "dollar-per-year" men, as they were known, while others served on unpaid boards and commissions.

Director of TEAC). Holloway was introduced to Clements' management style at these meetings. In his typical fashion he would toss out a topic for discussion, then go around the table and get opinions from everyone present. When everyone had had their say, he would make a decision on the matter and the meeting would end.

Third, Clements organized a meeting of Texas energy officials with the Texas delegation in Washington. The meeting was organized by the Office of State Federal Relations based in Washington. The Texas officials at the meeting included Clements, Hobby, Wallace, Vetter, Holloway and Bill Fisher of the Bureau of Economic Geology. Holloway remembers feeling somewhat uncomfortable from the very beginning of the meeting. The meeting was held in one of the House Committee rooms on Capitol Hill. The Texas officials, led by Clements, took the podium and the delegation (all 29 of them) sat in the audience chairs below. Somehow it did not seem appropriate. The symbolic message was opposite from the pecking order of politics. Holloway said he had never before looked a senior Senator in the eye from that position.

Clements proceeded to give the delegation a lecture on energy policy, and in short, told them they had not done a very good job. As summarized by Carolyn Barta "He jetted to Washington to meet with the Texas congressional delegation, peeving some members when he berated them over the lack of a national energy policy" (Barta, 1996, p. 223). Hobby told Holloway at the hotel that evening that Bentsen was really offended. Clements was surprised when Holloway told him later that the delegation was offended by the meeting. He said Bentsen was a good friend and he (Bentsen) had never said anything about being offended by the meeting (Holloway, 1999).

Governor Clements had many reasons to feel strongly about the direction of national energy policy, not the least of which was because he knew the oil and gas industry so well. He had built the largest drilling company in the world (SEDCO). But there were more personal and political reasons. The Congress had only recently spent the largest debate in history (18 months) over the various parts of President Carter's NEP. The Plan had been developed by a democratic president and its architect was none other than James Schlesinger. Schlesinger had been Secretary of Defense when Clements served as Under Secretary of Defense during the Nixon Administration.

Schlesinger and Clements did not get along well at all. They mostly stayed out of each other's way, and discord between them grew with time (Barta, 1996, p. 153). Nixon had organized a Cabinet-level Emergency Energy Action Group to deal with the energy crisis following the embargo in the fall of 1973; Clement was the Department of Defense representative on that White House group. But Schlesinger objected to Clements' involvement because of his oil industry interest and financial holdings, and because of his relationship with the shah of Iran. Clements was forced to remove himself from involvement in energy policy in January 1974. Schlesinger was responsible (Barta, 1996, p. 152). Clements was not alone in his dislike of the Carter NEP. The Democrats in Texas also thought it was very punitive toward Texas and did not solve the nation problem either. So Clements had plenty of support in his disdain for Jimmy Carter's energy policy.

Bill Clements made effective use of TENRAC. He was the first Republic governor of Texas since reconstruction and had to try and make the Governor's Office effective under very unfavorable political circumstances. Both the Senate and House were dominated by Democrats. Bill Hobby and Bill Clayton, both Democrats, were well established veterans in their leadership of the Senate and the House, respectively. Hobby and Clayton both had 6 years and three legislative sessions under their belts. The other statewide elected officials who served on the Council (the offices were named in the statute) were all Democrats—Railroad Commissioner Mac Wallace, Agriculture Commissioner Regan Brown, Land Commissioner Bob Armstrong, Attorney General Mark White, and Comptroller Bob Bullock all served on the Council of TENRAC. Other members of TENRAC included the Chairman of the PUC, Moak Rollins, a Clements appointee to the Commission, Chairman of the Water Development Board, Louis Beechrel, a Clements appointee to the Board, Health Commissioner Robert Bernstein, Chairman of the Air Control Board, John Blair, Parks and Wildlife Commissioner Perry Bass, and four citizen members appointed by Clements with concurrence by Hobby—James Russell of Abilene, Ed Cox of Dallas, Ed Vetter of Dallas, and Mike Halbouty of Houston. In all there were 12 Republicans and 10 Democrats on the Council. Clements' task was to get political consensus out of the group that would best serve Texas' interests.

Holloway was appointed Director of the Policy Analysis Division of GEAC by Governor Briscoe in January 1976 after serving as a staff member of the commission form of GEAC in 1973—74. He was appointed Executive Director of the second statutory form of the Council, known as the TEAC—appointed by Briscoe with recommendation of Lt. Governor Hobby with confirmation by the Senate. Holloway did not know Bill Clements when he (Clements) was elected Governor.

After an interim period of about 9 months Ed Vetter stepped down as Executive Director of TENRAC and became one of the Clements citizen appointees to the Council. Holloway was named Executive Director by agreement of the co-chairs, Clements and Hobby. Holloway was soon tested, Clements style, to see if he had the qualities Clements always looked for in people he worked with. He wanted to find out if Holloway had informed opinions to offer and if he understood who was calling the shots. Shortly after being appointed Holloway was called to Clements' office on short order for reasons not explained by his secretary, Janie Harris. Holloway arrived expecting to have some of Clements' personal staff in attendance—none were present. Holloway sat across the desk from Clements and eye-to-eye Clements asked bluntly if Holloway knew what his policy was on nuclear waste disposal. Holloway had not talked with Clements about nuclear waste but was very familiar with the topic—there had been considerable news coverage of a very messy low-level nuclear waste operation on Galveston Island and Clements was trying to get it shut down. Holloway had been directing the work of an advisory committee on low-level nuclear waste disposal which had been at work for several months and had drafted a report to the Council on policy recommendations that Clements had not yet seen. Holloway summarized what he thought Clements position was on the topic (which was not in conflict with the Advisory Committee report). Clements liked the direct, informed answer, and thereafter the two got along fine.

The nuclear waste event was Holloway's "litmus test." This was Clements' management style (Barta, 1996, p. 257). Holloway said later that he never had a communication problem with Clements—said he did his homework and the Governor an informed opinion. Clements always made sure he understood Holloway's reports and opinions, and he made decisions which gave clear direction. Holloway did not always agree with Governor Clements, but he did always know what "his policy was." One frustration with working with Clements was that one never got to revisit an issue. If the decision was wrong it just stayed wrong and one learned how to work with it. Clements did not like to change positions.

The changes of administrations in Washington D.C. became an important part of Clements's push to impact US energy policy which had been a complete failure under President Carter from the Texas perspective. Clements worked hard to help elect Ronald Regan and became a member of Reagan's "kitchen cabinet" for the 1980 Republican National Convention in Detroit. After Regan was elected in November 1980 Clements like other governors tried to get Texans appointed to leadership positions in the new administration. In the end he was only successful in getting one major appointment made—that of Allen Clark as deputy administrator of the Veterans Administration. Holloway was one of two finalists for the Administrator of the Energy Information Administration job recommended by Clements, but Senator Barry Goldwater of Arizona got his guy, Eric Evered finally selected for that role in the DOE.

Clements was an active member of the National Governor's Association, Southern Governors Association and the Republican Governor's Association, and he took his energy policy message to that forum. Ed Vetter and Walt Rostow (a University of Texas economics professor who served as foreign policy advisor to Presidents Kennedy and Johnson) wrote the first brief energy policy paper for Clements, which he got adopted in modified form by the NGA in the summer of 1979 (Barta, 1996, p. 237). A more extensive policy statement was drafted by a special team of TENRAC members in 1982. The team included Holloway, Ed Vetter, Moak Rollins, Bill Fisher and Walt Rostow. The policy paper was sent to President Reagan, Secretary of Energy, James Edwards, the Texas delegation and selected others.

TENRAC gained considerable recognition in Washington and among state and regional political groups and industry representatives who were interested in the energy problem. Engagement included personal visits to key congressmen, briefings, and congressional testimony. For example, natural gas policy became a major issue in the years following the enactment of the National Gas Policy Act of 1978. Several major attempts were made to modify the Act to provide a more complete deregulation of natural gas markets than that provided for in the Act. TENRAC provided analyses, briefing, and policy positions adopted by the Council and signed jointly by Clements and Hobby.

During the Carter administration TENRAC came out strongly against the WPT but supported the SFC. As a response from the SFC Exxon set up a large synfuels project in East Texas. Texas through TEAC and TENRAC implemented a Texas R&D effort of its own focused on coal gasification, in situ gasification of lignite, geothermal, enhanced oil recovery, renewable energy from wind and solar and conservation programs. The conservation programs were funded by pass-through of federal funds. Before TEAC was established there was no R&D entity focused on Texas energy. At the same time TEAC (followed by TENRAC) was created the major Texas universities (UT Austin, Texas A&M, University of Houston and Texas Tech) were granted R&D funding through newly created "energy centers," each with its uniquely allocated funding in university budgets. TEAC (and TENRAC) worked to coordinate the funding of high priority research projects among TENRAC and the universities, sometimes through jointly funded projects. TENRAC developed a Texas R&D long-range plan in 1982 that laid out priority research topics and projects (Texas Energy and Natural Resources Advisory Council, 1983a,b,c).

The pattern of providing analyses, briefings, and bipartisan policy positions was carried out on other major issues including nuclear power plant waste storage and disposal, the "windfall profits" tax on crude oil, restrictions on the use of natural gas as a boiler fuel, reorganization of the DOE, disposal of low-level (mainly hospital and research institution contaminated materials), import fees and quotes and synthetic fuels development.

Most of the major energy policy positions during and following the Iranian crisis were not partisan issues in Texas. Republicans and Democrats usually had the same basic views on matters, positions mainly dictated by the economics of energy in Texas. But there were disagreements that led to interesting interplay of politics and energy policy.

Bill Clements defeated Attorney General Hill in the general election of 1978 and Mark White won the race for Attorney General. White had been Governor Briscoe's General Council in the Governor's Office, and then Briscoe appointed him to be Secretary of State. From the Attorney General's position White was posed to run against Clements in the 1982 governor's race. As a result, every council meeting had the makings of some politicking by Clements and White. The politicking spilled over into energy policy, which at the time was the hottest topic in Austin. Holloway had to maneuver the Council meetings around and through this highly charged political setting; most of the time he was successful.

Holloway's routine was to have a brief meeting with Clements, Hobby and Clayton to set the agenda for quarterly Council meetings. He would then make personal visits to each of the principals on the Council. He tried to carry out the role as Executive Director in a nonpartisan fashion, letting the political officers work their politics around the issues and the objective staff work was brought to the Council.

There were two particular issues involving energy policy that became the focus of the Clements/White political battle that followed the 1978 election, and the quarterly council meetings. The first was the matter of a legal challenge to the Carter Administration's WPT, and the other was the Ixtoc oil spill. But interestingly it was the electricity price issue that won the Governor's race for White in 1982, not the WPT or the Ixtoc oil spill.

The electric pricing issue came as a surprise political issue, but was the direct result of the energy crisis of the 1970s, and the Natural Gas Policy Act of 1978. Rapidly rising natural gas prices in Texas came to the attention of consumers, not so much because of rising heating bills, but because of the indirect effect on rising electric bills. The Texas PUC had adopted a policy of allowing automatic pass-through of electric generation fuel cost to the consumer, an automatic fuel adjustment. During the 1978–82 period of the Iranian oil crisis and its aftermath, oil prices catapult upward to unparalleled levels; the oil market also affected natural gas prices since natural gas was a substitute for fuel oil in electric generation and in other energy markets.

Regulation of the electric utility industry at the state level did not exist prior to the Embargo. The Public Utility Regulatory Act (PURA) of 1975 as recommended by GEAC in the spring of 1975, created the state-level electric utility regulatory agency with original jurisdiction over investor-owned utilities, and appellate jurisdiction over municipally-owned utilities and coops. That structure remained until the Legislature began the restructuring of the industry in 1995. The restructured industry continues today under the PUC. The PUC helps oversee a restructured electric market with competitive generation and retail services supported by a regulated transmission and distribution component. Well spacing, well plugging, pipeline safety, and appellate jurisdiction over natural gas retail rates remain functions of the TRC.

Crude oil is refined into a number of fuels other than gasoline, including fuel oil for industrial and electric utility boiler fuel. So, as crude oil prices rose during the Iranian crisis, so did fuel oil, making it economical to use more gas and less fuel oil, thus driving up the price of natural gas.

The prices of natural gas, however, were, by this time, controlled by the Natural Gas Policy Act of 1978 (NGPA). This very complicated legislation defined some 22 separate categories of natural gas, some of which was decontrolled, and some that was allowed very high prices (by historical comparison) in order to encourage more production. In short, demand for natural gas and the NGPA allowed prices to rise rapidly as the market moved toward decontrol. Since electric utilities in Texas were fueled almost exclusively by natural gas, and due to the automatic fuel adjustment classes, high natural gas prices were immediately reflected in electric prices. The rapidly rising electric prices were more noticeable in Texas than gas utility prices because Texas has mild winters and hot summers.

The PUC policy of automatic pass-through of fuel costs to consumers meant that the Commission did not regulate this part of the electric bill. The basic rationale is that the PUC has no direct jurisdiction over natural gas prices, and so long as oil, gas, and coal markets are producing competitive fuel prices, there was no need for the Commission to try to regulate this part of the electric business, save that of making sure electric utilities were making reasonable contracts for fuel.

This issue found its way into the Governor's race of 1982 because the three Commissioners of the PUC are appointed by the Governor. Mark White recognized that consumers, most of whom are voters, were very upset by high air conditioning bills during the summer of 1982, the direct result of high natural gas prices, which, under PUC policy, was being passed on automatically to consumers without direct regulation by the Commissioners; the Commissioners by this time were all Clements' appointees. White easily convinced voters that high electric bills were Clements' fault, and that the cure was to change the automatic fuel pass-through policy, but more than that, the whole matter of the PUC regulatory framework needed an overhaul by the legislature—if he (White) were governor, he would fix the problem. White won the election, got the PUC Act rewritten, and in the process, the TENRAC was eliminated under the state's Sunset Law, (a story of its own developed below).

Among the issues that ultimately lead to TENRAC's demise was the issue of the legal challenge to the WPT on crude oil. This became a contentious matter between Clements and White. The WPT while purporting to be a tax on oil profits due to abnormally high oil prices under deregulation, as influenced by the Iranian crisis, was actually an excise tax (a percentage of the sales price of domestic oil), but with a major exception for Alaska. The justification was that high production and transportation cost required to get Alaskan oil to market in the lower 48 states put that production at such a disadvantage that the WPT would render it uneconomic to produce.

Mark White, as Attorney General, was responsible for bringing a law suit of Texas against the federal government and challenging the constitutionality of the WPT. White joined with several other Attorneys General to bring a law suit. Governor Clements thought General White was, by the Texas constitution, to represent Clements as Governor in the matter (Barta, 1996, p. 232). This issue became part of the TENRAC quarterly meetings during 1982. The other continuous matter between White and Clements was the Mexican "oil spill" from the Ixtoc offshore well that "blew out" in 1979 (Barta, 1996, p. 240). The well was drilled by one of SEDCO's drilling rigs, under contract with the Mexican government. The spill made quite a mess of the Texas beaches for some time, until the winds and tides finally pushed the oil out to sea. Clements made several statements about the relative importance of the oil spill—"it's no big deal," he said (Barta, 1996, p. 239). White made it a big deal, making much of the fact that it was Clements' firm (SEDCO) that was responsible. Clements pointed out that the Mexicans were operating the drilling rig, not SEDCO, and that the Mexicans had promised to pay for any damages on the Texas coast (Barta, 1996, p. 240). This issue, surprisingly, did not come up at the TENRAC regular meetings, White choosing to make use of the incident outside the TENRAC forum.

Mark White won the Governor's race in November, 1982, and took office in January 1983. One of his priorities for the legislature session was passing revisions to the Public Utility Regulatory Act (Handbook of Texas Online, 2018).[11] In the process of the 1983 legislative session, TENRAC had one meeting in February 1983, where White chaired the meeting and the new dynamics of the use of TENRAC as a political forum for newly elected officials became evident. The new members under TENRAC's statute included the newly elected Commission of the General Land Office, Gary Marrow, Commissioner of Agriculture, Jim Hightower and Attorney General, Jim Maddox. This proved to be TENRAC's last meeting. Holloway resigned, under pressure from White, in March 1983, and due to failure of White to support continuation of TENRAC under the Sunset Act (the support had to come from the House Energy Resources Committee), the renewal of the TENRAC statute died in committee and TENRAC was dissolved in August, 1983. During the House floor debate Representative Bill Hollowell of Grand Saline representing the district that was home of the East Texas Field of the 1930s in the deep east Texas oil country, stood and said "TENRAC couldn't find a quart of oil at an Exxon station." Such was the attitude of the House in 1983 when oil prices had come down, Jimmy Carter was no longer President, and the interest in any state-funded R&D on oil and gas alternatives was at a decade low. Ronald Reagan made a similar comment in the debate with

---

[11] The Texas Public Utility Commission was created in 1975 in part due to recommendations from the Governor's Energy Advisory Council growing out of the Arab Oil Embargo that was the impetus for creating the GEAC, but the pressure from public outcries over telephone utility issues was also a major reason for creating the new utility regulatory commission. After the first 8 years of PUC operations that included the state-wide regulation of electric utilities (original jurisdiction over investor-owned utilities and appellate jurisdiction over city regulation and coops) natural gas prices sky-rocketed causing electric rates to increase dramatically. The issue became a major part of the contest between Governor Clements and Mark White in the governor's race. Mark White promised to reform the PUC. The Texas Sunset Advisory Commission which reviewed both the PUC and TENRAC during 1982 recommended several changes to the PUC responsibilities, including limiting the automatic pass-through of fuel cost to electric consumers, a change which White supported. The Public Utilities Regulatory Act was amended by the Legislature in general session of 1983.

Carter in 1980 concerning the DOE spending on energy when he said "it hasn't produced a quart of oil or a lump of coal or anything else in the line of energy."

GEAC, TEAC and TENRAC all had a dual focus—policy and R&D. GEAC was made up of administrative and legislative branch leaders of state government supported by a large advisory committee from the private sector. GEAC's contributions to Texas and National energy sector innovation included policy and technological development that now spans four decades. The 40 specific projects of GEAC became a baseline characterization of the energy systems in Texas accompanied by recommended new directions. The studies included current and projected supply and demand for energy, the legal and regulatory framework for energy activities, environmental conditions related to energy production and consumption, and new technology, especially alternatives to oil and gas. The new technology studies included geothermal, lignite, solar, wind, enhanced oil and gas recovery, nuclear power, tidal, and Gulf-stream power and solid waste resources. These studies formed a base of information that launched policy directives and ongoing R&D programs for the next decade and beyond.

The policy directive of the GEAC resulted in ten significant legislative and state agency recommendations. These recommendations came from the Council's policy body of GEAC that was made up of leaders from state government (mostly statewide elected heads of state offices) supported by a large advisory committee from the private sector. The Council developed ten legislative and other recommendations, including; (1) creation of a statutory form of GEAC, (2) creation of an energy development fund for promotion of new energy technology to be managed by GEAC, (3) creation of the PUC for statewide regulation of the electric industry, (4) creation of a policy to phase-down the use of natural gas used for boiler fuel to be implemented by the TRC, (5) implementation of an energy efficiency labeling program by the Weights and Measures division of the Department of Agriculture, (6) clarifying the legal state of the mineral estate for geothermal resources by the TRC, (7) require state agencies to consult with and obtain comments and concurrence regarding environmental impacts associated with state permit applications, (8) require local governments to include energy conservation standards in building codes for residential and commercial buildings, (9) promote urban mass transit, and (10) request the Governor and other state officials to promote removal of the two-tiered pricing system on crude oil being implemented by the FERC, and to encourage advanced recovery from old reservoirs (Holloway et al., 1979, p. A-5).

All of the legislative policy recommendations of GEAC were passed in some form during the 1975 legislative session save one. Further, the state agencies mentioned in the recommendations did set policies in keeping with the other directives that targeted state agencies. The remaining recommendation, the formation of the Energy Development Fund, did not occur until the 1977 session of the legislature. TEAC carried out a Texas focused R&D program that helped form the basis for ongoing R&D that today includes wind and solar as well as conservation efforts for the residential and commercial sectors of the economy. TEAC also carried out a critic of the PIES modeling system that included both professional evaluations of the modeling and data structure but also the process of using the modeling system for evaluating national energy proposals. The outcome was to guide the development of procedures for the use and access of the modeling system that remains today. During the tenure of TEAC as the Texas energy policy and R&D office the Governor's Office continued to operate an energy conservation office funded by Federal energy conservation funding earmarked by Congress for cooperating states. This conservation office was housed in several state agency offices in the years following the Briscoe years, and exists today as the State Energy Conservation Office (SECO) as a division of the Comptroller's office. SECO partners with Texas local governments, county governments, public K–12 schools, public institutions of higher education and state agencies, to reduce utility costs and maximize efficiency, all funded by Federal conservation dollars from the US DOE.

TENRAC became the primary policy forum for developing Texas energy policy positions during the first Clements years of 1979–82 that helped change the national approach to the national energy challenge created by the two energy crises of the 1970s. TENRAC continued to manage and implement the Texas Energy Development Fund that funded a variety of R&D projects and wrote a 5-year plan for expanding the program of State government development of alternative energy technologies and policy models to support policy development (Texas Energy and Natural Resources Advisory Council, 1983a,b,c).

The final episode of Texas state governments attempt to reset energy policy came with the Ann Richards administration after she won the governor's race by defeating a West Texas oilman named Clayton Williams in 1990, serving one term from 1991 to 1994. Richards appointed Lena Guerro to fill a vacancy on the TRC who brought together a host of public and private sector leaders to define a Texas energy policy directive by an effort known as State of Texas Energy Policy Partnership (STEPP) and the Sustainable Energy Development Council. The group staffed by Carol Tombari, Director of Governor Richard's energy office, produced an energy policy document. Richard's attention focused on other matters but the effort was seen as a significant effort in support of energy conservation and renewable energy.

Interplay of policy, technology and markets following the energy crisis

This chapter reviews the structure and activities of the Federal and State government and private sector entities during the decade following the Embargo. These commissions and agencies are examples of institutional innovations. The commissions and agencies while being driven by duties defined in Executive Orders and statutes all had a fundamental institutional character—a means of generating new ideas. Measuring the effectiveness of such entities is a central issue in understanding how to generate institutional innovation in the future. The following summary discusses this challenge by way of examining the United States and Texas experience following the Embargo including an assessment of the successes of these institutions by way of measures and indicators of their *impacts* (Brooks, 2018).

The world energy crisis that followed the Arab Oil Embargo of August 1973 was clearly policy driven by key members of OPEC who made the decision to shut off production at a point in the dynamic shifting of world oil production that gave them the market power to achieve the price run-up that followed. But OPEC's ability to execute this run-up in price came after the oil markets of the world shifted the concentration of world production to OPEC countries from the United States and Texas during the previous decade. The Embargo event set off an ongoing struggle of the developed world countries to set new policy and redefine the role of governments, and the response of the private sector in order to assure adequate energy supplies primarily for national security purposes. But the struggle was also to develop policy to support economic growth and shield consumes from high prices. Oil shortages and high prices created recessions around the world. The policies that followed sometimes included attempts to directly control oil prices through statutory price ceilings. But oil prices in the United States rose anyway under the price ceiling system that was in place from the Nixon era until finally removed in the early Reagan administration. Nixon included oil price freezes along with economy-wide price controls intended to limit inflation which resulted from excessive Federal spending for the Vietnam War. The economy-wide price controls were subsequently removed since they were clearly creating a massive shortage problem, but controls were left on oil products. Price controls on oil at the wellhead distinguished "old" oil from "new" oil and set separate price ceilings for each to encourage new production but keep domestic producers from getting a windfall due to the high prices set by OPEC.

In addition to continuing oil price controls to limit the Embargo impacts on consumers, the Carter administration also proposed taxes on imported oil so as to encourage consumers to reduce consumption. The attempt failed and since the government had no control over imported oil prices, Texas average wellhead prices, which mirrored US prices, rose from $3.48/bbl in 1972 to $7.80/bbl in 1975 (domestic oil). Refiners paid $13.29/bbl for imported oil. During the Iranian crisis in 1979 that followed the Embargo, the imported price of oil rose to $40/bbl at the peak (approximately $100/bbl in 2016 dollars).

The embargo was a major disruption for the United States, and likewise for all of the developed world economies. The political attitude of laissez faire about the oil market obviously needed to change. Suddenly, national security, economic stability, and the international flow of crude oil supplies became different parts of the same problem. Governments were not organized to manage the problem; therefore, most of the early response to the Embargo was redefining the role of government and reorganizing government agencies to carry out new policies. Political attention to the problem involved shifting the power to respond from the private sector to government. An analogy of this effort is government's mobilization of forces needed to carry out a war. President Carter called the energy challenge "the moral equivalent of war."

Nixon's focus was on near-term price controls with rationing and reorganizing the government. But the markets responded as expected to the price ceilings—shortages of gasoline at the pump. So the government response to shortages was to set up a government allocation system to ration supplies—"share the shortage" burden—and tilt the allocation of oil to small refiners to keep them viable. To a large extent the system was like the rationing of oil products during World War II. The production-side focus was on the construction of the Alaska pipeline to encourage domestic production of the newly added reserves in Alaska, plus policy changes to encourage production of natural gas and to set reasonable standards for the surface mining of coal.

Nixon also had a long-term focus. His message to Congress in April 1973 called for a new ERDA incorporating the research functions of the AEC and, along with it, developing a new energy R&D effort of $10 billion for the next 5 years. Civilian nuclear regulation would be handled by a new NRC.

The markets responded. In the short term the government attempts to reduce imports by reducing consumption of gasoline and diesel through price controls and rationing created chaos with long lines at gasoline stations that ended in fights among consumers and even shootings by truck drivers. The long-term production responses eventually came but took decades to materialize. During the decade that followed the energy crises of the 1970s extensive exploration efforts got underway by the major oil companies in Alaska, the North Sea, the Caspian Basin, and Siberia. The exploration in Alaska was accompanied by the construction of the Trans Alaska Pipeline to move the oil to refineries in the lower forty-eight states.

The Trans Alaska Pipeline was designed to deliver up to 2.0 million barrels per day from Prudhoe Bay to the lower 48 states. Leasing in the other outer continental shelf areas (the east and west coasts and the Gulf) was also a part of the Nixon era plans. Implementation of the leasing programs became a major effort of the Ford administration that followed. The belief at the time was that the lower 48 domestic reserves would only decline—that new reserves could primarily be found in deep ocean locations.

The actions of the Nixon, Ford and Carter administrations did bear fruit by somewhat arresting the decline in US domestic oil production. Oil production was 9.2 mmbd in 1973 declining to 8.1 mmbd in 1976, then rising slightly to 8.6 mmbd in 1980 and 9.0 mmbd in 1985, mostly due to the addition of oil from the offshore Alaskan reserves. But net oil imports (imports of crude oil and products minus exports) continued to rise throughout the 1970s from 6.0 mmbd in 1973 to 8.0 mmbd in 1979. Net imports declined temporarily to 4.3 mmbd in 1983 and 1985 due to a combination of rising prices, government conservation

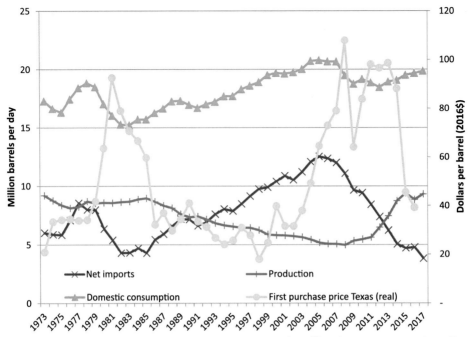

Figure 8.1 US oil production, imports, consumption and wellhead price: 1973—2017. *Energy Information Administration, Monthly Energy Review, January 2018.*

programs, an economic recession and increased production from Alaska. But imports rose again as the economy returned to normal growth and oil and gas prices declined during the mid-1980s and throughout the 1990s. Imports increased to 6.5 mmbd in 1988, 7.6 mmbd in 1993 and 9.9 mmbd in 1999 all as a result of declining US production and modest consumption growth. These trends continued as production declined to 5.1 mmbd in 2005 and net imports rose to 12.5 mmbd (see Fig. 8.1). The innovation that fundamentally changed this market dynamics was fracking technology, which finally produced dramatic results beginning in 2005 after three decades of development and demonstration. Fracking is the special topic of Chapter 9, Electric Industry Deregulation and Competitive Markets.

Texas and the United States spent the next decade reorganizing the government and trying to figure out the right combination of policy changes, market developments and R&D (to drive the long-term). In the months following the October 1973 Embargo, Nixon set up a policy office by Executive Order. The office was known as the FEO, which later became the FEA. FEO was responsible for putting together and administering allocation plans for crude oil and for all petroleum products as required by 1973 legislation. The Office was also responsible for putting together a plan for *energy independence by 1980.* The act also contained authority for gasoline rationing. FEO was charged in the executive order with advising the President "with respect to the establishment and integration of domestic and foreign policies relating to the production, conservation, use, distribution, and allocation of energy and with respect to all other energy matters" (Emergency Petroleum Allocation Act of, 1973, P. L. 93–159).

President Ford continued the focus on reorganizing the government, conservation at home and a focus on foreign policy. To address national security Ford set up the Strategic Petroleum Reserve in salt domes along the Texas and Louisiana Gulf Coast. The salt dome capacity was designed to hold 727 million barrels, the largest emergency supply in the world. This was another government action akin to actions in time of war. The markets settled down for a time. Market prices for oil, gas, and electricity were pretty stable and relatively low for a while, providing little incentive to add oil and gas reserves—the natural gas market was producing shortages in large part due to wellhead price controls in the interstate market. The approach to electric utilities was still dominated by the regulatory mind set and a focus on policy to make coal and nuclear replace oil and gas generation. The Ford Administration did propose legislation to create other institutional entities such as the Energy Independence Authority for long-term development. The Authority was to be a special corporation to provide Federal subsidies to companies for energy development projects. The Authority was to have $25 billion in equity and $75 billion in government-backed borrowing authority. Liberals in the Congress objected to the subsidy to big business and conservatives objected to the plan as further government interference in the marketplace (Energy Policy, 1981, p. 150).

Ford's 1975 State of the Union message set a goal of achieving a capacity of 1 million barrels a day of synthetic fuels by 1985 (Ford, 1975a,b, blc. 1, p. 42). The Energy Independence Authority legislation failed to get the support of the Congress. But the Energy Reorganization Act of 1974, signed by Ford on October 11, provided for a new entity, the Energy Resources Council. The Council was charged with developing "a single national energy policy and program" (Ford, 1975a,b, p. 231). The Council, however, turned out to have little influence and, in the end, the FEA became the chief energy policy agency of the government, evolving eventually into the DOE under President Carter.

President Carter continued the focus on reorganizing the government finally establishing the structure we have today. The legislation in 1978 created the DOE with a structure for management of the national labs, and especially important, created the Energy Information Administration. The several electric power administrations were transferred from Interior to DOE and the NRC was set up to regulate the civilian nuclear programs for electric power generation. Carter also established a long-term phase-out of the price controls set up by Nixon and continued under Ford. Finally, Carter tried to set up a new R&D entity—a joint public/private sector entity to produce an oil substitute from shale—the SFC.

The most important Carter-era contribution was the five energy bills that evolved from his much vocalized NEP. The legislation provided for the phase-out use of natural gas as a boiler fuel, phased out natural gas price ceilings, and deregulated "new oil" prices. The Iranian crisis sent oil prices to $40/bbl ($105 in 2016 dollars) resulting in a recession—the second oil shock recession of the decade. The overall market impact during the 1970s was a great incentive for adding oil and gas reserves, even though the 1980 oil price increase to $40 was temporary and prices declined to $10 per barrel in 1986—the decade-long trend was a 400% crease in the wellhead price since early in 1973 before the Embargo. The total oil and gas wells drilled increased from 27,000 in 1973 to 84,000 in 1982 following the run-up in prices from the Iranian crisis. Reserve additions increased from 1.1 billion barrels in 1977 to 3.0 billion in 1980. But production continued to exceed new additions to reserves so the level of proven reserves continued to decline from 35 billion barrels in 1973 to 28 billion in 1983. This pattern continued with only an occasional year where additions exceeded production, until the fracking generated additions began in 2005. By 2006 proven reserves had declined to 21 billion barrels (Energy Information Administration, 2018, 2000).

The natural gas market did not respond in lock-step with the oil markets, during and following the oil disruptions of the 1970s. Regulated natural gas wellhead prices increased during the 1970s, but the history and policy approach differed from that of the oil market. Imports, the major focus of oil policy, was not the focus since the only imports of natural gas came from Canada and Mexico. Gas imports were a small part of the US market. Rapid growth in natural gas demand in the late 1960s coupled with wellhead price controls in the interstate market produced shortages in the winters of 1970−76. Curtailments in the interstate market rose from 0.3 TCF in 1971 to 3.2 TCF in 1976 which amounted to 14% of demand (Holloway, 1977).

The very convoluted system of interstate wellhead regulated prices in the natural gas market evolved to include 22 distinct categories of natural gas (separate prices). Different prices were enforced based on vintage of the well and difficulty (cost) of recovery. "New gas" and "high cost gas" categories were created to encourage new reserve additions and production while preventing owners of "old gas" from receiving "windfalls." The issues of government regulation of the natural gas market were addressed by the Natural Gas Policy Act (one of the five bills of Carter's NEP) which eventually passed the Congress and was signed into law by President Carter. So continuous were the disagreements that the Act was debated for 18 months in the Congress before finally passing in late 1978.[12] The related Fuel Use Act placed severe constraints on natural use as a boiler fuel, which prevented new gas boilers from being deployed, and required the phase-out of existing

[12] The debate over natural gas regulation underlying the NGPA followed over 40 years of experience that began with the Natural Gas Act of 1938. Perhaps the most influential economist on the topic during the 1978 debate was Paul McAvoy of Yale (see MacAvoy, 2000).

unites by 1985 (Holloway, 1977). In short, the NEP legislation controlled the major use of natural gas for a boiler fuel in electric power generation because there were alternatives (coal and nuclear) and reserved natural gas supplies for commercial and residential use that were considered the highest priority. The combined effects of these national policies was offset in large part because the price of intrastate natural gas in Texas (about 40% of the US market) was not regulated, so reserves were added in the Texas intrastate market and shortages were eliminated.

The combination of United States and Texas natural gas policies in the 1970s added slightly more reserves than production during the 1980s pushing prices downward from $2.66/mcf in 1984 to an average of $1.67/mcf in 1987. The result was to only add reserves that approximated consumption, a condition that prevailed until fracking fundamentally changed the market beginning in 2005−06. Texas natural gas production declined from the peak at 9.5 trillion cubic feet (TCF) in 1972 to 5.5 TCF in 1999 (a 42% decline) before rising again to 8.8 TCF in 2015. The Carter-era Fuel Use Act that was intended to phase-out the use of natural gas for boiler fuel was repealed during the Regan era. The resulting growth in natural gas use that followed was concentrated in electric power generation in Texas where boiler fuel use in the electric power sector increased from 3.0 TCF in n1985 to 5.9 TCF in 2005. The dynamics of markets prevailed as natural gas became the fuel of choice in the Texas power industry. This electric market growth was in large part the cause for wellhead prices to increase to $10/mcf for a period in the summer of 2008 before the fracking contribution to production pushed prices back to the $3/mcf range in 2009. The other factor that drove prices down was the economic slowdown of 2008 following the financial crisis recession. The recession reduced demand across the board, including energy of all kinds (Energy Information Administration, 2017b).

The dynamics of the natural gas market that has evolved since the 1970s shortages and political action over pricing set up another innovation that has also changed as a result of the fracking technology and the deregulation of wellhead prices. Liquified natural gas (LNG) import facilities developed to import natural gas to fill the gap created by rising gas demand and declining production during the 1970−2005 period. The 20 plus year development of LNG import facilities have now become terminals for export of LNG. Some new ports are for export, others are import facilities that can also export. The improvements in LNG technology compared to the capabilities of the 1970s for the combination of liquefaction, transportation, and vaporization/compression changed the market prospects. The costs for moving LNG from several foreign sources into the US pipeline transmission system make the onshore delivered price of LNG competitive with wholesale natural gas in the deregulated US market. In short, the interest in LNG an import market was driven by (1) improved economics of tanker transfer and the recent volatile run-up in natural gas prices, that brought LNG economics in line with the US domestic market, (2) the outlook for significant

US natural gas market growth, and (3) a favorable regulatory climate for rapid approval of new terminal applications (Holloway, 2003).

The United States has four existing LNG terminals that have been in existence for 25 plus years, developed during the 1971−82 period. Two of the terminals— Cove Point and Elba Island—both of which opened in 1978 ceased operations in 1980. Cove Point reopened in 1995, but only for storage services to local utilities. Elba Island opened again for imports in 2001. Lake Charles opened in 1982 but operated only a short time before closing; it opened again in 1989. In 2001 three of the terminals—Lake Charles, Everett and Elba Island—received imports of LNG. Lake Charles received 61 tankers, Everett received 39 and Elba Island received one. All four terminals have since expanded their capacity (Gaul and Young, 2003).

Four LNG import facilities entered the permitting process in the 1990s and early 2000s to bring imported LNG to the Texas market t—one at Corpus Christi, a second one at Brownsville, a third at Freeport, and a fourth at Sabine Pass. As of January 2018 Freeport and Sabine Pass facilities were operating and Corpus Christi was approved and under construction. Eleven Other facilities were operational for the coastal areas of Louisiana, the southeast Atlantic, the northeast Atlantic and on the California coast. Two offshore Gulf of Mexico facilities were approved but not yet under construction.

The Reagan era changed the government approach from that of the prior three Presidents. Reagan's campaign promise was to remove of oil and gas price controls. He also promised to kill DOE, but in the end did not. Eventually price controls on oil and gas were removed, the mandated oil allocation among refineries was stopped, the gasoline allocation program was dismantled, the use of natural gas for boiler fuel use was left to the market place, and the SFC was abandoned. The United States and world recession in 1982 coupled with OPEC member cheating (exceeding their quotas) sent oil prices downward to $10 in 1986 ($22 in 2016 dollars). Reagan opened offshore (mainland outer continental shelf) and Alaska to drilling. Domestic energy prices settled downward to levels Nixon, Ford and Carter administrations would not have thought possible. But as a consequence drilling and reserve additions remained depressed resulting in rising imports. Net imports rose from 4.3 mmbd in 1985 to 12.5 mmbd in 2005.

But the DOE and all of its functions remained, including the very important primary source of reliable data on the energy sectors of the nation and the 50 states. DOE has an ongoing policy recommendation role, oversees the set of National Energy Labs and operates an ongoing R&D program focused more on early stage research and less on large-scale demonstration projects. The R&D program attempts to front load the basis for demonstration and commercialization by the private sector.

In a departure from that model DOE managed the implementation of the federal $130 million Stimulus program during the Oboma administration called the *Recovery Act—Smart Grid Demonstrations* during 2009—14. The funded projects were all about demonstration and commercialization of electric sector energy technologies.

The two energy crisis events of the 1970s set in motion several responses that have taken 45 years to fully develop. These crisis events, analogous to "hitting a wall" that often drives innovation set the pace of energy sector innovation that is evident today. The near half century of innovation has many parts but perhaps the most dramatic has been hydraulic fracking ("fracking") in oil and gas recovery and the restructuring of the electric industry. Texas has been a leader in the United States in deregulating the electric industry and developing and deploying the fracking technology, as well as many alternatives to oil and natural gas. The Texas policy and R&D entities also supported the early push of enhanced oil and gas recovery. These alternatives, first recognized in the policy and R&D programs of GEAC, TEAC and TENRAC include wind, solar, and lignite, geothermal, biomass, and several conservation technologies that have helped slow the dependence on foreign oil.

GEAC, TEAC and TENRAC were Texas government's attempts to redefine the role of government, drive a state and national energy policy and fund alternative energy resource development. As established state agencies (the TRC and the PUC) developed new approaches to oil, natural gas, lignite, geothermal, wind, solar and conservation technologies, and active markets developed, these advisory agencies closed down. TEAC and TENRAC funded Texas projects that included in situ gasification of lignite, fluidized bed technologies for converting lignite to a gas fuel, solar ponds in areas with high concentration of salt deposits, wind machines for stripper wells in isolated areas, and technologies for energy efficient buildings.

A unique Texas private sector organization developed in the aftermath of electric restructuring. Beginning in 2005 this Texas nonprofit called, the Center for the Commercialization of Electric Technologies (CCET), was formed to help guide the development of new technology in the Texas electric sector as it adjusted to the new era of completion, and a new world of technological innovation. After 10 years of joint government-private sector development and commercialization activities that added significantly to the innovation in the electric sector, CCET closed down in 2015. CCET existed for 10 years overseeing $30 million in demonstration and commercialization projects, which have helped push innovation in the Texas electric sector. The technologies and the innovation contributions are summarized in what follows in this chapter.

The private sector innovation, while supported by Texas government and private sector initiatives listed in the previous paragraph, came from the other major sources of innovation—the geological resources and the diverse and risk-taking energy sector leaders of Texas. The focus on the demand side of the oil market by three

Washington administrations could not change the long-term trend of rising imports of crude oil. A more powerful influence on rising oil imports was the continuing decline in US domestic oil production and the response of the private sector. The production declines included Texas with about 1/3 of the nation's production. Texas oil production peaked at 1.2 billion barrels in 1972 falling precipitously to 392 thousand barrels in 2006, a decline of 77% before the impacts of new technology (primarily fracking) turned the trend around bringing oil production back up to the 1972 peak of 1.2 billion barrels in 2017 (the topic of Chapter 9: Electric Industry Deregulation and Competitive Markets).

The fracking technology could not exist without the technology of horizontal drilling. The horizontal drilling first appeared in the Austin Chalk in 1934 to stop a Conroe, Texas oil field fire (Petroleum Technology, 1933). But the contribution that has recovered major quantities of oil found early success in the Austin Chalk formation southwest of San Antonio, Texas. Beginning in 1988 and extending through the 1990s horizontal drilling allowed major recovery of oil from unique reservoirs (Saucier, 2017). The technology was used importantly in the 1980s in Prudhoe Bay, Alaska and in the Bakken of North Dakota (Leibman, 1989). When combined with hydraulic fracturing supported by 3-D seismic data, horizontal drilling found its most timely and important contributions in the last decade in seven regions of the United States, including Permian, Niobrara, Haynesville, Eagle Ford, Bakken, Appalachia and Anadarko basins. The fracking technology innovation experience is the topic of Chapter 9, Electric Industry Deregulation and Competitive Markets.

"Hitting the wall" in the 1970s events sent both public and private sector participants in search of innovation—government policy, institutional changes, new markets, and new technology—eventually generated a technology and an industry restructuring that more than any other has turned the trends of the last 45 years in a truly new direction. Fracking and electric industry restructuring are the technologies and business model changes that are the subjects of following chapters. The new concern and focus of the energy debate is the impact of energy production and use on global warming and CO_2 emissions. This is the focus of Chapter 11, The Oil and Gas State Adds Renewable Wind and Solar.

Energy innovation growing out of the 1970s energy crises continues, even four decades later, to influence the geopolitical balance in the world economy. These innovations continue to impact overall economic growth and have shifted the market-driving marginal producer from the Middle East back to Texas for the first time since the 1960s prior to the Embargo. Such innovation also determines (in part) the distribution of economic activity and related environmental impacts among the nations of the world. The energy markets and energy innovation also are often the underlying triggers for military conflict that are evident especially in the Middle East and other parts of the world today. The energy challenge continues.

Measuring the impacts of entities created

The current chapter of this book is about the 1973—83 experience of a US and Texas government (*institutional innovation*) response to the challenges flowing out of the Arab Embargo of 1973 and the Iranian Crisis of 1979. The 2005—15 experience (summarized in Chapter 9) is about the Texas electric sector institutional innovation response to the challenges of restructuring the electric sector from a highly regulated industry to one of managed competition.

The challenge of assessing the successes and failures of both of the experiences is like the challenge Author Brooks faced when he became the President of the American Enterprise Institute (AEI)—measuring the success of nonprofit organizations. Nonprofits and governmental entities created with missions to advance innovation have the common challenge of reporting to their support groups so they can determine whether the effort is worthy of continued support. But measuring the achievement of success is a challenge. The temptation is to measure success by the *input* (level of effort) that goes into the enterprise to carry out the mission. Such things as size of the budget, members who join, number of employees working the problem, uniqueness, and longevity of the organization, etc. Such measurements are easy, but miss the mark. A second temptation is to measure the *output* (number of high-level meetings, number of significant publications, type, and significance of legislative recommendations, etc.). This easy measure also misses the mark. The more meaningful, but difficult measure is the *impact* relative to the mission (Brooks, 2018).

A significant contribution of nonprofits and special commissions of governments is the creation of ideas and, as discussed in Chapter 1, The Dynamics of Innovation and Technology in a Market Economy, ideas are the key drivers to innovation and are among the major contributions of the energy policy commissions formed in Texas following the Arab Embargo of 1973. The interplay of government officials and private sector leaders during the 10 years following the Embargo are evident in the projects funded and the policy formulations that grew out of the deliberations—everyone involved had a great interest in deciding how to go forward into a new era.[13]

The paragraphs below summarize the inputs, outputs, and impacts of the Texas Energy Policy organizations (GEAC, TEAC, and TENRAC). The inputs, outputs, and impacts of the electric industry nonprofit (CCET) are discussed in Chapter 9.

---

[13] An interesting and important contribution of inputs to a modern innovation process is "crowd-sourcing." This modern approach is an effective way of harnessing ideas from large groups of interested, engaged individuals and organizations. CCET used this process for some planning work on electric vehicle infrastructure development in 2010. One of CCET Board members, National Instruments was an avid supporter of crowd-sourcing.

GEAC (1973 commission) input

GEAC was organized to include all major statewide elected and appointed officials with major responsibilities in energy, and from a large group of private sector leaders of energy and chemical industries in Texas. Input to the Council came from this diverse source.

GEAC (1973 commission) output

1. GEAC adopted ten legislative recommendations.
2. GEAC competed 40 studies that made up a resource base for the future of energy R&D priorities.

   GEAC Impact: 7 of 10 legislative recommendations were adopted by 64th Legislature, including:

1. The establishment of an ongoing policy planning agency with a similar mission to that of GEAC, with State appropriations to support it. Other recommendations enacted into law included a geothermal production act setting up a regulatory process for the production of geothermal resources; establishing the Public Utilities Commission; establishing a requirement for environmental review of energy projects; encourage advanced recovery of oil; require urban building codes that promote energy conservation; and encouragement of mass transit for urban areas.
2. One recommendation that did not pass in the 64th Legislature passed in the following 65th Legislature. The recommendation was to create an energy development fund for identifying and funding alternative energy technologies. The result was the creation of the Texas Energy Development Fund to be managed by the TEAC.

GEAC (1975 Texas state agency) input

GEAC was organized to include all major statewide elected and appointed state agency officials with major responsibilities in energy, representatives of the House and Senate, and from a large group of private sector leaders of energy and chemical industries in Texas organized as an advisory committee. Input from the public was received in a 2-day conference on a draft of a Texas Energy Policy document which addressed all major energy topics including legislative and regulatory policies at both the Federal and State levels.

GEAC (1975 Texas state agency) output

1. A consensus-based policy document that laid out the Texas position on all of the legislative and regulatory energy issues of the time labeled Texas Energy Policy (GEAC, 1977b)

2. Creation of the first Texas comprehensive energy data base (GEAC, 1977c)
3. The first comprehensive Texas energy outlook to the year 2000 (GEAC, 1977a)
4. Recommendations to counter many of the provisions of the Federal National Plan of the Carter Administration.

GEAC (1975 Texas state agency) impact

The focus of this second state agency form of GEAC was organizing a number of energy producing states to influence Federal policy that was determined to be punitive of the energy producer states. The early forms of the Carter Administration NEP included the use of major energy taxes, price controls and a WPT on oil producers—the purpose was to equalize net energy costs to consumers regardless of the region of the country they resided in. The NEP also placed a heavy emphasis on energy conservation and provided funding for cooperating states to implement conservation programs. The GEAC leaders disagreed about taking Federal conservation funding and whether GEAC should focus on Federal policy rather than Texas policy. As a result GEAC expired at the end of its 2-year statutory life and a new, scaled-back agency was created to focus only on energy policy, both State and National.

Texas Energy Advisory Council (TEAC, 1977 Texas state agency) input

Texas leadership decided that an energy policy organization needed to continue to follow GEAC as it ended its 2-year statutory life. TEAC was formed in 1977 with much the same focus on State and Federal energy policy as GEAC, and with a similar membership structure, but without the Federal funded energy conservation office that created the internal conflicts in GEAC. TEAC was also given the responsibility of organizing and managing the Texas Energy Development Fund that passed the Legislature in 1977.

Texas Energy Advisory Council (TEAC, 1977 Texas state agency) output

1. A consensus-based policy document that laid out an updated version of the Texas position on all of the legislative and regulatory energy issues of the time labeled Texas Energy Policy 1978 Update (Texas Energy Advisory Council, 1979)
2. Maintenance of the Texas comprehensive energy data base (TEAC, August 1977)
3. A second comprehensive Texas energy outlook to the year 2000 (TEAC, June 1980)
4. Completion of a major challenge to the Federal government's energy modeling system used to defend the Carter Administration's NEP (Holloway, 1980)
5. Policy recommendations to counter many of the provisions of the Federal National Plan of the Carter Administration
6. Completion of several alternative energy R&D projects.

Texas Energy Advisory Council (TEAC, 1977 Texas state agency) impact

TEAC completed an evaluation of the economic modeling system developed and used to support the Carter National Energy Plan (PIES). The evaluation challenged the methods and processes used by the Federal government for evaluating national energy policy. The evaluation effort was supported by a nation-wide group of distinguished professionals. The result was to create professional respect at DOE for Texas analysis and policy development capabilities and to cause the DOE to develop an archiving system for their modeling systems and policy evaluations used in national policy initiatives so that their work could be accessed and evaluated by interested groups like TEAC or other national or regional groups.

TEAC provided analysis of State and Federal policy proposals and provided legislative testimony that helped Texas Legislative and Congressional members craft informed legislative proposals. The agency also developed and managed the Texas Energy Development Fund that funded a number of alternative energy projects including lignite, wind, solar, biomass, and geothermal energy sources, all of which added depth to the studies funded by commission form of GEAC. These research projects added to the knowledge base that continued to develop during the following decades.

TENRAC (1979 Texas State Agency) input

Texas leadership with the strong direction of the newly elected Republican Governor decided that an energy policy organization needed to continue and strengthen the energy policy work of TEAC as it ended its 2-year statutory life in August 1979. The TENRAC was formed in 1979 with much the same focus on State and Federal energy policy as GEAC and TEAC and with a similar, but expanded membership structure of 22 individuals representing Texas energy and natural resources agencies and legislative entity heads, and with several private energy company representatives. The new agency brought together the functions of TEAC, the federally funded energy conservation office from the Governor's Office and the Coastal Zone Management agency. The Texas State—Federal Office located in Washington D.C. was charged with coordinating closely with TENRAC to impact national legislation and Federal agency energy policy.

TENRAC was also given the ongoing responsibility of managing the Texas Energy Development Fund that had been created during the Legislative session of 1977—the Legislature in 1979 increased the State appropriations to fund the R&D work.

TENRAC (1979 Texas State Agency) output

During 1979—83 TENRAC adopted many resolutions representing a Texas consensus on energy policy focused on challenging the *Carter National Energy Plan* legislation that took the form of five separate bills:

The National Energy Act of 1978 (NEA78) It included the following statutes:

Public Utility Regulatory Policies Act (PURPA) (Pub.L. 95—617)

Energy Tax Act (Pub.L. 95—618)

National Energy Conservation Policy Act (NECPA) (Pub.L. 95—619)

Power Plant and Industrial Fuel Use Act (Pub.L. 95—620)

Natural Gas Policy Act (Pub.L. 95—621)

TENRAC produced several comprehensive analyses of the NEA supporting the Council's resolutions, testimony before Congressional committees and in an alternative to the Carter NEP by Governor Clements (Holloway, 1982).

Another major TENRAC policy initiative was to oppose the Carter Administration's WPT on crude oil (Crude Oil Windfall Profit Tax Act, P. L. 96—223). Several TENRAC coordinated studies examined the economic impacts of the several Carter-era legislative initiatives that were intended to counter the effects of and challenges flowing from the 1973 Arab Embargo and the 1979 Iranian Crisis. The TENRAC studies were all used to support congressional testimony as the legislation worked its way through the Congress.

Another major output of TENRAC was the development of a Texas policy on nuclear waste disposal. The results of TENRAC's efforts included a proposal for a low-level waste disposal facility in West Texas that would store medical and seismic technology nuclear wastes from the oil industry in Texas and similar waste from several other states. TENRAC also followed, and eventually opposed the location of a national high-level waste facility in subsurface salt beds in the High Plains of Texas (Holloway, 1982).

TENRAC funded dozens of R&D demonstration projects from the Texas Energy Development Fund with joint support from private and government entities that helped build a clearer idea of the future possible contributions from alternative energy and conservation efforts, and from enhanced oil and gas production (Texas Energy and Natural Resources Advisory Council, 1983a,b,c).

TENRAC impacts

TENRAC developed many strong consensuses-based positions on energy policy and actively participated in the national debate concerning the best ways to respond to the many challenges presented by the Arab Oil Embargo of 1973 and the Iranian Crisis of

1979—80. The agency followed the prior policy development begun in 1973—74 by GEAC and by TEAC in 1977—79. The agency was a unique formation that adequately represented, and was recognized as the most authoritative voice on Texas energy policy. This was so because the agency brought together the ideas and attitudes of the executive and legislative branches of State government, supported by academic and TENRAC staff analyses. The agency also included the input of the judicial branch because the structure (and the input on policy) came from the Attorney General who was a member of the TENRAC Board of Directors.

While it is impossible to know exactly how much influence TENRAC had on the National Energy Plan legislation, the final acts (1) included the full removal of crude oil and natural gas wellhead price ceilings that eventually came to pass in the mid to later part of the 1980s, (2) provided the means for the demise of the WPT which (unwittingly) expired of its own weight as the price of crude oil unexpectedly decreased from the high levels the bill's authors expected, (3) set up the structure of PURPA for the deregulation of wholesale electric prices sold in interstate commerce which later became the initiating influence leading to the Texas program to structure its own electric industry by a deregulation effort (the topic of Chapter 8), (4) provided for the rapid phase-out of natural gas use as a boiler fuel, but the high market prices that eventually came under the Natural Gas Policy Act reduced demand and led the industry expansion of supplies that made the Power Plant and Industrial Fuel Use Act useless or unnecessary, depending on your perspective, and (5) gave guidance to the structuring of energy conservation programs and provided ongoing funding to coop-erating states (including Texas) that remain today.

TENRAC carried out a study and consensus-building process to develop and site a Texas owned and operated low-level waste disposal facility. The agency developed the draft legislation for the site. TENRAC also successfully opposed the Federal government sitting of a high-level waste disposal site near Herford, Texas in the High Plains.

TENRAC's R&D program moved the ball forward in identifying the best options for adding alternative energy sources to the State's energy mix. Wind, solar, biomass, geothermal, enhanced oil recovery and conservation technologies were all a part of the TENRAC focus. One of the last efforts of the agency was to develop a 5-year energy research plan for Texas (Texas Energy and Natural Resources Advisory Council, 1983a,b,c).

The price of oil and gas has driven the on-again, off-again cycle of both policy formulation and R&D planning and project funding. The ebb and flow of politics and the accompanying attitudes about the role of government prevailed across the two decades following the Embargo of 1973. The decade of 1973—83 saw an evolving focus of governments, both in Washington and in Texas. As historically high prices and occasional shortages of oil and gas (the primary energy sources for well over a century) energized governments to get more involved in what had historically been a private sector activity. Then when the price of oil and gas declined to more normal levels, governments pulled back.

Mark White upended Bill Clements in his attempt at a second term as Governor in November 1982 and Ronald Regan beat Jimmy Carter who attempted a second term as President. The elections and oil market conditions changed the Texas and Federal energy policy dynamics. The price of oil fell to near $10/barrel for a short time in early 1983. Shortages were a thing of the past and electric and natural gas utilities were then under tight regulation. Following President Carter and the heavy-handed government action under the NEP, Ronald Reagan opted for reducing the role of the Federal government, including that of the DOE. TENRAC was at the end of a Sunset Review and the agency required a reauthorization to continue. Holloway, the long time Executive Director of TENRAC under Bill Clements resigned in the spring of 1983. The agency was dissolved in August 1983 under the Texas sunset act process.

CHAPTER 9

Electric industry deregulation and competitive markets

Contents

The context

Among the most important energy sector innovations of the Texas experience is that of the restructuring of the electric industry. The set of innovations, including legislative initiatives, public agency structural transitions, policy guidance, industry business model changes, consumer adaptation to change, and a set of accommodating technological developments took place over a period of about 10 years. And, this period from 1995 to 2005 brought the Texas electric sector from a traditional regulated rate-of-return model to a three-part structural paradigm including (1) a competitive generation market, (2) a regulated rate-of-return transmission and distribution sector, and (3) a competitive market for retail service.

The final outcome of the Texas move to a competitive electric market differs in important ways from deregulation of other sectors in the United States that began in the late 1970s. The deregulation of US industries has included railroads, airlines, oil and gas production and refining, financial markets (in part), and communications, and in the 1990s, the electric sector at the wholesale level. In most of the deregulation activity listed above the changes occurred at the Federal level since the US constitution lays the responsibility for protecting consumers in economic trades at the doorstep of the commerce clause. The commerce clause makes the federal government

Innovation Dynamics and Policy in the Energy Sector
DOI: https://doi.org/10.1016/B978-0-12-823813-4.00009-9

primarily responsible for enforcing the rules of competition for trade that crosses state boundaries. The Justice Department and the Federal Trade Commission (FTC) are chiefly responsible for enforcing antitrust laws guided by the courts in interpreting the commerce clause of the constitution. The Congress set up most of the antitrust laws during the last of the 19th century and early in the 20th century, laws that still dictate the rules of competition today.

An important exception to the federal dominance of the regulation-deregulation paradigm is where the limits of federal jurisdiction stop—commerce that does not operate across state lines. This distinction put Texas in the driver's seat for defining the deregulation structure and operation of the electric market since Texas has deliberately avoided interstate trading of electricity.

The 10-year-long process of restructuring the Texas electric market has recognized the potential benefits of deregulating this industry, but also gained an understanding that an active role must be maintained by public agencies and industry advisory groups to monitor and condition behavior. The conditioning includes prevention of excess market power by providers, ongoing monitoring of performance to guide rule-making and consumer education and engagement. The outcome has been a well-functioning electric market, but one that is a "managed" competitive market.

The current chapter of the book does not summarize the transition of the Texas electric industry from regulation to deregulation in all of its dynamics that would be needed to fully appreciate the enormity of the effort. Instead the chapter is focused on the key institutional, technological, and market innovations that emerged over this 10-year period and following. An excellent summary of the full detail of the electric sector deregulation is contained in a 2009 publication with contributions from 13 individuals involved in the restructuring effort and/or authoring of professional writing on the various aspects of the deregulation effort (Kiesling and Kleit, 2009).

The key elements of the Texas electric restructuring effort included (1) a competitive wholesale market design and implementation, (2) an assessment and maintenance of resource adequacy, (3) redirection and development of the modern regulated transmission sector, (4) allowance for and encouragement of distributed generation, (5) competitive retail market design and implementation, and (6) market monitoring to inform market participants and limit market power of providers. The process of restructuring began with statutory changes by the Texas Legislature in 1995 (SB 373, 1995). SB 373, among other things, permitted wholesale market participation of exempt wholesale generators, as defined by Federal law under the Public Utility Holding Company Act of 1935—generation facilities not previously able to sell power in the market place. SB 373 allowed nonutility wholesale market participants to offer market-based prices in Electric Reliability Council of Texas (ERCOT). In addition, SB 373 also allowed such generators to sell *only* wholesale electric power which exempted them from state utility regulation. Many other acts of the legislature, and policy changes by the Texas Public Utilities Commission (TPUC) would follow over the next 10 years finally completing all of the elements listed above in 2005.

One cannot fully understand the dynamics of the move to a deregulated electric market in Texas without connecting the lengthy process of 10 years with the developments in the natural gas industry. And, it is helpful to compare the Texas experience with the deregulation experience of California that proceeded Texas, and by recognizing the role of a key player in both markets—Enron corporation of Houston. The California experience served, unwittingly to inform Texas leaders so that Texas would not repeat their (California) mistakes.

The key connection of electric deregulation with the natural gas market developments began with industrial marketing programs for natural gas that developed in the early 1980s and continued until the early 1990s. The federal Natural Gas Policy Act of 1978 (NGPA) under the Carter Administration set in force a gradual deregulation of natural gas well head prices while also restricting the use of natural gas for industrial and electric generation (boiler fuel). There were many complicated sections of the legislation, and even more complexity added by regulatory modifications that followed over the period from 1979 until 1985. The NGPA was designed to complete decontrol of well head prices by 1985.[1] The end result, however, was to eventually produce somewhat of a glut in the gas market. At first there was a run-up of natural gas prices for high-cost of production gas near $9 per mcf at the margin, allowed under the NGPA, in 1981 to less than $2.00 in 1985 for spot-market gas in the years following (Holloway, 1986, pp. 9–10). The high prices for certain classes of gas produced more new gas than the Congress anticipated, and when combined with demand reduction rules for electric generators and industrial users (to phase out the use of gas for boiler fuel) created a glut of natural gas, rather than a continual shortage the Congressional leaders expected. So, the response to the glut was to allow direct sales of gas to industrial users at market prices, and to avoid the use of, and transportation charges for transportation services of the infrastructure pipeline.

The mechanisms for moving natural gas from producers to industrial users at competitive market prices included industrial marketing programs and so-called off-systems sales programs. These programs helped drive the market price of natural gas downward creating a new inter-fuel competition in the industrial and electric consumer markets which expanded the reach of natural gas in the energy markets of Texas, especially for use as a boiler fuel in electric generation where the alternatives included coal and nuclear plants with long led times for construction and increasing regulatory costs.

---

[1] NGPA defined some 29 categories of natural gas and placed evolving (rising) price ceilings for most of them. The idea was to decontrol "new" sources of gas in order to provide incentives for new discoveries driven by competition from alternate energy sources, limit excessive profits by owners of "old" gas reserves, provide extra price incentives for high-cost and deep reserves and to differentiate between gas trading in interstate markets from gas trading wholly within state boundaries (intrastate).

The success of these marketing programs helped move the federal policy toward deregulation of the wholesale market price of natural gas purchased by pipelines. This institutional/market-driven innovation radically changed the natural gas market across the United States. Natural gas producers were able to make direct sales of gas to industrial users without being subject to Texas Railroad Commission or Federal Energy Regulatory Commission (FERC) regulation that required payment of pipeline infrastructure costs routinely passed-on to retail consumers at the end of the pipeline infrastructure by consumers served by regulated pipeline distribution companies. This innovation that quickly spread to other states was led by Texas. But unfortunately, that was not the end of the story. The energy crisis in California that followed deregulation of electric markets led the way for a pullback from deregulation in many other states.

The leading company that helped to drive the market and policy changes resulting in the crisis of 2000 and 2001 was Enron. As developments occurred during the mid-1990s Enron developed its private company complex of subsidiaries and contract providers molded into a conglomeration ended in one of the largest and most far-reaching scandals in American history. At its peak growth Enron owned and operated a variety of assets including gas pipelines, electric generation plants, pulp and paper plants, water plants, and broadband services, operating both in the United States and abroad.

During the 1990s Enron and other market players took advantage of FERC policy deregulating wholesale electric sales across state lines in the electric market, coupled with deregulation efforts in California. The FERC policy led California to set up electric market deregulation that created an opportunity for market participants to manipulate the market price for electricity which in turn pushed up the price of natural gas for power generation. The result created the "California Energy Crisis" of 2000 and 2001. The fundamental vulnerability of markets for manipulation was the result of inapt California policies that decoupled investor-owned, vertically integrated electric companies where generation became competitive, but kept a regulatory cap on retail electric prices to protect consumers from expected rising prices (Sweeney, 2008). The other factors that fed the market manipulation included decontrolled electric prices for imported wholesale electricity (the result of national FERC policy) and weather conditions that limited the historic availability of hydro power imports from the Pacific Northwest.

Several methods were used by Enron and others to be able to use withholding of electric generation at peak demand times (to create supramarket electric prices), cross-border trading and other methods of market manipulation. Enron created false shortages by taking power plants offline thus pushing up the price said to be up to 20 times the real market value (Kiesling and Kleit, 2009). Due to the retail market price caps, the run-up in prices reduced profit margins of electric companies causing the bankruptcy of Pacific Gas and Electric Company and near bankruptcy of Southern California Edison. At least seven companies, including Enron, were taken to court and ultimately paid large fines, penalties, and refunds amounting to about $4.4 billion (Public Citizen's Energy Program www.citizen.org/cmep). Key individuals were convicted of criminal activity regarding the manipulation of prices. The FERC completed a major report on the matter in March of 2003 where they stated:

> *Staff found significant market manipulation, ... that significant supply shortfalls and a fatally flawed market design were the root causes of the California market meltdown.*

> *A key conclusion of this Report is that markets for natural gas and electricity in California are inextricably linked, and that dysfunctions in each fed off one another during the crisis. Spot gas prices rose to extraordinary levels, facilitating the unprecedented price increase in the electricity market. Dysfunctions in the natural gas market appear to stem, at least in part, from efforts to manipulate price indices compiled by trade publications. Reporting of false data and wash trading are examples of efforts to manipulate published price indices.*

> *And*

> *In a related finding, Staff concludes that large-volume, rapid-fire trading by a single company, in what was incorrectly assumed to be a liquid market, substantially increased natural gas prices in California.*

> *And*

> *Staff concludes that EnronOnline (EOL), which gave Enron proprietary knowledge of market conditions not available to other market participants, was a key enabler of wash trading. This created a false sense of market liquidity, which can cause artificial volatility and distort prices. Staff concludes that prices in the California spot markets were affected by economic withholding and inflated bidding. Staff finds this violated the anti-gaming provisions of the Cal ISO and Cal PX tariffs and recommends proceedings to require disgorgement of profits associated with these inflated prices.*

> **Federal Energy Regulatory Commission (2003)**

Enron filed bankruptcy on December 2, 2001, finally collapsing and ceasing all operations in 2007. The Enron stock market price went from $82/share down to $0.10/share in the end. Although it was not the sole factor the major contributor to Enron's decline developed as the company became a major participant in the California electric and gas markets. There are many parts to the California energy crisis which came to crisis levels for both energy and financial markets in 2000 and 2001. The failure of this dynamic event cost California residents and companies billions of unnecessary costs, and resulted in an ongoing failure of California and other states to miss out on the economic benefits of electric market deregulation—foregone economic benefits that continue today.

The institutional, policy, and market complexities of the California energy crisis was a "perfect storm" of institutional, policy, and market failures. Although the event was aggravated by drought that reduced hydro power production in the Pacific Northwest that California relies on, the combined innovation failure is perhaps unparalleled in US history. Enron took advantage of the market vulnerabilities created by institutional ineptness of the first order. The crisis followed partial deregulation of gas and electric industries allowing Enron to manipulate the market by withholding production and running up prices to artificial levels. One of the rules leading to shortages was allowing out-of-state electric producers to import electricity to California under deregulated prices. The imports were needed because California environmental rules made it difficult to add new generation in California, thus compounding the problem.

In contrast to the California experience the electric and gas markets in Texas continue to generate efficient market outcomes for Texans. The ongoing process of managing the restructured electric market requires periodic changes in the rules that make the market function efficiently. One of the latest adjustments still in progress is a plan to add marginal line losses to two managed market functions at ERCOT called "Real-Time Co-optimization." Line losses are unavoidable as electricity passes over transmission lines moving electricity from generators across the state to end users. These losses should be accounted for in efficient market operations beyond what the current market includes. A process is underway to properly include line losses in the market. As stated by ERCOT:

> In 2017, the PUC instructed ERCOT to assess the cost and time to implement Real-Time Co-optimization (RTC) and marginal line losses. RTC is in the process of procuring energy and Ancillary Services simultaneously in the Real-Time Market. With marginal losses, ERCOT would account for transmission line losses in its pricing mechanisms.
>
> **ERCOT, Inc. (2018)**

The ERCOT market currently operates by accommodating contracts for wholesale power that implement trades between competitive generator companies and competitive retail providers who sell power to end-users. These contracts are for defined quantities of power and periods of time beyond the very short-term. The short-term is a 15 minute adjustment market that is required to keep the grid running without outages (keeping the flow of electrons at approximately 60 hz). In order to allow the market to clear every 15 minutes while keeping the grid at the standard of 60 hz, ERCOT purchases "ancillary services" that supplement the market for power via long-term contracts among generators and retail providers. The market structure provides market-clearing prices for electricity delivered at each of several thousand "nodes" (geographically defined points on the electric grid) throughout the ERCOT region.

The dynamics of how the market has evolved since the beginning of restructuring at the wholesale level in 1995 is an ongoing process that initially took 10 years to complete with full implementation of retail competition in 2005. The managed competitive market requires ongoing adjustments, like the cooptimization mentioned earlier.

The 10-year march to restructuring of the electric market in Texas did not happen in isolation, however. To fully appreciate the achievement and the accompanying innovation contributions one must recognize the context in which it all took place. The context includes events in Texas, policy changes at the federal level, activities among certain other key states and the experience in Europe and Australia. The key context elements are summarized here before focusing on the key innovations that make up the major contributions of the Texas electric restructuring experience that now places Texas in a leadership position in electric sector innovation.

Perhaps the most important context that framed electric market restructuring in Texas is the influence of having a single regulatory authority. Texas is currently unique among the 50 states in having a well-defined and continually honored separation of federal and state regulatory jurisdiction, although it was somewhat of a bumpy road getting there. The federal government through the FERC does exercise some important, but limited jurisdiction as explained below, but when focusing on the development of competitive wholesale and retail markets under restructuring, having a single regulatory authority has been fundamental. Texas has avoided the restructuring failures of other states that, in part, grew out of poor attempts to coordinate federal and state actions. California is a prime example of the pitfalls of joint jurisdiction that, in large part was responsible for the chaos in the electric and natural gas markets in 2000–01 resulting in a pullback from retail electric competition. The experience resulted in consumer outrages, financial losses throughout the producer and consumer sectors and large fines and prison terms for key market players that exercised market power and violated federal and state law.

The path to a single regulatory jurisdiction in Texas began with the Federal Power Act of 1935 which established federal jurisdiction exercised by the Federal Power Commission (now the Federal Energy Regulatory Commission) over the interstate transmission and sale of wholesale power. The law, based on the commerce clause of the US Constitution, was the basis for federal regulation of prices for wholesale power transmitted across state lines (if such amounted to a substantial effect on interstate commerce), but it was also clear that such authority did not include electric power sold at retail or among entities where the power was solely transmitted and sold among entities within a state's boundaries. In response to the federal act Texas moved to avoid federal regulation by having Texas utilities isolated from interstate transmission and sales. The Texas structure was loosely maintained by informal agreements until an intrastate power pool was established in 1970 creating ERCOT (Spence and Buch, 2009, pp. 10–11).

ERCOT was at first only a regional electric reliability council overseen by the North American Electric Reliability Council and not acting by authority granted by

Texas statutes. It was a voluntary membership organization designed for interconnection among members for reliability purposes where members understood that they were obligated to notify other members if they planned to connect across state lines so that all members could disconnect from the to-be-interconnector thus preventing FERC jurisdiction on the rest of the group of members.

The Texas Public Utility Regulatory Act of 1975 (PURA) set up the TPUC as a state-wide regulatory body. It replaced the solely local regulation approach with the TPUC having original jurisdiction over investor-owned utilities (IOUs), with appellate jurisdiction over municipal systems and coops. ERCOT at this point was still a private-sector coordinating council (Spence and Buch, 2009, p. 14).[2]

The essential Federal policy which set the stage for deregulation in Texas, and for wholesale deregulation throughout all US states, was the Energy Policy Act of 1992 (EPAct). EPAct as passed was intended to open and expand the wholesale markets across the United States and to encourage new competitive generating companies to enter the market. This policy was implemented through policy of the FERC. EPAct passage was followed by FERC order 888 which mandated open-access wholesale wheeling at nondiscriminatory rates and the separation of wholesale sales from transmission services (called unbundling) but not reaching into the ERCOT market following agreements that had been incorporated into EPAct (Spence and Buch, 2009, p. 14).

Texas restructuring

In Texas the push to deregulation proceeded first by amending PURA to deregulate the wholesale market. The TPUC carried out the statute mandates in 1996 making ERCOT the first "independent system operator" (ISO) in the United States. Then the Legislature passed Senate Bill 7 in 1999 creating the structure for retail competition. The statue gave the TPUC the authority to make ERCOT an independent organization overseeing system reliability and wholesale and retail market operations. Specifically, ERCOT was assigned four primary responsibilities: (1) maintain system reliability, (2) facilitate a competitive wholesale market, (3) ensure open-access to transmission, and (4) facilitate a competitive retail market (Hunter, 2012).

---

[2] The expansion of the FERC jurisdiction among all states in the United States that depended on interstate transmission capabilities for power for commercial purposes and, importantly for reliability purposes, increased the interest in Texas for a state-level regulatory body with jurisdiction over both wholesale and retail markets. Further, the natural gas market turmoil discussed in Chapter 6, West Texas and the Permian Basin Early Innovations, created further incentives for state-wide regulation, which was a recommendation of the Governor's Energy Advisory Council in 1973 discussed in the Chapter 7, Panhandle Field and Natural Gas Flaring.

The transition to retail competition began in 1999 and opened the market to completion in 2002 and to full implementation in 2005 but only after executing several activities and rules intended to insure a smooth transition. As it turned out the transition was not so smooth. The volatile natural gas market made the transition challenging to say the least. Since natural gas was the fuel for a large part of electric generation the wholesale electric market swung erratically as natural gas prices sold for electric power generation went from $2.36/mcf in February 2002 to $10.72/mcf in December 2005, followed by retail electric power prices from 6.73/cents/kwh in January 2002 to 10.13 cents/kwh in September 2005 (US Energy Information Administration, 2017).

But the transition to competitive retail prices moved forward in spite of the unexpected volatile movement of natural gas prices. The Texas approach to smoothing the transition to retail competition stands in marked contrasts to other state regulatory bodies and legislative approaches. Senate Bill 7 assigned the TPUC and ERCOT the dual duty to oversee network reliability and retail operations. The approach of other states, for example California, was to follow the example of the UK by requiring retailers to purchase all of their power through a central clearinghouse; in California this was the California Power Exchange. The Exchange required buyers and sellers to clear daily bids for wholesale power through the central exchange. The states in the United States that followed the UK model then placed price caps on the transactions intended to shield consumers form radical price spikes. The outcome, of course, was to create shortages in the market, forcing purchasers to buy out-of-state power where prices were uncapped. This practice then resulted in California interest groups requesting the FERC to impose price caps on these interstate sales. Daily wholesale prices in California during this period ranged from a monthly average of about $50/MWh historically to $400/MWh in December 2000, upward to a daily peak of $1200/MWh. FERC's preference for letting wholesale prices fluctuate with the market conflicted with California's insistence on capping retail rates (Spence and Buch, 2009, p. 16).

The general conditions of gas market price extremes (which pushed wholesale electric prices up), dual regulatory jurisdiction, the California clearing market system and retail price ceiling policies created great opportunity for firms in the gas and electric markets to "game the system" by taking advantage of arbitrage opportunities. Such opportunities were created by differing rates (prices) in and out of state. Firms were able to exercise market power, the very conditions that have always worried opponents of deregulation. In the end the Commodities Futures Trade Commission, the US Justice Department, and the federal courts got involved and people went to prison and firms paid huge fines.

Texas by contrast took a different approach. Along with several other states, Texas allowed retailers to rely on long-term bilateral contracts with generators for wholesale power, and to supplement long-term contract commitments at the margin through a daily clearing market operated by ERCOT. This administered market-clearing operation would also balance the system load (a centralized balancing energy market). In addition, during the transition period Texas imposed a *retail price floor* on the incumbent IOUs to encourage market entry of competitors. The price floor was known as the "price to beat" (Spence and Buch, 2009, pp. 14–15). Incumbent utilities were allowed to recover stranded cost during the transition (stranded cost are capital costs not fully recovered at the time of movement from a regulated market to a competitive market). Also, during the 3-year move to retail competition "providers of last resort" were designated in all areas where choice by retail consumers was in effect. In the final implementation of competition cities and coops were allowed to opt-in to competition, but were not required to do so (Hunter, 2012).

Texas, uniquely in the United States, opted for a market structure known as an "energy-only" market. Such a market structure is expected to provide efficient energy at the moment but also provide the essential price signal to drive the addition of new capacity (the value of new generation). Other markets in the United States separate these two functions (Puller, 2009) and provide a separate market process for power (kilowatt hours) and addition of new capacity through a capacity market. All such market structures are faced by the same fundamental challenges.

The largest challenge to making an electric market function and produce benefits to consumers, while preventing the exercise of market power, is driven by the fundamental factor that is unique to the electric market. The short-term demand function is vertical. Without the help of a managed market overseer, consumers have neither the current information about energy use, the opportunity to exercise control over consumption or the accompanying marginal price that is fundamental to well-functioning markets. Furthermore, providers must provide power instantaneously to meet demand in order to keep the system up. Therefore, the "managed" market must provide education and timely information to consumers, a bidding structure for providers to compete on a fair basis and proper price signals to encourage the right amount of generation and transmission/distribution resources, and consumer consumption technologies—all for the long-term. Creating and maintaining a system to achieve these outcomes, along with a monitoring system to inform all market participants of the performance of the market is a tall order. But the wide-spread belief is that the TPUC and ERCOT, directed and mandated by the Legislature, are meeting these requirements.

The Texas market structure and operations

The restructured market has three primary components—generation which is competitive, the transmission and distribution component which is regulated and retail services which is competitive. Generators are required to register with TPUC in a process that assures TPUC and ERCOT that the generator can meet certain protocols for connection with the grid. Market participants are assured of access to the transmission system through a regulated "open-access" rule of the TPUC. The generators and retail providers then negotiate long-term bilateral contracts for providing the retail electric providers (REPs) with power. Bilateral contracts compromise 90%–95% of all power traded in the ERCOT market. REPs act to sign-up customers from residential, commercial, and industrial classes of users via short-term contracts (often no more than 3 years). ERCOT then operates a market-clearing operation to balance the load in the market, and to facilitate competitive prices. This "balancing" market accounts for 5%–10% of energy bought and sold. The certified generators are free to submit bids to ERCOT who runs the market-clearing operation—the Real-time Market and the Day-ahead Market. ERCOT also operates an "ancillary" service market to provide fast response capability for reliability purposes. Finally, consumers are free to switch REPs as consumers are able to do in other competitive markets, subject to conditions they agreed to with their current provider. Consumers have access to the full range of choice from all available REPs through an online portal at the TPUC called Power to Choose (Power to Choose http://www.powertochoose.org/).

The wholesale market-clearing operation at ERCOT first began by managing a market-clearing process at a large zone (geographical) level; the balancing market cleared every 15 minutes at three geographical "zone" levels. In 2015 ERCOT moved the market-clearing operation to a nodal system of more than 8000 "nodes." The wholesale market operation works as follows:

1. A voluntary Day-Ahead Market operates to help ERCOT and market participants prepare (anticipate) conditions for the Real-Time Market the next day. Market participates may submit offers to buy and sell energy on an hourly basis.

2. A Real-Time Market operates to clear the market and balance the load in real-time. This market provides an increment of power to balance the load outside of the long-term contract commitments of the participants. Market participants submit offers to provide generation output and bring generation online as needed; ERCOT may request additional generation as needed for reliability.

3. A Security-Constrained Economic Dispatch system selects the most efficient generation available every five minutes the meet customer demand within the limits of the transmission system.

4. Energy prices from each 5-minute interval reflect the marginal cost of the available resources on an "offer curve" made up of all offers received. There is an added component (adjustment) that reflects the value of energy during scarcity conditions.

5. Settlement in this Real-Time Market is made every 15 minutes where generators are paid a price that reflects conditions at the "node" level and load-serving entities (market participants who provide services—packaged power to consumers) are paid

load zone prices. The load zone prices include costs associated with transmission congestion—the cost of bottlenecks that invariably develop (ERCOT, Inc., 2017a,b).

An important tool used by ERCOT to balance the load in real time is an "ancillary service" operation where both generation resources and demand reduction users with approved telemetry provide competitive response capability. Further, the addition of advanced meters to the system (new digital technology) at consumer's locations provide capabilities for near real-time monitoring and demand management by consumers and their vendors. This advanced meter resource will become a more important resource for demand management in the future that promises to greatly improve the efficiency of the ERCOT electric grid, and therefore provide important benefits to consumers across Texas.

ERCOT also operates an emergency response service where loads (customers) and certain types of generators provide quick response during an emergency. This service helps ERCOT avoid implementing rolling blackouts. ERCOT procures emergency response service resources by selected qualified loads and generators for two different response times—30 minutes and 10 minutes. The procurements are made three times annually for 4-month periods (ERCOT, Inc., 2017a,b).

Performance

There are several means of measuring how well the Texas electric markets (wholesale and retail) have performed both during the 10-year transition period beginning in 1995 and, importantly, since full restructuring was completed in 2005. The key measures are (1) maintenance of reliability, (2) electric prices relative to those that would have prevailed in the absence of restructuring, (3) prevention of the exercise of market power, and (4) delivery of new innovations of retail services, and associated consumer satisfaction with electric service.

Reliability

The key indicator of electric system reliability watched by grid managers everywhere is the reserve margin. The focus of reliability assessment measures and operational policy is at the bulk power system level—the interconnected generators and transmission lines. The reserve margin measure is calculated as the available resources minus the system demand divided by the system demand at the peak hour in the year (or other reference period). The peak hour of the year in the southern United States is typically during August—the hottest time of the year when air conditioning load is the highest. And, there are several reserve margin measures used by grid managers that help translate such a concept into policy of the grid manager; the intent is to make sure managers can keep the grid up at all times. There is the "planning reserve margin" and the "reference margin level". Since resources (primarily generators connected to the grid) come and go as old units are retired and new ones are added, the use of reserve margins gets rather complicated. The planning reserve margin is used as a target for a future that typically runs annually 10 years out.[3] Since there is considerable uncertainty about both future load retirements and additions of new resources the use of reserve margins as a system reliability indicator is a probability conditioned measure (North American Reliability Corporation, 2017).

There are several other concepts, measures, and policies surrounding the reliability topic for the electric grid. Resiliency is an aspect of reliability defined as the ability of a system to absorb, survive, restore, and quickly recover from major adverse events (Silverstein, 2017). The changing mix of generation resources resulting from the addition of wind and solar generation require grid operators to adjust the resources available to fill the gaps in instantaneous response ability when the wind dies and clouds hide the sun. The challenge grows as the grid systems accommodate higher concentrations of renewable resources like wind and solar. The system must be able to respond quickly (ramp up or ramp down) to keep the system at 60 Hz. Since grid operators do not directly control generator resources they employ what is known as "ancillary services"—commitments to deploy at an agreed competitive bid price. In short, grid resources come in an entire array of generation resources by energy source type (coal, natural gas, oil, wind, solar, and hydro) and base-load and small quick response generators. Grid managers have to guide future changes to this mix, and make use of existing resources the balance the grid in real-time, all the time. The extent to which the grid is capable of responding to unfolding conditions with changing resource mixes is the resiliency measure.

---

[3] The typical definition of reserve margin for electric markets have several components as discussed by Wikipedia, "In electricity networks, the **operating reserve** is the generating capacity available to the system operator within a short interval of time to meet demand in case a generator goes down or there is another disruption to the supply. Most power systems are designed so that, under normal conditions, the operating reserve is always at least the capacity of the largest generator plus a fraction of the peak load. The operating reserve is made up of the **spinning reserve** as well as the **nonspinning or supplemental reserve**. The spinning reserve is the extra generating capacity that is available by increasing the power output of generators that are already connected to the power system. For most generators, this increase in power output is achieved by increasing the torque applied to the turbine's rotor. The nonspinning reserve or supplemental reserve is the extra generating capacity that is not currently connected to the system but can be brought online after a short delay. In isolated power systems, this typically equates to the power available from fast-start generators. However, in interconnected power systems, this may include the power available on short notice by importing power from other systems or retracting power that is currently being exported to other systems. Generators that intend to provide either spinning and/or nonspinning reserve should be able to reach their promised capacity within roughly ten minutes. Most power system guidelines require a significant fraction of their operating reserve to come from spinning reserve. This is because the spinning reserve is slightly more reliable (it doesn't suffer from start-up issues) and can respond immediately whereas with nonspinning reserve generators there is a delay as the generator starts-up offline."

The transition of the ERCOT system from a traditional regulated utility to deregulation did not suffer any significant threat of involuntary outages, and it proved to be resilient over some extreme conditions. In addition to rules adopted by TPUC and ERCOT the reserve margins left over from the regulated market, and favorable wholesale electric prices in the late 1990s and early 2000s, provided reserves well above those needed to meet any unusual temporary spike in demand and/or loss of capacity due to weather or other emergencies. The wholesale market price for electricity was, during the transition, primarily determined by the price of natural gas-based generation at the margin. This is still the case today.[4] A rising gas price from the gas market provided adequate incentives to bring on new generation; the somewhat separable incentives for wind generation added to the relatively high reserve margins (20% in 2006). But as the natural gas market prices fell rapidly from a peak in 2005 the events of 2011 threatened that record creating a heightened oversight through the proceedings of the TPUC. A very cold winter in February 2011 disabled generation and froze gas delivery equipment. These severe weather conditions lead to 8 hours of load shedding by ERCOT (Newell et al., 2012, p. 9). As the market conditions of low gas prices and therefore, low electric prices developed, scarce generation capacity additions raised concerns that an energy-only market would not maintain reliability for the system.

The February 2011 event focused the TPUC effort on the matter of supply scarcity pricing mechanism set up in the original wholesale market structure and its relevance to the reserve margin. The extreme weather event raised the question of whether the scarcity price limit of $3000 per megawatt-hour (MWh) was high enough to encourage new generation adequate to yield reasonable reserve margins. This $3000 price is known as the system-wide offer cap. For a short time (6 hours) the $3000 price ceiling was reached. This price compared with an average supply price in 2010 of $40 per MWh. During the event the cold weather caused over 50 generators to go offline with a loss of about 7000 MW with the a new winter load peak of 57,282 MW. ERCOT appealed to consumers across the state to reduce load but still called on utilities to shed 4000 MW or 8% of load through rolling outages and 10−45 minute interruptions of electric service (see Fig. 9.1 below).

Texas necessarily has very limited ability to meet extreme load conditions (e.g., when reserve margins fall) from out-of-ERCOT interconnected sources. By design, as described above, Texas pretty much has opted to manage its own load not relying on interstate connections to help in extreme conditions (But the ERCOT

---

[4] Base load plants (typically coal and nuclear) have relatively high fixed cost and low variable costs. Natural gas plants have relatively low fixed cost and high variable costs (primarily the price of natural gas). In competitive markets it is always the highest cost supply that is able to enter the market that sets the market price. The competitive electric grid clears the market on a very short-term basis, therefore the marginal cost in the market will be the high variable cost unit.

dollars per megawatt-hour ($/MWh)

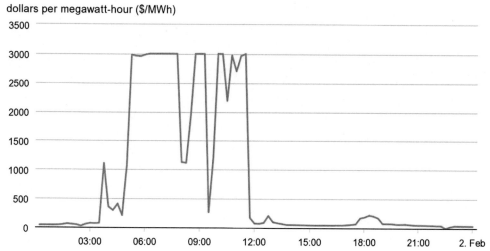

Figure 9.1 ERCOT real-time electric power price, Wednesday, February 2, 2011. *Derived from ERCOT data, Published by: U.S. Energy Information Administration.*

market is a large and diverse system providing the capability of management under extreme conditions). Therefore, the February 2011 event caused considerable attention to the issue of whether the target reserve margin of 13.75% is adequate, and whether the system-wide offer cap needed to be raised to encourage new resources to keep the reserve margin at or above the target.

The downward trend in system reserve margins and the February 2011 event gave the TPUC leadership concern as to whether adequate price signals from the energy-only market was providing enough incentive, soon enough to avoid an unacceptable risk of rolling blackouts. The circumstances that led to this heightened concern was that almost 10 years had passed from the initial opening of the wholesale market to full opening of the retail market in 2005, during which period there had been little concern about resource adequacy. The unique conditions during the wholesale completion era had maintained a reserve capacity of near 20%, but this condition had very little relation to the restructuring process. The excess capacity resulted from other conditions that were somewhat independent of restructuring. Namely, the unusually hot summer of 2012, following the February 2011 event provided the first real test of whether the energy-only market would provide enough new capacity soon enough to avoid rolling blackouts. In the end the severe heat subsided and new capacity addition came online which relieved the pressure on the TPUC to change the policies. Some "tweets" were made, but fundamentally it has been the consensus that marginal prices in the wholesale market are adequate to bring additional capacity to ERCOT (Newell et al., 2012).

The Commission in 2012 and following maintained an ongoing process to focus on "resource adequacy" commonly referencing the system reserve margin in

ERCOT. The target acceptable reserve margin was set as 13.75% in earlier proceedings—a target believed to be adequate to cover the prospect of "one load-shed event in 10 years." As the events of 2012 unfolded the Commission's 13.75% reliability target became the focus of intense debate. The debate was whether the restructured market provides adequate net additions of generation capacity and demand response (DR). During the peak summer season of 2012—an unusually hot summer—some reserve margin projections were far below the target level of 13.75%. The period that brought intensive focus on resource adequacy, however, was actually in the spring when, although the system peak was far below the summer peak that followed, the resources serving that spring demand were limited by units being out for repairs and the high penetration of wind units that provided power mainly in the night time hours leaving base-load units and natural gas turbines responsible for meeting the peak during the afternoon hours.

Since the prospect of the reserve margin falling to levels that portend rolling blackouts is a probability matter, the Commission set a target of "one load-shed event in 10 years," as an acceptable level of risk.[5] While high market prices during scarcity situations are needed to encourage investment, constraints are needed to prevent investors from using market power to run up prices. The solution was for the Commission to enforce offer caps when a load-shed event occurs. The offer cap going in to 2012 had been set at $3000/MWh. But concerns grew with the absence of new generation investments. Conditions threatened the ERCOT system with a reserve margin much lower than the target 13.75%. Changes were made to raise the system-wide offer cap to $4500/MWh in 2012 and potentially increase it in future years up to $9000/MWh. The actual cap following the 2012 policy change is subject to market conditions tied to the economic net margins of peaker generation units.[6] A high net margin will keep the system-wide offer cap at a low level; a low net margin will set the cap higher per the schedule set in the TPUC order §25.505. As of 2017 the system-wide offer cap has only been reached for a short time in 2012 so the effective offer cap remains at $4500/MWh.

During a review of the resource adequacy standard in 2014 the TPUC decided to replace the "one load-shed event in 10 years" standard with a three-part metric consisting of (1) the market equilibrium reserve margin, (2) the economically-optimal reserve margin, and (3) the associated levels of expected unserved energy (Public Utility Commission of Texas §25.505, 2012). The actual reserve margin as of 2017 did not exceed the target level of 13.75%. Since the intense debate of 2012

---

[5] Rolling blackouts are the actions taken by the grid operator to avoid cascading to system collapse.

[6] "Peakers" are power plants that generally run only at system peak demand. Because they supply power only occasionally, the power supplied commands a much higher price per kilowatt hour than base load plants.

when the Commission expected the reserve margin to drop to 13.14% in 2013 it stood at 18.2% in 2017 and was expected to rise to 25.4% in 2018. The TPPUC has projected reserve margins well above the 13.75% target for the next 10 years.[7] Under the TPUC order §25.505 the risk of major disruptions in power service (in the extreme, rolling blackouts) will be assessed on an ongoing basis. The managed energy-only market seems to be capable of providing adequate incentives for new generation and demand-side investments to keep the reserve margin at acceptable levels (Public Utility Commission of Texas Report to the 85th Texas Legislature, 2017).

The rise in the concentration of renewable on the Texas grid is continuing to complicate and challenge the approach to mechanisms for maintaining system reliability. The ongoing discussion includes the provision of essential reliability services, which includes frequency response and increased system flexibility. A recent NERC 2017 Long-Term Reliability Assessment is an ongoing focus on meeting these characteristics and other reliability of the grid challenges. Among other challenges the rising concentrations of renewable are causing more cycling and ramping (North American Reliability Corporation, 2017).

Prices

An important measure of the restructured market performance is the prices of electricity with restructuring compared to prices if regulation had continued. The price of electricity has clearly followed the price of natural gas since the wholesale market restructuring was completed. The correspondence between the natural gas price paid by electric generators and wholesale electric prices from 2002 through 2016 as shown in Fig. 9.2.

Fig. 9.3 below contrasts the retail prices of electricity and the natural gas price faced by electric generators. The trends since 2009 show that electric prices (corrected for inflation) have declined and are highly correlated with the decline in natural gas prices paid by power generators. There are, of course, other influences on the retail price in addition to natural gas generation prices, such as congestion charges from locations with transmission bottlenecks. The retail electric price includes market-based generation cost plus regulated transmission and distribution costs and market-based retail service provider fees. While the rise and fall of natural gas prices primarily drives the price of electricity—natural gas generation is the marginal generation in the market—the relationship is not one to one between a change in the price of natural gas and the change in electric prices but the trends are directly correlated in real dollar terms.

In the ERCOT market the regulated rates for transmission and distribution (T&D) are "socialized," meaning they are spread across all retail consumers. The policy in the

---

[7] Recent announced closings for a number of aging coal plants have challenged this 2018 expectation (see ERCOT, Inc., 2017a,b).

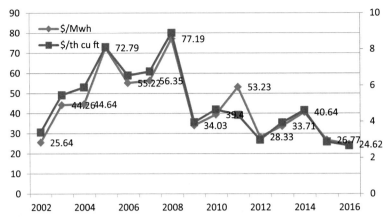

Figure 9.2 ERCOT wholesale electric and natural gas prices. *U.S. Energy Information Administration, Average retail price of electricity monthly and Natural Gas Electric Power Price at: https://www.eia.gov/ electricity/data/ and https://www.eia.gov/dnav/ng/hist/n3045us3m.htm.*

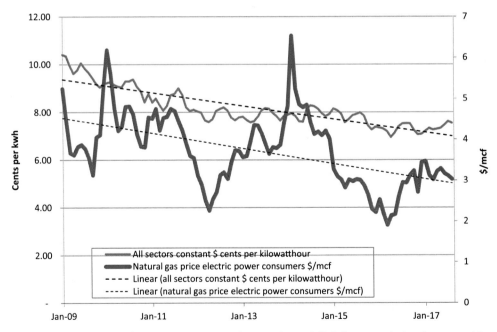

Figure 9.3 Texas retail electric prices (2009 inflation adjusted $) following wholesale competition and fracking influence. *U.S. Energy Information Administration, Average retail price of electricity monthly and Natural Gas Electric Power Price at: https://www.eia.gov/electricity/data/ and https:// www.eia.gov/dnav/ng/hist/n3045us3m.htm.*

ERCOT market has been to expand the transmission part of T&D over the 2002—17 year period, mostly to bring wind power in West Texas to the market that is concentrated in the large metropolitan areas of Dallas-Fort Worth, Houston, San Antonio, and Austin. The expansion of the transmission system was initiated by the Texas Legislature in 2005 in a proactive initiative to alleviate grid congestion (bottlenecks to get power from generation to consumers). This was followed by an ERCOT study of expected wind projects in defined geographical areas. The TPUC set up planning process designed to add transmission capacity to the defined Competitive Renewable Energy Zone (CREZ). As a result transmission projects amounting to $7 billion of investment by seven transmission companies were added to the system. The CREZ plan supported the addition of 18,000 MW of wind generation to the ERCOT market.

The TPUC made a study in 2006 to determine the comparison between prices that have prevailed since 2002 compared to what would have existed without deregulation. Their finding for the Houston area was that prices would have been from 18% to 26% higher under continued regulation; in the TXU area around Dallas the finding was similar—11%—18% (Kiesling and Kleit, 2009, p. 6).

Retail service and consumer satisfaction

The restructured market has been a major break from the past for final consumers, especially the residential class. Without a choice of their electric service provider the consumer pretty much just flipped the light switch and paid the bill when it came at the end of the month. Consumers did not need to know much about alternatives or to have much control over their energy use and cost. Retail competition changed that paradigm. Suddenly consumers could choose among several providers and a multitude of bundled programs that put them in control rather than a one-service-fits-all of the past.

Choice is fundamental to making competitive markets work, so upon restructuring in ERCOT consumers needed to educate themselves in order to make good choices as they commonly do in other markets. But consumers needed some help in the form of an introduction to this new world of choice, and better access to information. So, a significant part of the transition from the fully regulated electric world to completion was a consumer education program designed and implemented by the TPUC. The program resulted in the creation of a web portal at the TPUC where consumers have access to all of the program offers by REPs and consumers can switch providers and programs in short order. SB 7 also set up the process to support the consumer education and assistance for low-income consumers with budget problems. The statute authorized the TPUC to establish a system benefit fund (SBF) for the purpose. The SBF was set up to provide rate discounts for low-income consumers, TPUC consumer education activities, a weatherization fund targeting low-income consumers and lost property tax revenue. Funding of the SBF was by a per kwh charge included in the "wires" charge just as for T&D paid by all customer classes (Kiesling, 2009).

The TPUC hired J.D. Power and Associates in 2008 to study consumer satisfaction with the service provided under the restructured market and has continued the studies ever since. The 2010 study showed that 41% of customers have been with their current provider for at least 3 years compared with 49% in the prior year. Ten percent of customers indicating they intended to stay with their REP and 25% saying they definitely would stay with their REP. The J.D. Power surveys have continued each year since and show increased satisfaction. Satisfaction is highest in the enrollment/renewal factor. The statement by J.D. Power from the 2013 study is that customers who are satisfied with their REP are more loyal with their brand and more likely to renew their service. The main reason for customers selecting their current REP was price.

When customers are satisfied with their REP, they are more loyal to the brand, more likely to renew and more likely to recommend the REP to family and friends. When J.D. Power compared Texas to eight other states in the 2016 survey they found that Texas had the highest overall score in their (Power's) satisfaction index (J.D. Power and Associates, 2016).

A measure of the performance in the retail market, in addition to the price decreases for power relative to that which would have occurred in a continuation of the historical regulated rate-of-return world, is the number of REPs and program offers that have entered the market. By 2007 over 95 products from more than 25 REPs were active in each region, and enough switching had occurred that REPs provided the majority of energy sold in the ERCOT region. As of September 2016, 109 REPs were operating in ERCOT, providing 440 total unique products, 97 of which solely support electricity generated from 100% renewable sources. As of March 2016, in the portion of the state that is open to customer choice, 92% of all customers had exercised their ability to switch providers (Public Utility Commission of Texas Report to the 85th Texas Legislature, 2017).

The role of the market monitor regarding market power

A key to maintaining confidence that electric service providers (primarily large generators) are not exercising market power to set prices above competitive levels is to keep an independent monitor engaged to monitor market conditions and behavior. Actions by the TPUC to condition behavior and to promote competition are often based on the information and analysis of the market monitor.

The conditions in most markets of the economy create competitive market outcomes in the United States (and other free-market countries) that need no government oversight to insure good outcomes for consumers. Competition drives market outcomes to prices near the marginal cost of production accompanied by the most efficient technology and business organization—the combination of which guarantees efficient outcomes for consumers, while providing a reasonable rate-of-return for business investment.

But some markets do not support a large number of competitors (or potential competitors) and are therefore dominated by only a very few large players. Such industries are disciplined by federal and state antitrust laws and enforcement. The principle public agencies with responsibility for antitrust oversight include US FTC, the Justice Department and federal courts. The US antitrust laws grew out of the market extremes of the 1890s as a consequence of the exercise of market power (e.g., the ability to set prices above competitive levels and/or collusion agreements among suppliers to divide up the market shares. Such concentrations are also present on the demand-side of markets where a dominant few customers agree to collude). The US antitrust laws that are still the basis today for disciplining such market behavior grew out of the scandals in manufacturing and transportation industries—oil, steel, and railroads of the 1890s and early in the 20th century.

The primary laws that are designed to discipline markets subject to anticompetitive behavior are the Interstate Commerce Act of 1887, the Sherman Act of 1890, the Clayton Act of 1914 and the FTC Act of 1914. This basic structure was expanded in the 20th century by the Robinson—Patman Act of 1936, and the Celler—Kefauver Act of 1950. These antitrust laws are designed to guard against collusion resulting in the restraint of trade and to limit mergers and acquisitions that lesson competition. A very powerful part of the antitrust laws is the provision of "treble" damages (triple of actual economic damages) resulting from conviction of antitrust law violation in a court of law.

The fundamental economic conditions of competitive markets that discipline even monopoly and oligopoly industries do not exist in restructured electric markets. Therefore, the restructured electric markets are set up as "managed" competition, meaning there is continual oversight and regulatory rules that are designed to simulate outcomes of a natural competitive market. Perhaps the most important competitive market fundamental that is missing in the electric market is the short-term consumer demand. In economic jargon the ***short-term demand function is vertical***. That is, the consumer will not—cannot—respond to price signals to decide on how much electricity he is willing to buy at the going market price. Under the regulated rate-of-return paradigm the consumer does not know how much electricity he is buying, has little control over the level of use and does not have the ability to "shop" among providers for a better deal. The consumer just gets a bill at the end of the month for the electricity consumed (as measured by the electric provider), billed at the price previously allowed by the latest rate case. Therefore a restructured electric market must be managed to change that paradigm.

A second key characteristic of competitive markets that is missing in the traditional rate-of-return paradigm has to do with entry of market players into the market. In competitive markets entry is not constrained. There are conditions that have to be met, of course, such as environmental laws and many forms of government reporting. But fundamentally, entry is open to any firm that wishes to enter the market. The regulated rate-of-return paradigm sets out geographical service areas and ordains the only service provider able to provide services in the domain. The restructured electric market has to provide the open-access but also must set the technical conditions for a market participant to connect to the grid to provide, take or sell power, and to do so in a way that promotes fair competition.

On the generator end of a restructured electric market there are fundamentals of electric grid systems that have to be managed—conditions that do not exist in other markets. The grid must provide a continual flow of power from generation to the consumer—demand must be met on an instantaneous basis; failure to do so will bring down the grid. Further, the probability of maintaining the power flow is impacted by a number of uncontrollable factors. In short, the electron flow must be managed for reliability purposes—keep the grid up at all times, or under extreme conditions provide limited outages (rolling blackouts).

The managed competitive electric market in Texas is designed to meet all of the conditions one would expect to see in well-functioning competitive markets—the product is always there for consumers, there is adequate information for consumers to make choices about managing their use and selecting a provider, and investors have the ability, and opportunity to enter and leave the market for generation and retail service provision. Finally, since the transportation and distribution function (much like highways for transportation) is not competitive, it is planned and provided by regulation and is guaranteed to be "open-access" to all providers (Kleit, 2009).

The role of new technology and institutional changes

The fundamentals of the electric grid until the last two decades have not changed much since the first grids transitioned from DC (direct current) to AC (alternating current) in the late 1800s.[8] The traditional system that has provided reliable electric power for over 120 years includes generation power plants connected to a transmission (high-voltage) system to deliver electrons across relatively long distances. Transformers at substations then reduce the voltage and deliver electrons to consumers via lower voltage distribution lines. Until recent changes the electric grid was a one-way system designed to deliver power from the generator to the end-user instantaneously. There was no significant storage of electrons on the large systems across the United States and there was no customer-owned generators allowed to inject power into the system from the end-users on the distribution system in these fully regulated utility systems.[9]

---

[8] There are significant advantages of AC (alternating current) over DC (direct current) in using transformers to raise and lower voltages to allow transmission of power over long distances. Therefore, direct current was replaced by alternating current in power delivery.

[9] There are few exceptions to the general system design described in this paragraph. First, hydro-power damns perform a storage function by storing water which is released as needed to flow through turbines to generate power. Storage is not electrons directly but of the water that drives the electric generator. Second, residential and rural wind-charging systems were developed in the 1930s which stored power in battery systems that discharged power as needed for simple home lighting uses.

The traditional systems were built on the assumption that the industry was fundamentally a natural monopoly that had to be regulated with a rate-of-return style regulatory agency.

But things began to change in the 1990s with deregulation where market participants could expect to get rewarded by competitive rates of return on investments by innovating in the electric industry. In 2005 upon the eve of full retail and wholesale deregulation in Texas the major electric T&D companies and high tech entities formed a nonprofit company to share research, development, demonstration, and commercialization RDD&C efforts that are inherently risky ventures. The institutional innovation growing out of these conditions created the Center for the Commercialization of Electric Technologies (CCET) based in Austin, Texas, in September 2005.

The institutional innovation (CCET) amounted to organizing a forum for the interchange of ideas, and joint investment in RDD&C projects growing out of new technology development interests. The historically regulated electric utilities decided to address the changing technology landscape together with the regular interchange of ideas and joint funding of projects with high tech companies. The effort increased the collaboration among electric companies and high tech companies in Texas, and ultimately with participation from across the United States and some participation from European based companies. The high tech companies have a history of, and an ongoing process for, investment in risky technology ventures. But the high tech companies needed direct access to the electric companies that manage the flow of electrons to understand the context for innovation, and electric companies needed to understand the venture process of the tech world.

The status of the changes underway in the electric grid, and CCET's 10-year effort to advance innovation in the Texas market was summarized on CCET's webpage in the summer of 2015.[10]

The electric grid in Texas and elsewhere across the U.S. and the developed world is undergoing a technology transition. This developing modern grid joins IT technology with the traditional engineering approach for management of the flow of electrons. The transition from a one-way system of electron flow—from central station generation to the consumer—is giving way to a new system of two-way information communication and flexible generation at several places on the grid. This modern grid includes, among other things, renewable generation by the end-use customer, and the active participation of industrial, commercial and residential consumers in managing their load. The promise is that such diverse generation and load management will benefit both the consumer and the grid manager while supporting the

---

[10] CCET began operations in September 2005 and voted to end the effort in the summer of 2015. During CCET's 10-year long effort the organization carried a dozen or so projects amounting to a $30 million RDD&C effort.

transition to a more efficient grid. But this vision of a modern grid has many challenges that are the focus of experts around the world.

While there is much yet to be completed in this transition, much has been accomplished. Texas, in many ways has led the nation in this electric grid transition that is taking place in a Texas-style deregulated market. The Center for the Commercialization of Electric Technologies (CCET) played a significant role in these achievements during the last decade. I led this Texas non-profit organization as it carried out several important technology demonstration projects during 2005–2015. At its peak CCET included 22 utility and technology companies.

Holloway (2005–2015)

Since the transition of the electric grid began seriously in the 1990s key technologies have changed the landscape of the industry in Texas and elsewhere, but the Texas efforts have led much of the change, and CCET was a significant contributor to the changes. The technology changes, in addition to the fundamental restructuring of the market, have included (1) full implementation of advanced, digital meters across Texas (called Smart Meters), (2) development and deployment of synchrophasor systems to monitor grid conditions at 30 times per second, (3) deployment of ERCOT systems to reliably manage the grid with wind power capacity rising from near zero in year 2000 to over 19,000 MW in 2017, (4) development of 1000 MW of utility-scale solar systems, (5) development and deployment of three utility-scale battery systems in Texas to aid in accommodating more renewable generation and better management of peak system loads, (6) advanced restoration and recovery technologies to quickly recover from storm damage, and (7) deployment of hundreds of REP programs to allow a wide range of choice for consumers to manage their load and avail themselves of services to fit individual needs. To accompany the deployment of advanced electric meters the TPUC with participation by the ERCOT, T&D companies developed a smart meter portal to make maximum use of 15-minute digital meter data to support development of consumer programs for more efficient load management through enablement of secure communications with customer in-home devices.

CCET carried out demonstration projects that moved new technology forward in the Texas electric markets. The testing and deployment of new technologies included (1) syncrophasor technology deployment, (2) testing the ability of electric vehicles to provide ancillary services to ERCOT, (3) experimentation with pricing experiments in a residential community to test consumer's responsiveness to electric prices, (4) the design and deployment of a utility-scale battery system operated in conjunction with a wind farm, (5) testing the use of digital software and communication technologies to activate reliable consumer DR via the advanced metering system joined with the smart meter portal, and (6) addition and testing of technologies to closely monitor the electric behavior of concentrated solar communities.

Many other technological and institutional changes are in the RDD&C pipeline in Texas including the exploration of large-scale integration of electronic and digital systems at a municipality-level called the Internet of Things. Also, advanced monitoring and communications systems are coming to monitor grid behavior at the distribution end of the grid—an expansion of synchrophasor systems that are now monitoring the transmission end of the grid. This is a capability that does not now exist. Such systems are needed as a counterpart to developing DR capabilities and consumer connected generation that that will challenge grid reliability. Also, storage technologies (in addition to batteries) are entering the market testing phase including compressed-air storage that makes use of underground geologic formations to store compressed air that drives above ground generators. Microgrids are forming, primarily at US military bases where operations have to remain viable if the larger grid they ordinarily rely on goes down. But hospitals and other critical operations are also experimenting with microgrids for the same reason. Robots and drones are also coming of age to replace human labor in ways that allow better detection of power line failures and performance of routine repairs. Automation of all kinds is changing the job landscape and presenting regulators with a major challenge to adjust. Oversight becomes a new game when automation replaces human decision-making that is the focus of much regulation. Finally, there is a rising interest in data analytics (sophisticated algorithms and large data sets) to improve efficiencies of virtually all large energy systems operations, including the electric grid.

The interplay of policy, markets, and technology

The restructuring of the Texas electric market developed in what is arguably the most integrated combination of markets, policy, and technology of all the key Texas energy events summarized in this book. The 10-year-long process is a great example of the ongoing interplay of the three drivers of innovation: policy, markets, and technology, all driven by various means of a continual generation of new ideas. As the process unfolded the federal policy to deregulate wholesale electric markets created the interest to deregulate wholesale markets in Texas, but importantly without Federal jurisdiction. The single regulator condition made the process much more efficient than that in other states. The Legislature revisited the issues of policy in each session of the legislature that occurred every 2 years. The institutional innovation that followed created a very open process of hearings and rule-making at the TPUC that was continual over the period as the commissioners developed the rules of the market and reported back to the Legislature. ERCOT has operated an ongoing, open process of market participants to develop and execute the balancing market and an ancillary service market, provided ongoing market information, organized market advisory committees, held open meetings for market participants, and operated the grid reliably.

The market for coal, natural gas, wind, and solar generation has been an extremely important part of the march to a deregulated electric market in Texas. Deregulation took place even as the underlying energy markets for coal, natural gas, nuclear, and wind and solar technologies adjusted to the upheaval driven by fracking that changed everything (see the following Chapter 10). Texas instituted wholesale restructuring following the 1995 legislation the price of natural gas for electric generation was about $2.50/mmbtu compared with delivered Western coal prices of $1.25/mmbtu. Three years after the 2005 full retail deregulation the price of natural gas to electric generators was $8.71 compared to delivered coal prices of $1.88 (average 2008 prices). By 2014 the price of oil and natural gas fell by two-thirds making coal and nuclear uneconomic in the long term (Energy Information Administration SEDS database). And, in the beginning of the 10-year deregulation process wind generation had only begun to dot the landscape in West Texas and both rooftop and generation-scale solar were mostly future ideas. As of the 2005 full implementation of retail restructuring wind generation reached 10,000 MW of capacity and utility-scale solar was on the horizon. The ERCOT system now accommodates 19,000 MW of wind and 1000 MW of utility-scale solar. Over the decade of deregulation natural gas-based generation became the marginal generation that set electric prices in the market; rising electric prices encouraged innovation across the spectrum, helping to make wind and solar more attractive, and enhanced interests in storage technologies and consumer DR. With the addition of Federal subsidies for wind and solar the rising marginal electric prices help drive technological innovation.

In short, the primary fuel markets have been quite turbulent during the 10-year deregulation period, subjecting the process of deregulation to high attention of market participants and consumer groups that challenged the support for completing deregulation. But policy changes have continued to evolve, and technological advances have changed the landscape. Texas policy changes began by leveraging the Federal initiative to deregulate wholesale markets but maintained a clear demarcation of state/federal regulatory authority that allowed Texas to do it the Texas way, simplifying the dynamics of change. Texas then took the initiative to use the legislative and regulative powers to help drive innovation.

The two-part move to competition first in the wholesale market, then the retail market, all proceeding with care over 10 years to complete the market restructuring. In the meantime the Texas market participants, with leadership from the TPUC, and the market participant process at ERCOT followed the market restructuring with new technology, especially the state-wide implementation of digital meters that became a major building block for the creation of effective DR capabilities. DR is fundamental to consumer participation to drive the market—a base characteristic for all well-functioning competitive markets. And, the TPUC and ERCOT supported the rapid development of wind power by planning and overseeing a competitive process for extending the transmission system into the West Texas wind generation zones to allow delivery of wind (and eventually, utility-scale solar) into the large municipal load centers of central and south/southeast Texas. The dynamics of the drive to deploy wind technology that leads the nation could not have happened without the policy of ERCOT-driven transmission planning and the policy of "socializing" the transmission expansion cost among all consumers, matched with the power of the Federal subsidies.

Texas did not invent all of the technology that is now in place in the evolving ERCOT market. Instead, Texas took advantage of wind technology developed in Europe early on and set up the market incentive to bring the wind industry companies to Texas through a Renewable Portfolio Standard that targeted penetration levels in the future. CCET was one mechanism for helping to introduce wind companies to the Texas market. This same process was carried out in the advanced meter technology deployment. By 2006—07 CCET had membership from all four of the key advanced meter companies (Itron, GE, Landis-Gyr, and eMeter) as members while the T&D companies finally choose one of these companies to deploy advanced meters on their systems. The communication technology companies that helped develop the communication systems to deploy and use advanced meter data joined CCET to increase their prospects for participation in the Texas market. These companies included Current Communications and GridPoint. The DR focused company, Comverge helped drive DR technology as a CCET member. Battery companies including Younicos (a German company) and Samsung SDI (a South Korea company) were CCET members that developed and deployed the CCET-sponsored utility-scale battery at Reese Technology Center in Lubbock. The three largest REPs were active CCET members as the retail market developed, including TXU Energy, Direct Energy, and Reliant Energy and helped shape the early programs offers to the Texas retail market.

The success of CCET in winning a $27 million American Recovery and Reinvestment Act project supported important technology projects during 2009 through 2014 that remain as major contributors to innovation in the Texas market, including (1) the key role in developing and deploying synchrophasor technology for the real-time monitoring of grid events/conditions at 30 times per second, (2) the third utility-scale battery that operates in conjunction with a wind farm that demonstrated effective battery use for five grid functions, (3) pricing experiments that help inform the market about consumer response to several pricing programs, (4) technology for real-time monitoring of solar communities, (5) demonstration that electric vehicles can provide ancillary services to ERCOT to help manage the grid, and (6) the testing of advanced communication systems for supporting residential DR systems.

In summary, the 10-year transformation of the Texas electric grid from regulation to restructured deregulation has been a success in methodically generating new ideas, and integrating policy, markets, and technology. The contributions have included important institutional innovations through ongoing rule-making and oversight at the TPUC, the grid management by ERCOT and the operation of a nonprofit called

CCET that helped develop, test, and integrate policy, markets, and technology in the Texas market during the full market transition period of 2005 through 2014. The key policies to deregulate the wholesale and retail markets directed by the Legislature and carried out by the rule-making by TPUC and operations of ERCOT were first of all founded in the separation of Texas and Federal regulatory jurisdiction that left Texas in control of the deregulation process and oversight that followed. The key implementation that followed the Texas statutory changes included (1) the orderly spin-off of generation and retail from transmission and retail services, (2) the development of a real-time and day-ahead markets, augmented by an active ancillary service market operated by ERCOT, all to supplement bilateral market agreements between generators and REPs, (3) the guarantee of open-access to transmission and distribution services, (4) the use of price floors (the Price to Beat) during the transition stage rather than price ceilings that failed the retail market attempts in California and other states, (5) the provision of a consumer portal for consumers to choose REPs, (6) the establishment and funding of consumer information programs to educate consumers, and (7) the ongoing employment of a qualified market monitor to enlighten market participants, and to guard against abuse of market power by generators.

The move to deregulated electric markets, in addition to a well-developed institutional innovation at the TPUC and ERCOT has required a process by diverse group efforts interested in technological innovation, some of which was necessary to make the electric system function under deregulation, and some of which is designed to move the grid to a modern form generally called the "Modern Grid" or "Smart Grid." Many of the technologies that have been developed, or are still in the pipeline are not directly required for the deregulated market to function, but are very much an outgrowth of the process of deregulation that has created opportunities to provide valuable market additions and consumer benefits. This development of a technological focus has also been driven by Federal and Texas policy to support and integrate renewable technologies (wind and solar) and storage (batteries and other forms of storage such as compressed-air storage and flywheels).

The key technological innovations required to make the deregulated Texas electric market function has been the advanced meter technology. It is fundamental to making the retail market function. The Power to Choose portal developed and maintained at the TPUC website is also fundamental to deregulation where informed consumers choose their REP. The market system software and data at ERCOT that supports the real-time and day-ahead markets (the balancing market), and the supplemental ancillary service markets are also fundamental. Finally, the planning and deployment of modern transmission system additions make the (now) nodal market operational. And, while not required for a deregulated market per se, the transmission system additions to bring wind power from West Texas to load centers in the rest of the state were required for meeting Federal and Texas renewable energy goals.

The other key technologies that are a part of the ongoing development of the Modern Grid include grid restoration and recovery from storms, storage technologies, synchrophasor systems to monitor the grid in real-time, technologies (both communications and devices) at the end-user level for integrating and managing load in response to price signals and emergency conditions like controlled temporary outages. And, as the number of devices at the consumer end of the grid continue to multiply, the functioning of the larger Internet of Things that will certainly be a part of future communities. The advent of electric vehicles will add requirements for technological innovation to support wide-spread charging stations, and the advent of large numbers of electric vehicles will present an opportunity for the supporting battery systems to provide grid ancillary services. CCET was an institutional innovation to support collaborative efforts to explore new ideas, research, develop, test, demonstrate, and commercialize new technologies. CCET participated in, and/or funded projects in all of the above areas during its 10-year life. CCET was responsible for introducing the synchrophasor technology to ERCOT participants, and through the work of a company leader, Electric Power Group of California, organized, developed, and deployed an ERCOT-wide synchrophasor system that continues to monitor the ERCOT grid and provide valuable operator warnings in time for operators to prevent expensive outages.

Impacts of the restructured electric market innovations

The restructuring of the Texas electric market has created benefits to the consumers of Texas and enhanced economic growth. It continues to provide valuable results that give guidance to other regions who undertake the task of deregulating the industry—it is a grand experiment with great impact. The experience has been an example for the country, and is also the singular most complex development of competitive market design and implementation in modern times carried out with great success. While competition in the newly created generation and retail markets has kept investments in new capacity adequate to prevent blackouts in periods of peak system demand it has also tested the ability of an energy-only market[11] to incentivize investment in new generation adequate to meet unusual peak demand. This test of "resource adequacy" was being played out for a second time during this writing in the hot season of 2018 just as low natural gas prices made older coal plants uneconomic and forced them to schedule shut downs.

The Texas restructured electric market has been the only US experience with a successful energy-only market design, and has so far met all of the significant expectations of the market design leaders. It has provided enormous consumer choice and gone through

[11] An energy-only market for competitive electric systems relies totally on actual, short-term wholesale prices to incentivize additions and contractions of generation capacity for the market. A key alternative instituted elsewhere, and more commonly, is to separate the power component from capacity and set up the market to generate both while maintaining a design that successfully coordinates the two.

two cycles of testing as to whether energy-only markets can provide adequate price signals to generators in order to attract investment in new generation when extreme weather, combined with periodic shutdown of large base-load plants for repairs, can still provide power to meet peak system demand. So far, the answer is "yes." Another measure of good performance that has produced positive impacts is the range of choice available to consumers at the retail end of the grid. The restructured market has produced over 116 active REPs competing for markets throughout ERCOT, and these companies have created dozens of products to appeal to consumers' preferences at acceptable prices—the clearest indicator of a successful retail market.

Another impact with positive benefits will be played out over a long period of the future as Texas has become a grand experiment that will continue to provide guidance to other markets in the United States and around the world. The experimental value is in part because of the diversity of the industry serving the full Texas market. The structure finally put in place by the Texas legislature for restructuring the industry did not mandate restructuring for all entities providing power to Texas consumers. The mandate for restructuring only included the privately owned utilities—not coops and municaple utilities (MUNIs). These publicly owned entities had (and still have) the option to opt-in to the restructured system. So far, only one coop out of 75 and not any MUNI has opted-in. The result is that the Texas provides an ongoing experiment with competitive versus traditionally regulated electric systems producing results of price, reliability, and consumer satisfaction throughout the state. The impacts of this grand experiment will continue to provide impacts on future deregulation experience—yet another example of the benefits of diversification.

In short, the ERCOT restructured market has proven that a managed, competitive market can provide a great range of choice for consumers at an acceptable system-wide performance of reliable power when compared with the coops and MUNIs in this the smallest of three US reliability systems (ERCOT, Eastern Interconnect and Western Interconnect). It is reliable and provides electric service at prices equal or less than traditional regulated electric systems. And, the Texas experiment has, at the same time, delivered on these key impacts of deregulation by accommodating the largest concentration of intermittent wind generation in the United States—now large-scale solar is being added to the system further adding to the complexity of intermittent and traditional generation that the Texas market continues to integrate.

CHAPTER 10

Hydraulic fracturing: the permian basin challenges organization of petroleum exporting countries leadership

Contents

The beginning

Perhaps the most important energy technological innovation of the last 50 years is hydraulic fracturing (fracking) in recovery of oil and gas. The technology found its entry into the market place in a serious way beginning in 2005 in the Barnett shale north of Ft. Worth, Texas and quickly spread to a half dozen other plays in the United States. In a remarkable explosion of drilling and production since 2005, US and Texas oil production has recovered from a near 50% decline off of the peak of 1970 to return to a production level of 10 mmbbl/d in 2017. The record for natural gas has been remarkable also, and has been the earlier contributor of the two in launching the fracking "revolution."

The fracking technology has turned the energy world upside down. OPEC is no longer able to set the price of oil in the world market because their market share has significantly declined and fracking has made US production much more responsive in the short term than anywhere else in the world. Once wells are completed in shale formations with horizontal drilling that taps large quantities by one well, operators can shut down and start up much faster than has ever been true with traditional vertically drilled wells. As a result of fracking, the US has become a significant exporter of oil for the first time in decades. And, natural gas production from fracking has led to exports expanding, primarily via conversion to liquefied natural gas (LNG). The US electric market has switched a major portion of generation to natural gas, and CO_2 emissions in the United States have declined. Finally, the consumer prices for oil and gas products, especially gasoline and residential natural gas, have declined significantly

Innovation Dynamics and Policy in the Energy Sector
DOI: https://doi.org/10.1016/B978-0-12-823813-4.00010-5

235

from the peaks of 2008. There are many other economic and geopolitical impacts that have followed.

The fracking revolution is built upon the important technologies developed in a major way by work of the University of Texas Bureau of Economic Geology (BEG) research and a following industry education program in partnership with the Texas Independent Producers and Royalty Owners Association. The effort grew out of an ongoing research program in "unconventional" oil prospects beginning in the late 1980s and continuing into the 1990s. BEG's program of economic geology focused on a strategy of oil and gas recovery by use of enhanced recovery techniques used to unlock previously unrecoverable oil. The focus was on recovery from existing plays rather than from wildcats. This emphasis was continually voiced by Dr. Bill Fisher, Director of the BEG on many occasions at industry conferences. Eventually the Bureau and Texas Independent Producers and Royalty Owners Association hosted a series of educational seminars all across the state. In a follow up Department of Energy (DOE) engaged with funding and along with the Bureau helped inform the rest of the nation, an action that did a great deal to advance enhanced oil recovery and then fracking.

Natural gas from shale

The fracking revolution, which became economic in the Barnett shale in 2004, was the final stage of a lengthy development spanning several decades focusing on natural gas recovery. In 2004 there were roughly 400 producing horizontal wells in the Barnett shale; by 2010 there were more than 10,000. But much like other technologies fracking did not develop overnight. Natural gas was first extracted from shale in Fredonia, New York in 1821. In 1947 napalm was used to extract natural gas from shale in limestone formations in Kansas. Water and gelling agents replaced petroleum products to complete fracking in 1953. The Energy Research and Development Agency launched the Eastern Gas Shales Program and the Western Gas Sands Program in 1976. The Natural Gas Policy Act of 1978 allowed higher price ceilings for unconventional gas. Mitchell Energy and Development conducted the largest yet massive hydraulic fracturing demonstration with DOE assistance in 1978. A federal tax credit is provided for unconventional gas. In 1991 Mitchell Energy experimented with mapping and drilling with DOE and Gas Research Institute (GRI) support. In 1998 Mitchell adapted Union Pacific Resources' slick-water fracking in the Barnett shale. In 2003 Southwest Energy used Mitchell's slick-water technique to extract oil (King et al., 2015).

The fact that oil and gas lay unrecoverable in large quantities in so-called tight sand and in scale rock formations was well known decades before the Barnett shale development. But attempts to exploit oil and natural gas from these sources eluded developers until Mitchell Energy and Development finally made the right combination of technologies that leveraged the work of others, combined three major technologies, joined the resources of both public and private sectors, and finally attracted large funding for commercialization through a sale of the company to Devon Energy.

Shale deposits of resources occur throughout the known petroleum basins of Texas and the US. It was well known that some process of cracking (fracturing) the shale rock formation would release much of the trapped oil and gas, and other processes could recover oil and gas from tight sand formations (A North Texans for Natural Gas Special Report, 2016). The challenge was finding methods and technologies that could be used down-hole to fracture the rock and open pathways through tight formations and then remove the resource economically. George Mitchell and his team at Mitchell Energy and Development at the Woodlands north of Houston worked the problem for natural gas in the shale formation of the Barnett play for the better part of a decade before making enough headway to attract the necessary capital to prove that the technology they developed could produce gas competitive with existing techniques. The challenge was met by combining the fracking process with three-dimensional (3D) seismic imaging to find the richest concentrations of shale natural gas found thousands of feet underground. Further, the team used the developing horizontal drilling technology to penetrate the formation on a large enough scale to recover economic quantities from a single well.

Mitchell and his team managed to fund and develop their technology in the Barnett shale in the Fort Worth area where the company had leases producing natural gas by traditional methods and serving customers in the Chicago area via sales to interstate pipelines. Early attempts to fracture the shale using various combinations of foams, gels, and carbon dioxide (CO_2) with water proved promising, but were not productive enough to be economic. After several years of effort and dozens of wells the team decided they had to try another approach. The pay-off came from a combination of water, sand and a gelling agent derived from the guar bean. The mixture they called "slick-water frac." The gas production from injecting the mix under high pressure was surprisingly large, but still at a cost that would not compete with existing technologies.

The Mitchell team eventually prevailed but the dedication finally paid off with the help of research results from the DOE and the GRI via the Eastern Gas Shales Project (EGSP), and high price incentives from the Natural Gas Policy Act for unconventional gas. Another influence at a timely moment was information about the use of water-based fracking by a competitor, Union Pacific Resources. The Mitchell effort eventually received funding support from the EGSP. The early results from use of the "slick-water" trials attracted a larger company, Devon Energy who acquired Mitchell Energy and Development at year-end 2002 for $3.5 billion (Breakthrough Institute, 2015).

The result of the decades-long effort, driven in large part by the tenacity of George Mitchell, has been truly remarkable. The contribution from fracking in the Texas and US gas markets was first brought to market by the Mitchell team in the Barnett shale. While market prices clearly drive competitive markets the complexities of several factors responsible for bringing a new technology to market cannot be determined by immediate response to current market prices. The quantity contribution from fracking in the gas industry has occurred in a market where the average wellhead price of natural gas rose from near $2 per mcf in 1999 when the Mitchell team was struggling to make the technology work, to $9 per mcf in 2009 and back to about $2.50 in 2016.

The fracking revolution that began in the Barnett shale quickly spread to other natural gas and oil plays (shale and tight sands formations) principally in the regions of the Permian (Texas and New Mexico), the Eagle Ford (Texas), the Haynesville (Texas and Louisiana), the Anadarko (Oklahoma), the Niobrara (Colorado and Wyoming), the Bakken (North Dakota and Montana), and the Appalachia (Ohio, Pennsylvania, West Virginia, and New York; see Fig. 10.1). Today the largest gas production from fracking is from the Marcellus formation.

The fracking contribution in the Marcellus is largely attributable to Bill Zagorski, a geologist at Ranger Resources. The successes in the Marcellus came in 2004 following the successful economic production by Mitchell in the Barnett shale. The Zagorski story is much like that of other innovation. He had a failed natural gas well and a fellow geologist encouraged him to try a combination of horizontal drilling and hydraulic

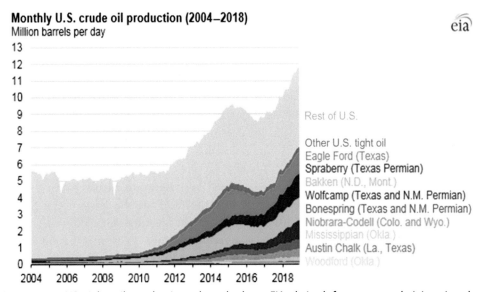

Figure 10.1 US tight oil production-selected plays. *EIA derived from state administrative data collected by Drilling Info Inc. Data are through January 2018 and represent EIA's official shale gas estimates, but are not survey dat. State abbreviations indicate primary state(s).*

fracking. The effort paid off much like that of Mitchell in the Barnett shale. It turns out that the Marcellus is sandwiched between two other formations, the Upper Devonian, and the Utica. This fact increases the prospects for long-term major production that may dwarf the discoveries and production made so far (Zagorski et al., 2012).

The impacts of fracking on US natural gas production is evident in Fig. 10.2. Reserves have doubled, and production has risen by more than 50%. Development of ports that were originally planned to import LNG have switched to ports for exports, or become ports that do both imports and exports. Exports of LNG did not exist before the fracking impact that began in 2005; exports are now at about 1.3 TCF per year and expected to rise to 9.5 TCF by the end of 2019 (Fig. 10.3). While the export share of total natural gas production is quite small (about ½ of 1%) exports are bound to be an important part of US participation in world oil and gas trade in the future. Texas LNG ports at Sabine Pass (actually in Louisiana near the border of Texas and Louisiana), Freeport LNG (southwest of Houston), and Corpus Christi LNG (Cheniere Energy subsidiary at Corpus Christi) are now, and will continue to be, leaders in the United States that also includes Cameron LNG in Louisiana, Elba Island LNG in Georgia, Cove Point in Maryland, and others planned or in the permitting process on all three US coasts.

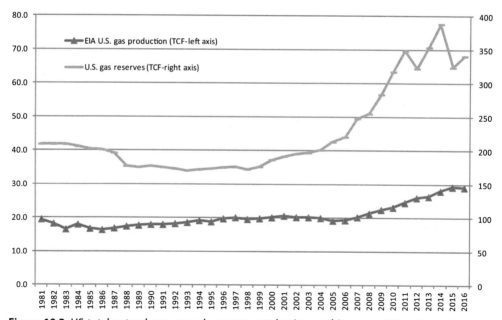

Figure 10.2 US total natural gas proved reserves, production, and imports, 1983—2016. *U.S. Energy Information Administration, From EIA-23L, Annual Report of Domestic Oil and Gas Reserves; U.S. Department of Energy, Office of Fossil Energy, Natural Gas Imports and Exports; Natural Gas Annual: 2016, DOE/EIA-0131(16).*

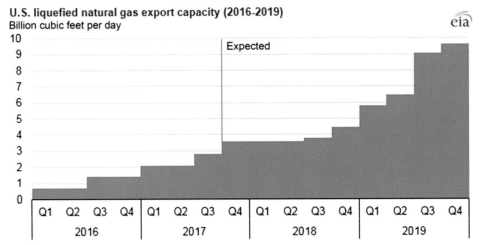

U.S. liquefied natural gas export capacity (2016-2019)
Billion cubic feet per day

Figure 10.3 US liquefied natural gas export capacity (2016—19). *U.S. Energy Information Administration, based on IHS and trade press.*

While imports still exceed exports that balance is bound to change (net imports have fallen from 10 billion cubic feet per day in 2006 to about 2 billion cubic feet per day in 2017). China has become a major importer of LNG as the country struggled to reduce the severity of the air pollution that has become one of the most severe in the world. China followed the dictates of a long-term plan to reduce air pollution in 2016—17 that has become a national emergency. The plan calls for a major switch to distributed clean-coal but the immediate means of improving emissions is the import of LNG. The strong move in the direction of power plants fired with natural gas was so dramatic in 2017 as to create a major natural gas shortage in China. Imports of LNG increased by 45% in 2017—all of which pushed natural gas prices to a historic high of $12/MMBtu. Natural gas has become the short to medium term solution for reducing CO_2 emissions around the globe (CERAWEEK by IHS Markit, 2018, p. A8).

Energy Information Administration (EIA) expects the US LNG export capacity to rise to 9 BCF/day by the end of 2019 (Fig. 10.3). This world trade development is the joint result of rising world natural gas use to drive down CO_2 emissions and the US fracking revolution.

Crude oil from shale and tight sands

The fracking revolution that began in the natural gas industry in the Barnett shale spread quickly to crude oil recovery in contrast with the early industry view that oil production from fracking would not be significant. But the impact on Texas and US oil production and the world market has been much more profound in the oil markets than in the gas markets. The US oil reserves have risen from a four-decade low of 20 billion barrels in 2009 doubling to 40 billion in 2015. Production has risen from a four-decade low of 1.8 billion barrels in 2008 to a recent high of 3.4 billion in 2015. The impact of fracking has turned the US oil import market from an apparent ever-increasing upward trend into a rapid decline, shifted the balance of market power (the marginal world producer) from Saudi Arabia to the United States (Figs. 10.4 and 10.5) and it has happened even as crude oil prices have fluctuated from a low of $10/bbl in 1986 to $110/bbl in 2008, and a recent low of $25/bbl for a short time in 2016.

The truly surprising result of the fracking revolution on OPEC's behavior is that Saudi Arabia, the prior marginal world producer, has pressed the other OPEC members and Russia to reduce production in order to raise world prices. The strategy is dependent on the price elasticity of demand—that the percentage increase in prices from production limits will be greater than the percentage decline in production, thus increasing revenues. The Saudi's and other OPEC members have greatly increased

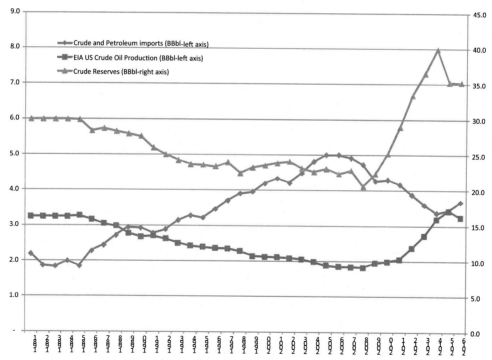

Figure 10.4 US crude oil and lease condensate proved reserves, production, and imports, 1983–2016. *U.S. Energy Information Administration, Form EIA-23L, Annual Report of Domestic Oil and Gas Reserves; Form EIA-814, Monthly Imports Report; Petroleum Supply Annual 2016, DOE/EIA-0340(16).*

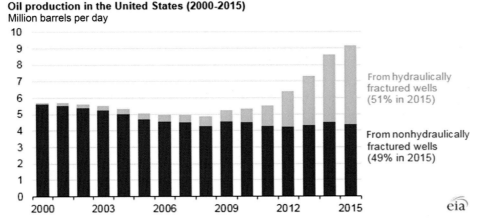

Figure 10.5 Oil production in the United States (2000—2015). *U.S. Energy Information Administration, IHS Global Insight, and Drilling Info.*

national expenditures over the last four decades that makes them highly dependent on oil revenues. But the quick response capacity of US fracking is continually challenging the OPEC strategy—US oil production rises to challenge OPEC's market share keeping prices below Saudi expectations.

According to the EIA oil production from fracking wells now exceeds that from traditional wells (Fig. 10.5). EIA expects that trend to continue, and the level of total US oil production will continue to rise from near 10 million barrels per day in 2017 to 12—14 million by 2025 and beyond, with reserves additions continuing to support that level of production for the foreseeable future (U.S. Energy Information Administration, 2018a). A common narrative from conference participants and related sources is that the US will surpass Russia as the world's largest oil producer by the end of 2018 (Alessi, 2018).

The major contribution to oil production from fracking in the United States has come from Texas regions including the Permian basin in west Texas and eastern New Mexico, and the Eagle Ford in south Texas. The Austin Chalk fields in east Texas and the Haynesville in Texas and Louisiana add some oil production but produce mostly natural gas from fracking. The fracking technology with horizontal drilling produces natural gas, natural gas liquids, condensate and crude oil. The Texas regions combined produce approximately 4.2 mmbbl/d of the total US production from fracking of 5.2 mmbbl/d.

The Permian is a 300 mile square area lying above a complex over lay of geologic formations which began production soon after World War I is now the largest producing basin in the United States, large share of it from fracking wells. The region's deposits may grow in discovered oil and gas to match or exceed that of the Ghawar field in Saudi Arabia, now the largest field in the world. The Permian is made up of oil and gas formations stacked one over the other, a structure called "multiple stacked plays." About 80% of the plays are within 10,000 ft. of the surface but also includes multiple formations reaching more than 3 miles below the surface. There are a dozen or more formations and some experts think the Permian may eventually contain reserves greater than the Ghawar field in Saudi Arabia which has reserves of 70 billion barrels. One estimate is that the Permian may have more than 160 billion barrels with 75 billion barrels in the Spraberry and Wolfcamp plays alone. There is an estimated 40 billion barrels in the Delaware formations. Several of the formations are made up of tight oil shales and will yield economic production from hydraulic fracking. EIA estimates current production from five Permian plays of about 3 mmbbl/d and rising (Fig. 10.1).

The second largest current oil production from fracking is from the Eagle Ford in south Texas which is currently producing over 1.2 million barrels per day of oil and over 6 billion cubic feet per day of natural gas. Oil production was 1.7 mmbbl/d in 2015 (U.S. Energy Information Administration, 2018b). The play produces from depths between 4000 and 14,000 ft. which began in 2008 and has followed a $30 billion in investment. The Eagle Ford formation is directly beneath the Austin Chalk and may be up to 400 ft. thick (Eagle Ford Shale News, Market Place and Jobs, 2009—2017).

The Haynesville Region of Texas and Louisiana currently produces mostly natural gas at about 8 BCF/day primarily from shale gas formations (U.S. Energy Information Administration, 2018b). The USGS estimates the play holds 174.6 TCF of recoverable shale gas resources, second in the US only to the Appalachia region (Hammes et al., 2011). An advantage of the region in the natural gas market is its location near the large complex of natural gas pipelines connecting the Gulf Coast resources to markets from Texas to the northeast US.

The Anadarko Basin in Oklahoma and the northeastern part of the Texas Panhandle, southeastern Colorado and western Kansas produces about 0.5 mmbbl/d of oil and 6 BCF/day of natural gas (U.S. Energy Information Administration, 2018b). Over 4000 horizontal wells were completed as of 2012 at vertical depths as great as 15,000 ft. There were four major horizontal plays in progress in 2012, including the Cleveland, the Granite Wash of Pennsylvanian, the Mississippian limestones and the Upper Devonian Woodford Shale (Mitchell, 2012).

The Niobrara Region of Colorado and Wyoming currently produces approximately 0.6 mmbbl/d of oil and 5 TCF/day of natural gas (U.S. Energy Information Administration, 2018b). The formation of the Niobrara is located between 7000 and 11,000 ft. below the surface. Geographically the location is north of Denver, Colorado extending into the Wyoming Powder River Basin. The Denver Julesburg and Powder River Basins produce mostly oil while the Greater River and Piceance basins produce mostly natural gas (DiLallo, 2016).

The Bakken Region of North Dakota and Montana currently produces 1.2 mmbbl/d of oil and over 2 BCF/day of natural gas. The region covers about 200,000 square miles from the Late Devonian to Early Mississippian age. Most of the production is in North Dakota. Oil was first produced from the Bakken in 1951 but production faced major hurdles until fracking and horizontal drilling changed the way oil is recovered. This region is one of the top tight oil production regions of the US all made possible by the fracking revolution and the current North Dakota production is second only to that in Texas (U.S. Energy Information Administration, 2018b).

Interplay of policy, technology, and markets

The complexity of the interplay of markets, technology and policy that resulted in the shale revolution is perhaps the most dramatic of all of the events reviewed in this book—events spanning a period of 120 years in the Texas and US energy sector beginning with Spindle Top in 1901. The various drivers of innovation—the policy dynamic, the changes in markets, evolutions of the business model, and the technology responses built off of prior and cross-industry technology deployments—makes the fracking revolution somewhat of a template for future energy sector innovation.

A significant and ongoing concern of government energy policy is the energy−national security link. The 1973 Embargo and the following Iranian Crisis of 1979 that grew out of World War II governmental structures in the Middle East, and the ongoing US support of Israel, set the stage for the fracking revolution that followed 30 years later in the United States. The Allied Forces' focus on oil supplies to support the war effort targeted the Middle East even as the US supplied much of the oil for the Allied Forces. During the War the European countries looked to the Middle East for future national security as the oil discoveries in Iraq, Iran and Saudi Arabia showed enormous reserves—Europe had little. OPEC developed in the years following World War II and finally gained enough market penetration to influence world oil prices in 1973. The Israeli Palestinian conflict that resulted in the Six-Day War of 1967, with the United States supporting Israel and the Arabian countries supporting Egypt, was followed by a Saudi-led attempt to use oil as a weapon. The effort failed. But conditions changed by the start of the Israeli-Egyptian conflict of 1973 and OPEC, led by Saudi Arabia, imposed an oil embargo on the United States and a number of US ally countries.

By 1973 the OPEC market concentration condition changed as supplies rose rapidly in the Middle East, and reserves were in decline in the US. The Embargo followed as an Arab nation response to the US support of Israel with major impacts felt by the developed economies of the world. In the United States, government policy in response to the Embargo took the front seat in trying, clumsily to chart a new course. The new course was a change in the role of government that was heretofore a laissez-faire attitude evolving to that of direct government interference in energy markets. The primary rationale was that of federal government war-time command and control for national security purposes. The government attempts in this new world evolved from ill-conceived direct control of markets with quotas, price regulation, government allocations of energy among consumers, and downstream suppliers. These were the initial policy responses of the Nixon Administration which evolved painfully to a more price and R&D incentive and tax-based approach by the Carter Administration. But during the post-Embargo period the private sector was not just setting on the side lines waiting for government action. The industry recognized that the world of oil-driven markets previously dominated by the Texas Railroad Commission (TRC) policy of oil market stability had been over-turned by the Embargo. Thus the efforts of industry leaders like George Mitchell went to work in the 1980s, 1990s, and early in the decade of 2000—10—the fracking revolution followed, albeit 30 years after the Embargo.

The 1970s through the 1980s were filled with OPEC-led changes in world energy prices created by production target agreements among OPEC members. In the US, changes in government policy, and the influence of energy markets and policy on economic growth also resulted in price volatility. Compared to the prior 30 years of TRC policy of stable oil prices, this was indeed a new world. But in spite of periods of high prices ($40 per barrel during the Iranian Crisis of 1979) to quite low oil prices ($10 per barrel for a short time in 1986) the general expectation of both the public and private sectors was that the long-term outlook was that of rising real energy prices (net of inflation) as the United States and much of the world faced declining reserves of economically recoverable oil supplies.

Against the backdrop of expected long-term rising oil prices, it was common knowledge that oil and gas lay unrecoverable in large quantities in so-called tight sand and in shale rock formations. This recognition existed decades before the Barnett shale development. But attempts to exploit oil and natural gas from these sources eluded developers until Mitchell Energy and Development finally made the right combination of technologies that leveraged the work of others, combined three major technologies, joined the resources of both public and private sectors, and finally attracted large funding for commercialization through a sale of the company to Devon Energy.

The Mitchell team experience tracks our best understanding of how successful innovations unfold. Scott Berkun concludes that there are five categories that form successful innovation patterns he has observed from studying many innovation success stories: (1) hard work in a specific direction, (2) hard work with direction change, (3) curiosity, (4) wealth and money, and (5) necessity. In addition to patterns there are major challenges innovators must overcome. Berkun's list of eight challenges include (1) find an idea, (2) develop a solution, (3) identify a sponsor and funding, (4) reproduce (scale) the process, (5) reach a customer, (6) beat the competitors, (7) be timely with the idea and development, and (8) keep the lights on (Berkun, 2007, pp. 45–46).

Virtually all of the challenges identified by Berkun were encountered by George Mitchell and his team operating in the Barnett shale for more than two decades before fracking entered the market as a viable innovation. Mitchell needed to replace the dwindling reserves of natural gas serving pipeline customers in the Midwest that constituted his customer base. The team knew natural gas deposits existed under their leases but there was not a technology that would recover them economically. The ability to release gas by fracturing the shale had been around for a long time but the technology did not exist to fracture the shale deep below the surface and recover natural gas economically.

The beginning idea of the Mitchell team was to make existing natural gas leases more productive by fracturing the shale deep below the primary production zones using various combinations of foams, gels, and carbon dioxide (CO_2) with water. While promising, these attempts were not productive enough to be economic. After several years of effort and dozens of wells the team decided they had to try another approach. The pay-off finally came from a combination of water, sand and a gelling agent derived from the guar bean. This mixture they called "slick–water frac." The gas production from injecting the mix under high pressure was surprisingly large, but still at a cost that would not compete with existing technologies. The Mitchell team eventually succeeded with the help of research results from the DOE and the GRI via the EGSP. But even with this help the team would not have succeeded without high price incentives from the Natural Gas Policy Act of 1978 for unconventional gas. Another influence at a timely moment was information about the use of water-based fracking by a competitor, Union Pacific Resources. The Mitchell effort also received funding support from the EGSP. Then the early results from use of the "slick-water" trials attracted a larger company, Devon Energy who acquired Mitchell Energy and Development at year-end 2002 for $3.5 billion (Breakthrough Institute, 2015).

Impacts of the fracking innovations

It is difficult to overstate the importance of the hydraulic fracturing (fracking). As in other innovations the best measure of success of an innovation is the *impacts*. This technological innovation has, in 12 years of time since commercialization began in the Barnett shale of North Texas, upended the geopolitical balance in the energy world, including changes in US foreign policy (especially in the Middle East) and returned the US and Texas oil and gas production levels not seen since 1972. No one, not even the most optimistic forecasters expected that production levels (that had fallen more than 50% by the 1990s from the high of 1972) could return to 1972 levels in ten short years. The resulting production impacts have been so great that US policy makers have allowed crude oil to be exported from the US for the first time since the Arab Embargo of 1973. The oil production from the US led by the Permian Basin of Texas has increased to levels that place it in third place in the world behind, or near that of Saudi Arabia and Russia (Bloomberg News reported in January 2018 shows US average expected in 2018 at 10.3 mmbd; Saudi Arabia, 10 mmbd; and Russia 11.2 mmbd; Summers, 2018).

The rising production of natural gas drove the market price down to 1980s levels near \$2.00 per mcf at the wellhead that have made coal and nuclear power plants uneconomic, and as a result significantly decreased the CO_2 emissions significantly below expected levels, and kept electric prices for consumers from rising. Natural gas power plants can respond quickly to keep the power grid up when wind and solar power fail when the wind drops and clouds reduce solar unit outputs.

Finally, the fracking technology has raised employment levels and the economic viability of communities and states with an oil and gas economic base where fracking is economic, especially, Texas, Louisiana, New Mexico, Kansas, Oklahoma, North Dakota, and Pennsylvania. Local property taxes have remained strong to support local economies, and state severance taxes have supported state government public sector services.

CHAPTER 11

The oil and gas state adds renewable wind and solar

Contents

Renewable energy definition

Renewable energy is normally defined as an energy source that is inexhaustible, that is will not be depleted by our continued use. The list of renewable includes solar, wind, biomass, tidal, wave, and hydropower.

The national move toward clean, renewable energy that promises to make the United States and the rest of the world a more environmentally friendly place has been increasing for more than three decades. The US participation in the 2016 UN Paris Agreement resulted in signing of the Paris Agreement on climate change by the Obama administration. The Trump administration has announced the intent to withdraw from the Accord, because the administration is doubtful about the contribution of fossil fuel use to global warming, and because the United States has done more to restrict CO_2 growth than fairness among nations would dictate. But there is a long-held belief by scientists that CO_2 emissions from the burning of fossil fuels is a primary, or perhaps the most of the cause of the accumulation of greenhouse gasses in the upper atmosphere that is resulting in global warming. The substitution of renewable wind and solar power generation for fossil fuel-based generation is being promoted by a number of federal, local, and state incentives and has become the primary focus of national, state, local, and international policy to address climate change. Texas is no exception. There are, of course, other policy and technology alternatives for slowing or reversing the global warming trend. These alternatives cannot be adequately discussed here as an adequate treatment of the topic is a book all its own dealing with global warming, its causes and policy alternatives.

Innovation Dynamics and Policy in the Energy Sector
DOI: https://doi.org/10.1016/B978-0-12-823813-4.00011-7

Wind energy electric generation

The wind energy experience in Texas is an interesting study of the interplay of technology, policy, and markets leading to the largest concentration of wind generation in the nation. That is, the large wind generation is somewhat surprising in the state most known in the energy world as the US leader in oil and gas production. The story of Texas wind energy would not exist, of course, except for the plentiful wind resource to generate electric power. But, that fact alone is not enough to explain the dramatic rise in wind generation capacity in Texas—a capacity that has risen from 2 GW in year 2000 to 20 GW by the end of 2017 (Fig. 11.1). During a low total-demand day on a Sunday afternoon in January, and accompanied by high winds, the contribution of wind to electric power reached 44.71% of instantaneous total generation on the Electric Reliability Council of Texas (ERCOT) system. Total wind output over the October month in 2016 averaged about 17% of load (ERCOT, Inc., 2017a,b).

The wind resource is concentrated in West Texas and the High Plains. It is highly valued by wind farm developers, but mostly cursed by residents because of its destructive power in storms that cross the region in the spring, and because of its impact on soil erosion. Residents have many "bad hair days" every year. While the wind is a valuable resource for the electric grid its contribution and management by the grid operator is complicated because wind is highly variable across the seasons, and across the 24 hour day while the

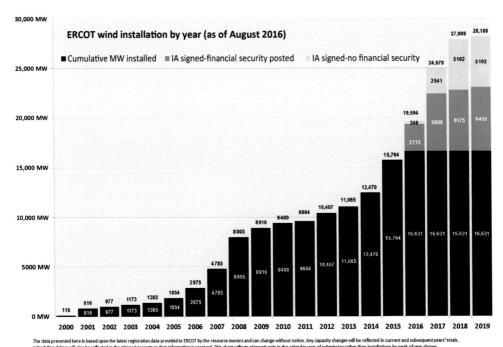

Figure 11.1 ERCOT wind installations by year.

demand for electricity highest in the early afternoon most days and highest of all in the summer when total demand is about twice that of the winter season due to the summer air conditioning load. To most everyone's surprise, the wind blows most productively for wind power machines during the late evening and early night time hours of the day, and in the spring rather than in the early afternoon of the day and during the summer when it is most needed. The wind resource is also valued by wind farm developers along the Texas Gulf Coast, both on and off shore. The resource here along the coast is more constant throughout the 24 hour day than in West Texas and the High Plains, and therefore has this advantage over the West Texas and High Plains resource.

The wind as a source of power did not begin with the 1970s energy crisis focus on alternatives to oil and gas. Windmills have been an essential part of the rural Texas farm and ranch economy for pumping water from underground aquifers for 150 years. And "wind chargers" (small wind turbines) charged battery banks to provide electricity for lighting of rural Texas households as early as the 1930s before the Rural Electrification Authority brought central source electric power to rural Texas.

Interests in wind generation for the electric grid intensified following the Embargo of 1973 as mentioned above with the work of Governor's Energy Advisory Council (GEAC), Texas Energy Advisory Council (TEAC), Texas Energy & Natural Resources Advisory Council (TENRAC), and State Energy Conservation Office (SECO) as well as US Department of Energy (DOE) R&D funding and private sector initiatives. Michael Osborne is created with setting up the first wind farm in Texas in the summer of 1981 when he installed five wind turbines near Pampa, Texas and sold the power to Southwestern Public Service for 2.7 cents/kwh. The project only lasted about 4 years with frequent repairs and damage from lightning strikes taking their toll on cost, and because the price of natural gas that was the main source of electric generation soon declined; but it was a beginning in the part of the state with a good wind resource (Galbraith and Price, 2013, pp. 22–23 and 33–35).[1]

---

[1] The Galbraith and Price book entitled *The Great Texas Wind Rush* is a well-written book on the Texas wind energy story that reviews the industry technology changes since the beginning of small wind machines for water pumping in the 1800s and continuing to the modern large turbines that now dominate the landscape in the High Plains and West Texas. The book is also a review of the industry leaders who lead the way. The development of the Texas wind industry followed the work of the important Texas energy policy agency that existed for ten years first as a special commission, then in three statutory forms following the Arab Oil Embargo of 1973. The agency contributions came from the development of energy policy leading to legislation and through energy projects funded by the agency's program that managed the Texas Energy Development Fund. The current writing does not try to duplicate the many contributions of *The Great Texas Wind Rush*—199 pages in length—but relies on it for much of the following summary. I recommend the book—it is a good read. The current writing, following the structure of the previous chapters, focuses on the important innovations (institutional, technological, market, and policy influences) that have brought Texas to the position of the leading wind energy state in the Nation.

The first commercial wind farm in Texas called the Texas Wind Power Project, was developed by way of a Lower Colorado River Authority (LCRA) contract to purchase power from a wind turbine company known as Kenetech. The project came online in 1995. Land was leased to the state by landowners who in turn leased the site to Kenetech who set up the wind farm. A transmission line was added to connect to a nearby existing transmission line, and a legal arrangement from the Texas legislature allowing "wheeling" of the power to get it to users on the Texas electric grid.[2] The story of how this first wind farm with 112 turbines on a leased land area near the Delaware Mountains in far West Texas is a complicated affair. A company named US Windpower, renamed Kenetech, when it reorganized in the late 1980s had a wind-machine manufacturing plant in Livermore, California where they began manufacturing wind machines based primarily on the developing California market in the 1970s and early 1980s. But Kenetech was struggling after the poor decade for renewable energy in the late 1980s which saw low natural gas prices, removal of federal subsidies and failure of a California rule that amounted to a quota system requiring utilities to buy power from independent producers like wind farms.

US Windpower, operating from location in Massachusetts, had developed the first wind farm in the United States—a 20 machine operation in New Hampshire in the mid 1970s (Galbraith and Price, 2013, p. 87). The company already had projects in Spain and India. When California created incentives for renewable energy in that state, which had good wind resources along the coastal areas adjacent to the coastal mountain range, US Windpower set up a manufacturing plant in California which made them attractive to wind development interests in the Texas market. After winning a competitive bid from LCRA, US Windpower (now Kenetech) supplied the wind turbines for the Texas Wind Power Project near the Delaware Mountains in West Texas.

A second West Texas wind farm, called Southwest Mesa was developed by FPL Energy (a subsidiary of Florida Power and Light) consisting of 107 Danish machines which were twice the size of the Kenetech machines at the Texas Wind Power Project. A third project was developed by Vistas and located near Big Springs, Texas. Also, Enron bought a US wind machine company located in Thachapi, California called Zond. The company built a West Texas experimental wind farm near Fort Davis consisting of 12 Zond machines. At the time in 1997 Zond had become the largest wind power company in the US Zond then became a major player in the developing Texas market (Galbraith and Price, 2013, p. 114).

The modern era focus on wind as a source of electric power in Texas, however, began with the Texas response to the Arab Oil Embargo of 1973. The GEAC identified promising alternatives to conventional oil and gas and made policy recommendations to

---

[2] Wheeling is the transportation of electricity from within one utilities service area to another service area in order to reach consumers.

the Governor and the Legislature. Among the GEAC reports completed was the first ever assessment of the potential for commercial grade electric power generated from wind resources in all regions of the state—research conducted by Vaughn Nelson of West Texas State University and Earl Gilmore of Amarillo College (Governor's Energy Advisory Council, 1975). TEAC and TENRAC, the successor forms of the Council continued the early GEAC work on wind energy in Texas that included testing and demonstration of wind turbines, investigating storage systems to pare with wind generation and completing early testing of wind-assisted irrigation applications for agricultural use of wind power as a remote source for electric motors for irrigation wells (Texas Energy Advisory Council, 1983).

An important driver to the creation of the Texas energy council (institutional entities that lasted for 10 years and beyond) was the expectation that Texas oil and gas was in permanent decline and the state needed to explore all of the alternatives—a "necessity" of sorts. It seemed necessary to develop alternatives during the decade following the Embargo. Following the idea that necessity is the mother of invention the state tried policy and development of alternatives to oil and gas. But invention had many "fathers" that led to the current wind energy technology in place today. Such developments primarily generated by policies, some of which were not so different from policies and markets that first helped generate the Embargo in the first place.

The reactions to the Embargo, were somewhat independent of natural gas market developments—a resource that at the time was not a player in the international markets for energy. The somewhat independent gas market developed shortages in the 1970s, high natural gas prices in the 1980s, mandatory reduction of natural gas for powering electric generators and industrial boiler fuel users (off-gas provisions Texas Railroad Commission Order 600 and off-gas provisions of the federal policy on boiler fuel use of scarce natural gas (Powerplant and Industrial Fuel Use Act, 1978). Also, they were independent of problems with evolving regulatory rules that were impacting the development of nuclear power plants following the Three-Mile Island power plant melt-down in 1979 (Nuclear Regulatory Commission, 2018). Finally, Texas researchers and entrepreneurs had a pivotal role in creating the wind industry in Texas—a development that leads the nation in wind energy production. The leaders included Vaughn Nelson of West Texas State, Earl Gilmore of Amarillo, Michael Osborne, businesswoman and investor, Jay Carter of Carter Wind Energy, Mark Rose of LCRA, Ed Winkler, businessman and Austin City Council member of Austin, Gary Marrow Commissioner of the General Land Office, and Bob King of Good Company).

Solar energy

The renewable solar energy resource, like its companion wind energy, is huge in the Lone Star state. Dr. Al Hildebrandt, Director of the Solar Energy Institute at the

University of Texas estimated the btu quantity of solar that could be generated from large utility-scale solar towers that could be constructed in Texas at 3 quadrillion btu or 45% of Texas 1972 total energy consumption. Hildebrandt's study was one of 40 studies published by GEAC in 1975—studies that formed the Texas R&D and policy basis for the state's response to the Arab Embargo of 1973. More recently, Frontier Associates used data from a Texas Solar Radiation Database (TSRDB) in 2008 to estimate the total Texas btus of energy from the sun in Texas, which they estimate to be 250 QBTUs, where 1.0 QBTU is the annual consumption of 3.0 million Texans (Frontier Associates, LLC, 2008, pp. 3–8).[3]

The total QBTU of solar, of course, could never be available for power generation because one could never collect it all. The number just measures the total resource base of energy from sunlight. The practicable use of the measurements and data base summarized by Frontier is to estimate the amount of energy from the sun that is available at a particular location where one might locate a solar collector like a set of PV panels, a parabolic dish or a power tower.

TEAC and TENRAC funded a number of solar energy projects during 1977–82 out of funds from the Texas Energy Development Fund, including residential applications, solar assist application in agriculture, industrial applications and solar distillation of alcohol fuels for the transportation sector. Also included was a project on optimizing solar repowering of retired generators in electric utility applications. Repowering allows the utility to switch to an alternative generation source and benefit from having the infrastructure like transmission hookups already in place and to avoid much of the usual regulatory approval requirements that would accompany the creation of a new generation site. SECO has funded many more solar projects since that time using federal conservation funding and leveraging other funds.

There are two types of solar power applications in the modern economy. There is the so-called rooftop application that generates power for residential and commercial appliances in buildings, and industrial applications for heat or power—all on the distribution end of the grid. Then there is utility-scale solar that generates power as companions to other power generators including natural gas, coal, nuclear, hydro, and wind. Like other states, Texas has developed both.

The early solar applications in Texas grew out of demonstration projects like those that TEAC and TENRAC funded that were mostly small scale application demonstrations and development programs on the distribution end of the grid, some of them leveraging federal R&D and conservation programs. Implementation of solar installations usually

---

[3] The TSRDB is derived from 89 National Solar Radiation Database (NSRDB) collection sites. The NSRDB was developed and maintained by the National Renewable Energy Laboratory in Denver, Colorado (Wilcox, 2007). The TSRDB available at http://www.me.utexas.edu/∼solarlab/tsrdb/tsrdb.html.

allows sponsors to qualify for federal, state, and local tax subsidies. Some of the demonstration projects focused on distribution-end applications in remote areas that had either high cost utility-provided electricity or no available electric utility service at all. For example, there are many small solar cell devices such as deer feeders on hunting leases across the state. Traffic signals to monitor the speed of vehicles and display the speed on a monitor in plain sight to drivers are common applications on Texas streets and highways today.

By far the most recognizable application of distribution-end solar today is rooftop solar photovoltaic (PV) cells that substitute for utility power in residential and commercial buildings by converting sunlight to electricity. State, local, and federal subsidies and favorable Public Utility Commission (PUC) regulation encourages these applications. Controversial rule making arises over the technical conditions required of such solar generated power units for putting power into the grid—power that exceeds the appliance direct use. There are both economic issues and safety concerns at the heart of the matter. Transformers that reduce the voltage of a utility distribution line from 69 kV to 110 and 220 volts for residential and commercial power are not designed to have power flow in the opposite direction, so controls are in place to protect the utility assets. The traditional electric grid was not designed to handle generation sources coming from the distribution end of the grid. Functions like voltage regulation, over-current protection, switching techniques and maintenance and power restoration practices were all designed for managing the grid with generation coming from the upstream source over the high voltage transmission lines which is then stepped down to distribution voltage. Managing the grid for all of the above functions for the combination of generation from both ends requires special new systems and practices in order to safely and effectively operate and maintain high performance of the grid. This is not saying that management of the grid from generation at both ends of the grid cannot be done, just that it will require time and resources. An example of an extreme case is that of San Diego Gas and Electric (SDG&E) where concentrations of up to 18% of the distribution load is from solar has become a major challenge. In 2015 35% of SDG&E generation was wind and solar—all of which is intermittent power.

A chief challenge for the SDG&E utility in managing high concentrations of renewable is that without system improvements, the renewable units cause voltage "swings," and can even cause an outage. Such voltage swings also cause equipment (appliance) damage for homeowners and commercial industries. SDG&E actively supported a state requirement for new solar systems to install "smart inverters" needed to address the voltage swing problem. An additional challenge for the utility that has a responsibility to keep the grid reliable and stable is that grid operators cannot call on these solar units to produce power when a sudden need arises. The grid manager often has to add generation at a moment's notice which can only be provided by other generation types like gas-fueled units. That is, high concentration of renewable means the utility has fewer options for providing quick-start units or power from "spinning reserves" when needed.

The permission for a distribution-level customer to put power into the grid is controlled by an "interconnection" agreement for larger generators, or registration with ERCOT for small generators of less than 10 MW of capacity. Generators smaller than 1 MW, are not required to register or to have an interconnection agreement. The interconnection agreement/process is the built-in technical protection for the grid owner. An exception for small generators avoids imposing major regulation on small direct generation (DG) owners. The typical case is a generator with less than 1 MW capacity like a rooftop solar or small wind machines. Units smaller than 1 MW in ERCOT are monitored and solar in all the relevant load zones amounted to 157 MW as of the fourth quarter 2017. Small wind units amounted to 4.6 MW (ERCOT, Inc., 2018).

The chief economic issue for small generators at a distribution location is whether the home or commercial end user gets credit for the power injected into the grid at the wholesale, or the retail price of electricity. In Texas there are "feed-in tariffs" provided by municipal utilities like Austin Energy and CPS that credit a user's generated power put into the grid at the retail price, given to the customer as a credit on his monthly bill. In the ERCOT restructured market the price provided for customer-generated power put into the grid is a matter of negotiation between the consumer and the Retail Electric Provider. Energy from registered DG units is settled at the Zone settlement point price.

The applications of solar utility-scale generation on the generation end of the electric grid have several generation options, each with its own economics and regulatory issues. There are *solar-thermal* power stations like that studied and promoted by Dr. Al Hildebrandt during GEAC's tenure in the 1970s. These units have a set of mirrors in a field that concentrate sun light on a receptor at the top of a tower that heats oil and generates steam for power generation. There is one solar power tower operating today in the United States in San Bernardino, California rated at 392 MW. There is a planned solar-tower unit rated at 1600 MW in Nye County, Nevada. The other more common technology for "concentrating" solar heat uses mirrors in a parabolic trough shape for concentrating sunlight on a focal line (receptor) rather than a tower. The focal line collects the heat and uses it to generate electric power like other steam generators. There are currently three of these power plants operating in California and one in Arizona. These generators are rated at 280 MW for the California projects to 361 MW for the Arizona unit.

Utility-scale solar in the United States (both PV and thermal technologies) have grown rapidly during the last 5 years. These sources made up 2% of total US utility-scale generation capacity in 2016—growing at an average 72% per year between 2010 and 2016. The total US capacity of utility-scale solar was 21.5 gigawatts (GW) in 2016. California had 9.8 GW of the total; Texas ranked seventh with 0.6 GW (Fig. 11.2; Energy Information Administration, 2017a).

Austin Energy, among all the Texas utilities, has the largest commitment to utility-scale solar generation in Texas. It currently has commitments to purchase solar from a 120 MW

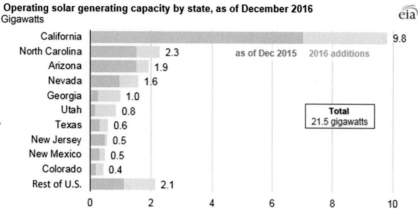

Figure 11.2 Solar generation capacity by state. *U.S. Energy Information Administration, Preliminary Monthly Electric Generator Inventory.*

plant called the East Pacos Solar Facility in Pacos County beginning in 2017. It is a second Pecos County project. A 2016 solar unit called Roserock Solar is a 150-megawatt (MW) solar power plant. A 2011 plant at Webberville, called the Webberville Solar Project, is a 30 MW facility. In December 2017 Austin City Council approved a contract to purchase of 150 MW of solar from Intersect Power. The Council has a goal of 50% of Austin's power from renewables by 2020 (Austin Energy News Release, 2017).

At this writing DOE Solar Energy Technologies Office issued a funding opportunity announcement intending to fund 70 projects with $105.5 million in research dollars supporting projects on four solar topics: (1) Advanced Solar Systems Integration Technologies (up to $46 million, ~14 projects), (2) Concentrating Solar Power Research and Development (up to $24 million, ~21 projects), (3) Improving and Expanding the Solar Industry through Workforce Initiatives (up to $8.5 million, ~4 projects), and (4) Improving and Expanding the Solar Industry through Workforce Initiatives (up to $8.5 million, ~4 projects) (U.S. Department of Energy, Solar Energy Technologies Office, 2018). So, the innovation in solar technologies that are intended to ramp up the contribution of solar power for the electric grid is not done. Texas innovators will be an ongoing part of the development.

The interest in wind, solar, biomass, in-situ gasification of lignite, fluidized-bed technology for lignite, geothermal production, the full range of conservation technologies, and other alternatives to existing oil and gas products waned in the late 1980s and 1990s. This was the result of market prices of oil and gas products declining back to near historic low levels (prices adjusted for inflation). Under such conditions alternatives energy sources were for the most part "priced out of the market". The drive for deregulating the electric market as a response to federal policy for wholesale power, the wind technology development in Europe, and the return of rapidly escalating market prices for oil and gas products reignited the interest in wind and solar.

The political agreement to restructure the Texas electric sector needed the support of renewable energy and energy conservation proponents. When joined with the federal mandates applicable to all other states in the nation, interest in meeting renewable targets in Texas became a focus of the 1999 legislation. The outcomes of the restructuring initiative (Senate Bill 7 in 1999) included a Renewable Portfolio Standard (RPS). Texas currently has the largest installed wind generation capacity in the United States. With a growing electric load, the region served by the ERCOT grid reached an overall peak load of 71,093 MW in August 2016, and reached a new winter record at 62,855 MW in January 2018. Increasing energy demand coupled with legislated RPSs has focused efforts on inclusion of renewable—especially wind—since 1999. As of November 2017, ERCOT had an installed wind capacity of 19,800 MW (see Fig. 11.2) which provides about 10% of the ERCOT electricity generation; wind generation capacity is 38.4% of the total. For a short time in the winter of 2016 wind generation equaled 45% of the total generation. The construction of new transmission lines into West Texas and the Panhandle area of Texas were initially set to accommodate 19.5 GW of wind but the plan has been revised adding to the initial transmission capacity. It is estimated that the capacity of wind generation could grow to 28,000 MW by 2020. Wind farms are being added in the coastal areas of South Texas as well as the High Plains and West Texas, presenting somewhat different power profiles to system operators than the wind from West Texas and the Pan Handle. All wind farms are distant from the main load centers, and present grid management challenges for ERCOT's system operators in accommodating both grid reliability and generation dispatch related to wind variability.

The challenges to the Texas electric grid posed by this additional growth in wind generation, magnified by rapidly growing solar capacity in ERCOT underscores the need to establish new response mechanisms to wind and solar variability. Center for the Commercialization of Electric Technologies (CCET)'s $27 million stimulus grant project in 2009–14 represented a multifaceted synergistic approach to managing fluctuations in wind power in the large ERCOT transmission grid through better system monitoring capabilities, enhanced operator visualization, improved load management, and consumer response to pricing incentives (Center for the Commercialization of Electric Technologies, 2015).

There are two overriding concerns with high levels of renewables on the electric grid. First, transmission capacity to move wind power from the remote areas of Texas where the wind resource is best to the large urban areas of concentrated demand has to be provided. Second, the grid manager must have the capacity and legal authority to manage the grid by bringing up other generation when intermittent renewable are suddenly not available, and to maintain reliability of the system while doing so. These issues are summarized below as currently operating in the ERCOT market of Texas.

Transmission capacity

Traditionally transmission capacity additions under the old regulatory system was added incrementally by individual vertically integrated electric companies, and the additions could not be put in the company's rate base to earn a return until the addition was "used and useful," meaning investments had to be made "on the come." The challenge of adding large capacity wind farms to the Texas grid under the restructured market to meet to goals of the RPS meant that this approach had to be changed.

Texas chose to address the transmission capacity additions to support wind generation by setting up a transmission capacity addition process called Competitive Renewable Energy Zones. The assignment was given to ERCOT as the grid manager to define transmission corridors through a participatory planning process. Once the zones were defined electric companies (usually joint ventures of several firms) bid on projects to construct and own transmission systems to support up to 18.5 GW. The winning teams then constructed the transmission lines and received regulatory approval of transmission rates to be paid by retail electric providers who would then pass on transmission costs to consumers obtained competitively as defined by the PUC with guidelines from the Legislature via Senate Bill 7. The unique rules for beginning to receive a return on investment approved by the PUC allowed sunk cost to be added to transmission providers as the investment was incurred, all of which had to be adjusted by a follow up proceeding when the transmission capacity went into operation. This was key to getting the transmission capacity added in record time, and the catalyst for driving wind companies to invest in wind farms.

Reliability

Intermittent wind generation creates management problems for the grid manager because sudden drops or increases in wind currents and thus generation (ramp up or ramp down) that have to be met quickly to keep the grid up and power flowing. So-called "spinning reserves" must be available as backup and other tools activated to compensate (e.g., reducing load). Also, wind machines are a source of instability—the grid has to be maintained at very near 60 hz at all times. In technical terms the grid manager must monitor the grid for what is called "regulation up" (frequency above 60 hz) or "regulation down" (frequency below 60 hz) and quickly adjust resources to move the grid back toward 60 hz. There are a number of tools that the grid manager has to work this ongoing challenge—one is real-time monitoring with a system called synchrophasors. Synchrophasors from a network of sensors obtains grid condition readings at 30 times per second.[4]

---

[4] CCET developed a synchrophasor network for the ERCOT system during the 2009−14 time period of a Federal grant under the stimulus program funded by Congress in 2009 (Center for the Commercialization of Electric Technologies, 2015).

Interplay of policy, technology, and markets

Innovation has many parents leading one to ask "if necessity is the mother of invention, who is the father?" One father that was a stimulus for wind machines was the deployment of the newly invented barbed wire designed to protect farmers from the herds of cattle that, in the early days of ranching roamed the plains on large ranches. As farms developed—which proved to be a better return on investment than large ranches—farmers needed to protect their crops. The answer was barbed wire. But barbed wire cut off cattle from springs and rivers that were their natural source of water. Hence, the stimulus for more windmills emerged. In the 1930s rural farmers and ranchers developed wind charges matched with battery banks to provide the first electricity for many residents. The Oil Embargo set in motion another phase of wind power development.

Early studies of alternative energy resources, engineering estimates of energy outputs from a set of emerging technologies, market assessments, economic benefit-cost studies, and outcomes of demonstration projects were carried out by organizations like GEAC, TEAC, TENRAC, and SECO. These institutions helped create a knowledge base in the decade or so following the Oil Embargo of 1973. That world-wide event was a wakeup call for the developed world economies, and for Texas in particular.

The first significant sized wind farm in Texas was developed by with a contract between LCRA and a wind turbine company, Kenetech coming online in 1995. The story of how this first wind farm with 112 turbines on a leased land area near the Delaware Mountains in far West Texas is a complicated affair. The effort joined both Texas and federal policy, private sector investment, a loose coalition of state agency and private sector actors with a vision of renewable energy, institutional maturity (especially ERCOT and the Texas PUC, and favorable relative energy market prices. Combined these factors chipped away at the oil and gas dominance of the energy industry in Texas. The generation capacity for wind farms has risen from the start-up projects in West Texas in the mid 1990s that got the development going, to a measured 2 GW of capacity in year 2000 rising to 20 GW today. This growth has averaged 14.5% per year—a remarkable development by any measure.

Institutional organizations form important development and demonstration capabilities by matching university talent with research and development entities like the Center for the Commercialization of Electric Technologies of Austin, the Alternative Energy Center at Canyon, the Texas Tech research facility at Bushland, the Texas Tech Wind Science and Engineering Research Center, the Scaled Wind Farm Technology Facility at the former Reese Air Force Base site neat Lubbock and Group NIRE of Lubbock have been highly important in advancing wind technology in Texas.

There has also been a continual evolution of the technology of wind machines since the beginning of wind power capture of the late 19th century. This progression has included the transition from barbered wire to windmills to wind chargers, to early wind machine tests of "egg beater" blade structures, gear-box to direct-drive generators, two-blade to three-blade machines, dynamic controls to face into the wind, and shut-down capabilities when winds reach 50 mph or severe storms arise.

So what made this development of wind power generation work? Probably the most important single factor was the federal production tax credit policy. But the rapid wind installation also benefited enormously from local property tax abatements, the Texas RPS, high natural gas prices early on, problems with continued development of nuclear plants suffering from a large regulatory burden and safety concerns, coal prices that included coal train transportation cost from Wyoming, and ongoing concerns about CO_2 emissions from burning coal. Not to be overlooked was the collaboration among state and local leaders, private investors willing to risk their capital, and the sheer tenacity of optimistic wind technology "believers" like Vaughn Nelson of West Texas State, Earl Gilmore of Amarillo, Michael Osborne, businesswoman and investor, Jay Carter of Carter Wind Energy, Mark Rose of LCRA, Ed Winkler, businessman and Austin City Council member of Austin, Gary Marrow, Commissioner of the General Land Office, and Bob King of Good Company.

Another highly important factor was the ability to leverage the cutting edge technology developed in the European countries—countries that were much more dependent on imported oil and gas than the United States at the time of the Oil Embargo. These energy resource deficits lead the Europeans to become leaders in wind machine technology. And, we benefited from those developments. Leading European companies have installed a large share of the 20 GW of wind capacity in the Texas market today. Finally, all of this effort, favorable market influences, and public policy would have produced little except for the very large wind resource base in Texas.

Solar development has benefited from some of the factors that drove the wind industry in Texas, but the set of influences are somewhat different and some of the important factors in wind development are missing for solar. The result is an obvious trailing of contribution from solar even though the base resource is similarly large.

The important production tax credit that has been most important for wind has also been available for solar, and the RPS also includes solar. There is also the ability in ERCOT to provide power at both the distribution and transmission ends of the grid which supports both utility-scale wind and solar, as well as distribution-level locations like community-based units and rooftop solar.

Impacts of renewables innovation

The impacts of wind and solar in Texas include mostly their contribution to clean technologies that have reduced the use of natural gas and coal for power generation, compared to the concentration of, especially coal-fired power plants that would have continued to dominate the industry growth without wind and solar. The wind companies that have benefited by the growth of 20 GW of wind machines have mostly benefited out-of-state and foreign based companies who brought their technology to Texas. The primary manufacturing of wind systems is in the production of the pedestals and blades that are sometimes made in Texas. The same is generally true for solar companies, although there have been a number of small start-up companies develop in Texas.

The substitution of wind and solar for coal and natural gas power plants has reduced the level of CO_2 emissions to the atmosphere, which is an important impact that helps meet US goals to reduce CO_2 emissions. While the level of CO_2 emissions reductions have been estimated it is difficult to know or measure any reduction in global warming that may result, but the impact is no doubt positive.

Perhaps one of the most important impacts of the rapid growth of wind farms in Texas has been the development of grid control technologies and industry protocols required for managing the dynamics of integrating intermittent generation with quick-start gas units to keep the grid up and operating at 60 hz when wind and solar suddenly go offline.

Another impact has been the local area lease payments to land owners where wind and solar farms are located. These rural communities derive as much income from this source as from use of the land for crop and livestock economic returns.

CHAPTER 12

Capture and global warming: the technology and regulation debate

Contents

Global warming and CO_2 emissions

Climate change is in the news daily as US and world leaders try to forge agreements to limit human contributions to the warming of the planet. The national debate continues as disagreement ensues not only about the human contribution to global warming, but also about the extent of warming over the long haul. More narrowly, there is an ongoing debate specifically about whether CO_2 is the major contributor to warming. The debate over CO_2 emission controls is front and center when the topic makes the news. The threats of long-term warming that portends melting of the polar ice caps, rising sea levels, weather pattern modifications that result in episodes of severe flooding and unusual heat waves, and the health effects of warming on plant, animal, and sea life are being monitored and analyzed by scientists across the globe. The recent trend of the world's global temperature is not much a part of the debate, although some experts question whether we are seeing a long-term trend or simply the early stages of a long-term natural cycle. It is clear, however, that a significant rise in temperatures has continued since the beginning of the 1970s. The focus of the debate therefore is on what to do about it.

The global warming challenge cannot be adequately covered in this short chapter. The current and developing literature on the topic has produced many books and now dominates much public policy debate on energy topics. The subject matter in the physical science realm is complex because there are several gases that are included in the known sources of greenhouse gases that have been determined to contribute to

Innovation Dynamics and Policy in the Energy Sector
DOI: https://doi.org/10.1016/B978-0-12-823813-4.00012-9

the global warming phenomena. Furthermore, the span of time for the individual gases to dissipate from harmful accumulations the upper atmosphere varies from a few years for methane gas, for example, to more than a 1000 years for CO_2. Then on the policy solution prospects part of the debate there are a host of policies that are been discussed to address the greenhouse gas issue. The solution to the problem obviously requires participation by many nations around the world, so governance issues abound when global policy approaches are considered. Finally, there are many technology possibilities for addressing the challenge, some of which are in late stages of the R&D spectrum and others that are mere ideas not yet receiving significant attention.

Because of the above complexity there is no attempt here to present a comprehensive review of the global warming topic. The focus is on the current outstanding efforts in Texas among selected efforts elsewhere. There are several reasons why Texas will make a significant contribution to the ongoing global warming solutions. The beginning of the discussion is with the evidence of global temperature rises.

The chart below (Fig. 12.1) shows the gradual rise in world surface temperature since the beginning of the industrial age near the turn of the last century. The measured temperature rise has been roughly 1.5°F (0.83°C). The chart shows temperature differences (called anomalies) from the average temperature of 1901−2000. There have been times in the history of the world when temperatures were higher than today as a result of natural conditions and cycles surrounding that period are evident in the data over the long term. But the trends since the early 1970s are clearly on the rise as the world dynamic systems push the averages up. Since the temperature measurements are estimates rather than absolute measures they are subject to error so the chart also shows the 95% confidence level band, meaning there is a statistical probability of about 5% that the real (unknown) actual number could be outside the range.

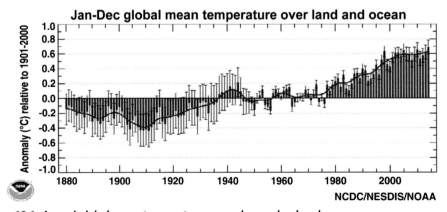

Figure 12.1 *Annual global mean temperature anomaly over land and sea.*

The data for Fig. 12.1 mostly comes from monitors stationed at key places around the globe, but of course the coverage is not everywhere so methodologies are used to estimate the temperature at each spot and aggregated to approximate global averages. Other methods are deployed to estimate temperatures in earlier times. In short, there are actual readings and then there are methodologies for estimating the aggregates, and extending such estimates over long periods of time.

Scientists have developed a consensus view that the recent world temperature rise is primarily due to human actions, especially the CO_2 emissions from the burning of fossil fuels. The common belief is that the cumulative level of greenhouse gases in the upper atmosphere, about 63% of which is CO_2, causes a rise in world temperature because they form a blanket around the globe holding in heat from the sun, some of which would otherwise escape. The normal condition is that the heat escape is the balancing that prevents the globe from warming. The oceans absorb CO_2 from the atmosphere and cause the sea level to rise. The scientific community consensus is that if warming continues its rise to 3.6°F above the 1901–2000 average, we will be in great trouble. We are about 40% of being there today. The 3.6°F rise (equal to 2°C) is the highest extreme in the historical record, a time when oceans were a larger percent of the surface area of the globe, and polar ice caps were considerably smaller than today. So, a 3.6 degrees rise above the preindustrial levels has become the target for policy objectives. That is, the policy target is that the world should prevent such a severe rise from happening. We are already at about 1.5°F above the preindustrial age level, and it does not seem possible to stop the rise completely, but slow it we must. More than a 3.6 degrees rise would take the climate outside of the range of historical observations reaching back several hundred thousand years.

The challenge presented by climate change, primarily due to warming of the globe, is complex; there are many factors at play. The comprehensive measure of global warming is the combination of the atmospheric, land, and ocean temperature increases. The combined measure developed by the scientific community is an index that allows us to monitor global temperature, to use statistical methods to identify causes of warming, and to forecast future changes. But measurement and monitoring that is fundamental to identifying causes of temperature change is a challenge. The analysis of causes is the other fundamental. The challenges are enlightened by the following chart that represents the earth systems as they exchange carbon by both natural and artificial activities (see Fig. 12.2).

The chart shows both sources and sinks. The complexity of this carbon exchange system gives rise to a number of policy prescriptions to stop the rise in world temperatures by (1) preventing CO_2 emissions through capture and sequester, (2) redirecting and storing the CO_2, and/or (3) developing technologies to off-set the impacts of CO_2 concentrations in the atmosphere, on land and in the sea.

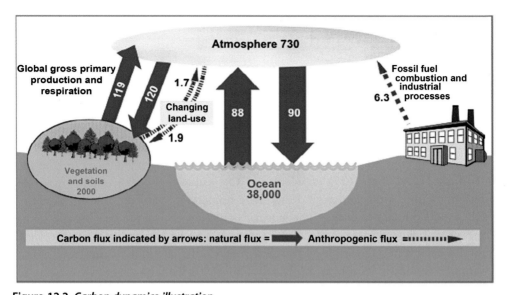

Figure 12.2 *Carbon dynamics illustration.*
Intergovernmental Panel on Climate Change, 2001. Climate Change 2001: The Scientific Basis. IPCC, UK.

The first requirement for analyzing the global warming causation is to construct data systems that quantify temperature, CO_2 and other variables needed for a thorough analysis. The temperature concentration of CO_2 in the atmosphere over a 400,000 year period is shown in Fig. 12.3 below. CO_2 concentrations have risen rapidly since the beginning of the industrial revolution at the end of the 1800s, but rapidly since 1950. Fig. 12.3 shows that the increased concentration has accelerated since 1950, and is correlated with global warming. Fig. 12.4, however, shows that the movements of temperature have not been a steady rise since 1880. There have been periods of rises, a period of no general rise and then the rapid rise since 1950. But the annual movements have been up and down throughout while the concentration of CO_2 continues to rise steadily, making it obvious that more is going on than CO_2 concentration. That is, the analysis needs to identify multiple factors explaining global warming.

The debate about the extent to which CO_2 emissions from the burning of fossil fuels is the cause of global warming continues because the ability of physical and behavioral scientists to separate the influence of human activity from natural system cycles is subject to uncertainty. The uncertainty comes from many sources including the theories of physical and behavioral scientists, the quality of long-term climate data, the measurement of CO_2 emissions and its distribution around the globe, and from methods of statistical analysis, especially the ability to model climate change and its causes over long periods of time in the past when actual measurements did not exist.

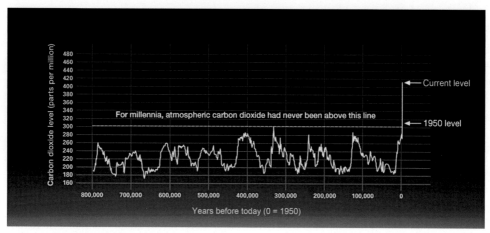

Figure 12.3 *Global CO$_2$ levels.*
http://climate.nasa.gov/evidence/. This graph, based on the comparison of atmospheric samples contained in ice cores and more recent direct measurements, provides evidence that atmospheric CO$_2$ has increased since the Industrial Revolution. (Credit: Vostok ice core data/J.R. Petit et al.; NOAA Mauna Loa CO$_2$ record.)

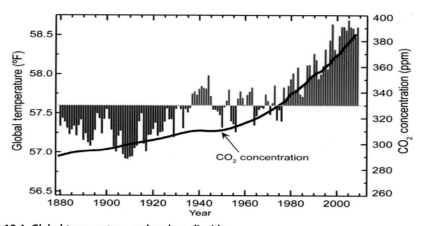

Figure 12.4 *Global temperature and carbon dioxide.*

There are a few scientists who do not accept the aforementioned summary of CO$_2$ influence, and the debate goes on. There are several reasons for questioning the commonly accepted narrative summarized in the prior paragraphs. First, the measurements of actual temperatures via monitoring stations around the globe have not been in place over the long period of the world's history so other methods are used to estimate temperatures during the long-term past. Also, the current monitors only provide measures of specific spots on the globe which are then aggregated to a global measure. In short, the temperatures from the long historical period can only be approximated by the combination of actual measurements, use of temperature indicators from the geological record, and other sources. And, then there is additional uncertainty surrounding the narrative when it comes to determining causation rather than relying on simple correlations. Both physical and social science theories provide the basis for the analysis of causation, and the discipline of statistics provides the methods for assessing the probability that the estimates of causation, given the theoretical model, are reliable—that is, statistics provides the means for reaching the conclusion that estimates of the extent to which the parameters of tested causes is likely true because the hypotheses of causation cannot be rejected at a high level of statistical probability. But the mathematical model tested with statistical methods depends on the model specification which comes from theory—theory that is always subject to debate among the professionals within a discipline. But the more times we fail to reject the theoretical model, and/or the statistical tests from repeated sampling of a population (the full domain of the subject matter), the more confident we are that the theory and statistical tests can be accepted for public policy or other uses.

Critiques of statistical analysis results are often based on a number of limitations inherent in debates about causation in all of science, and such limitations certainly exist in the climate debate about CO_2 and global warming. There is the matter of measurement errors—are the temperature and CO_2 measures and other variables accurate? There is the matter of specification errors in setting up mathematical representation of the hypothesized relationships for testing. Is the CO_2 concentration linear or some form of nonlinear relationship to global temperature increases? Are there variables left out of the analysis that may cause the analyst to attribute causation estimates to the wrong variable? For example, the analysis needs to separate the CO_2 effects from that of other greenhouse gases. There are additional challenges of statistical analyses of time-series data and causation. In analyses of causation over long periods of time an analyst must account for the lag period effect. For example, how quickly does increased CO_2 concentration in the atmosphere find its way into the ocean environment? And, finally, there is the matter of assigning a probability. The practice in statistical analysis is to determine if the hypothesis should be rejected—a test of significance. If the tested relationship is not rejected at say a 95% level of confidence, then the analyst has a reasonable basis for concluding that the relationship is actually true—at least until ongoing tests conclude otherwise. Such a test in the current case is whether one can conclude that CO_2 is likely a cause of global warming, at least until further testing proves us wrong. Further, are we confident that the parameter being tested in the analysis adequately distinguishes this cause from causation due to other sources specified in the model.

The importance of statistical analysis on the climate change debate cannot be overstated. Analysis of the human contribution to global warming includes the use of statistical models that recognize the difference between correlation and causation. Methods for estimating causality in the climate change controversy include both physical and behavioral science disciplines. Statistical methods provide a means of assessing the probability that a measured causation parameter of a mathematical model is reliable for explaining, modeling, and forecasting. In the climate change debate statistical methods are means of determining whether human activities, especially CO_2 releases from the burning of fossil fuels are a cause of global warming, as distinct from natural sources, and if so to estimate the contribution of CO_2 among all sources and natural cycles that contribute to warming in a defined period of time. The statistical method used must separate the human induced CO_2 causation from other sources of CO_2 releases into the atmosphere, all of which, when taken together causes warming. The question is examined through testing of a hypothesis based on mathematical equation representing the theory. Once we are confident that we know the relationships, and can do a reasonable job of forecasting world temperature rises, the next set of questions arise: What is to be done about it?

Since human behavior is directly involved in emissions of CO_2 a thorough analysis should recognize the human activity that removes CO_2 from the environment as well as the various activities that emit CO_2 to the environment. Perhaps most, but not all of that CO_2, emissions go into the atmosphere. Since humans cannot destroy physical elements, we know taking CO_2 from one system means it has to go somewhere else in global systems. That is, if the proposition is that human activity is the cause of global warming through CO_2 emissions into the atmosphere, the analysis should recognize the human activity that changes CO_2 concentration elsewhere in the environment. This recognition opens the public policy debate to a much larger set of policies for addressing climate change than simply implementing regulatory policies to limit emissions from burning of fossil fuels.

Having said all of the above, however, there is currently a consensus among leaders in the scientific community that global warming is happening and that human activity, especially CO_2 emissions from the burning of fossil fuels is a major contributor, perhaps the major contributor. There is much less of a consensus about what should be done about it. Major differences of opinion exist among analysts concerning reasonable solutions. The differences are present among professionals within the various disciplines and, among the scientific community, citizens, and the politicians and their public agency regulators. The public policy choices include a regulatory approach to limiting CO_2 emissions by the chief actors—namely the electric utility and manufacturing industries, and automobile manufacturers who sell gasoline and diesel-powered vehicles. Other approaches focus on a set of technological innovations to capture CO_2 and store it in deep geological formations, or make valuable products from it. There is also a host of technologies that could counter the global warming effects of CO_2 by off-setting the CO_2 effects.

Interplay of policy, technology, and markets

The global warming/CO_2 event is just now unfolding in Texas, across the United States, and around the world. There are well-developed markets for CO_2 but significant policy changes are still being debated and technology changes are still very much in the "pipeline" of development. There are only a few examples of policy implementation available for analysis to guide future developments. As the results of emerging technologies become available, policy direction will come better into focus. The markets for CO_2, current policy developments and maturing technology ideas are discussed below. Texas is an active participate in these unfolding developments.

Markets

Markets drive the production and consumption of fossil fuels and their substitutes, determine the mix of new energy sources entering the market, change energy consumption patterns and drive investment in new technology. There is a well-developed market for CO_2 in the oil business known as enhanced oil recovery (EOR).[1] The use of CO_2 injected into oil reservoirs has been underway for decades, especially in the Permian Basin of West Texas. A 2006 study estimated the Permian Basin to have original oil in place of 95.4 billion barrels, of which 35% (33.7 billion barrels) have been recovered. The 61.7 billion barrels remaining is called "stranded" oil, a large portion of which is being recovered using CO_2 injection (Advanced Resources International, 2006, p. 8). The remaining oil is the target of fracking efforts.

The first CO_2 EOR recovery project in the US was in Surry County Texas, a project known as the SACROC Unit (Scurry Area Canyon Reef Operators Committee). The project relied on CO_2 produced from a natural gas processing plant 220 miles to the south, both locations being in the Permian Basin. The project was economic because oil prices were relatively high during the 1970s and early 1980s,

---

[1] Stages of oil recovery include primary recovery that is the traditional recovery using the natural pressure of reservoirs to push oil to the well bore. Secondary recovery uses the injection of water (called water-flooding) into the reservoir to help release oil from the formation rocks and push it to the well bore. Tertiary recovery primarily consists of injecting CO_2 under high pressure into the reservoir (CO_2 flooding called Enhanced Oil Recovery or CO_2 EOR) which is miscible (mixes) with the oil; there are several approaches to CO_2 recovery. Tertiary may also include chemical and thermal methods. The typical progression is that water-flooding for secondary recovery begins after production from primary recovery has about run out. Tertiary then follows secondary. But each of these stages depends heavily on the reservoir conditions and the chemical compassion of the oil in place, so EOR doesn't work everywhere following secondary recovery. Hydraulic fracturing is now adding a fourth dimension to oil and gas recovery (the topic of Chapter 9: Electric Industry Deregulation and Competitive Markets).

and because the cost of CO_2 delivered to the production site was relatively inexpensive. This project opened the door to CO_2 injection in the Permian which then spread to other oil production areas of the US that have the right conditions. The Permian Basin became the primary CO_2 EOR production area of the country producing approximately 80% of US CO_2 EOR oil recovery. The US production rose slowly at first but began to accelerate in the mid-1980s reaching 87 million barrels per year by 2008.

The full development of CO_2 oil recovery in the Permian relied on CO_2 from natural underground deposits in southern Colorado and northern New Mexico. There are three sources with pipelines delivering CO_2 to EOR projects in the Permian—Sheep Mountain and McElmo Dome in Colorado and Brovo Dome in northeastern New Mexico. In recent years 1.6 billion cubic feet per day of naturally occurring CO_2 are produced resulting in about 170,000 barrels per day of incremental oil production from dozens of projects in the Permian. The successes in the Permian have led to major projects in Louisiana and Mississippi, Oklahoma, North Dakota, Wyoming, and Utah (National Energy Technology Laboratory, U.S. Department of Energy, 2010).

CO_2 is also produced from nonnatural sources called anthropogenic sources. For example, CO_2 is a byproduct of a natural gas processing plant like the one that served the original SACROC project in 1972. Another example is CO_2 from a lignite-fired Dakota Gasification Company synthetic fuels plant in North Dakota that provides CO_2 for an oil recovery project across the border in Canada. Other anthropogenic sources of CO_2 can easily come from coal-fired power plants, ethanol plants and other industrial processes. Such projects can ship the CO_2 directly to oil recovery projects, permanently store the CO_2 in underground formations (known as sequestration), or temporarily store the gas to be produced and reinjected into an oil recovery project. The full development of the CO_2 industry will likely include all of these alternatives, depending on the economics peculiar to the situation. A rough idea of a CO_2 EOR project that would pay a reasonable rate of return on investment at April 2018 oil prices is summarized below (Fig. 12.5).

Tax incentives, which have greatly determined the economics of CO_2 recovery, date back to federal legislation in 1979, a time when oil price ceiling controls were still in place for ordinary oil classes. The National Energy Policy Act during the Carter Administration provided incentives for EOR. The incentives were continued by the US Federal EOR Tax Incentive of 1986 after ordinary oil prices were decontrolled. The incentive provided a 15% tax credit that applied to all costs associated with installing CO_2 equipment, the purchase price of CO_2 and CO_2 injection costs (National Energy Technology Laboratory, U.S. Department of Energy, 2010).

There are other markets that embody CO_2, either directly or indirectly, since consumer and industrial products of many kinds contain or are aided in effectiveness by CO_2. For example, it is an inert gas in welding and fire extinguishers and used as a

Oil price ($/barrel)	$70
Gravity/basis differentials, royalties and production taxes	($15)
Net wellhead revenues ($/barrel)	**$55**
Capital cost amortization	($5–$10)
CO_2 costs (@ $2/Mcf for purchase; $0.70/Mcf for recycle)	($15)
Well/lease operations and maintenance	($10–$15)
Economic margin, pre-tax ($/barrel)	**$15–$25**

Figure 12.5 *Illustrative costs and economics of CO_2 EOR project.*
National Energy Technology Laboratory, U.S. Department of Energy, 2010. Untapped Domestic Energy Supply and Long Term Carbon Storage Solution. Available at: www.netl.doe.gov.

pressurizing gas in air guns; it is a chemical feedstock and used as a supercritical fluid in decaffeination of coffee. It is added to drinking water and carbonated beverages (beer and sparkling wine). A frozen form is a refrigerant called "dry ice" which is often used in the transport of medical supplies and organs for transplant, and for keeping food and drinks refrigerated for travelers on the road.

CO_2 is produced by animal and aquatic life when they metabolize carbohydrates and lipids to produce energy by respiration. It is returned to the oceans by fish gills and to the atmosphere by the lungs of air-breathing land animals. It is removed from the atmosphere by plants during photosynthesis. Deforestation therefore reduces the natural system's capacity to remove CO_2 from the atmosphere. Concentrations of CO_2 in the oceans result in acidification because it dissolves in water to form carbonic acid. Coral reefs are damaged or destroyed by high levels of carbonic acid.

In short, CO_2 is a key, directly or indirectly to all life forms, and it is an ingredient in many production and consumption items found in everyday use. A natural balance in the dynamics of earth systems keeps problems of CO_2 concentrations in the atmosphere at bay, but when the natural balance is interrupted by human activity, severe challenges may be created, like warming of the globe. When CO_2 is released into the atmosphere at abnormal rates the natural systems are interrupted. CO_2 released by activities such as the burning of fossil fuels and captured must then go somewhere. So one solution to CO_2 emissions from industrial and electric utility power plants fired by fossil fuels is to capture the gas and store it underground (like the naturally occurring CO_2 deposits). Or, store it temporally and then use it for a beneficial purpose like EOR. CO_2 emitted into the atmosphere concentrates in the upper atmosphere and some of it is absorbed by the sea. The challenges and options for controlling and eliminating these concentrations are complex. Fundamental to designing solutions requires understanding the long-term dynamics of all impacted earth systems—not an easy task.

In addition to CO_2 markets for EOR other markets influence and help or hinder efforts to devise policy solutions for reducing CO_2 emissions. For example, market systems impact the demand for energy (mainly consumption of transportation fuels and electric power) and therefore the level of CO_2 emissions into the atmosphere. So, reducing the use of fossil fuels for transportation (gasoline, propane, and diesel) reduces CO_2 emissions by vehicles. Changed consumer behavior like ride-sharing and toll roads could reduce the vehicle emissions. However, if the means of such emission reduction is to substitute electricity for fossil fuels and the electricity is produced by fossil-fueled power plants, the net reduction may disappear and only change the location from urban areas where the vehicles are concentrated to rural areas where power plants are located. In that case we might only get a net reduction in CO_2 emissions if the electric power increase by switching transportation to electricity is produced by wind and solar or other renewable power sources like hydropower and geothermal. Increasing housing demand increases the harvesting of trees for lumber adding to deforestation and a reduction of CO_2 absorption from the atmosphere.

As a general matter the data show that CO_2 concentration in the atmosphere began to rise with the advent of the industrial revolution beginning about year 1880. One way or another, the industrial revolution, for all the good that it has brought to the world community, has complicated life by the emissions of CO_2 to the atmosphere. The future version of this revolution needs to reduce CO_2 emissions, or use markets and technology to off-set the negative effects that are driving global warming. Such solutions will likely be a combination of (1) substituting renewable for fossil fuel—based power generation, (2) the capture and sequestration of CO_2, (3) diverting CO_2 from emissions to useful produces, and (4) development of new technologies that off-set the effects of CO_2 concentrations in the atmosphere, in the seas, and on land. All of these approaches can be driven or aided by markets. Effective policy changes must account for and make use of markets, leading most economists to favor the straight forward taxation of CO_2 emissions or alternatively, taxation of the carbon content in fossil fuels.

Policy

The public policy surrounding oil markets is inexorably linked to national defense. This national defense focus started with World War I but mostly grew out of World

War II when oil from the United States became a major factor in winning the War. Hitler ran short of oil during the Russian invasion and the lack of fuel became and major factor in the rest of the war years. The fuel for the Allied forces was underwritten by the US oil industry. Thus, national defense became an integral part of US foreign policy that remains today. National defense is often a key part of US foreign policy in the Middle East, in dealing with Russia and North Africa, as well as other international relationships. The national defense concerns are almost always a part of the public debate in the US Congress when energy policy solutions are offered. Changes in energy policy have been a part of presidential campaigns ever since World War II.

US energy policy and the resulting market adjustments were reset following the removal of federal price ceilings on domestic crude oil in the mid-1980s. Since then, a fundamental characteristic of the market for crude oil is one of price volatility and future price uncertainty. The development of oil markets in the Middle East shifted the balance of market power beginning with the Embargo, and with such changes, new price volatility. The world market and the location of discovered reserves at the time had shifted the production concentration from the United States to the Middle East to such an extent that Organization of the Petroleum Exporting Countries became the market influence leader. The United States responded by changing policy during a politically tumultuous time. For the next 10 years the United States controlled the US oil price by federal price ceilings, differentiating between "old" oil, "new" oil and high-cost-of-production oil, like CO_2 EOR recovery.

The removal of price ceilings was finally achieved in the mid-1980s and prices were allowed to float with world market conditions. At this point, producers had to learn new ways of dealing with price uncertainty. This was achieved by the creation of futures markets for crude oil and major oil products. Futures markets reduced the impacts of price uncertainty through active participation by producers and consumers of oil products, and their essential counterbalance—the oil market speculators.[2] The futures market for crude oil and key oil products like fuel oil has since been traded daily at the New York Stock Exchange, as well as in the Tokyo Commodity Exchange (TOCOM) futures markets.

Oil prices in the United States have ranged from $10 per barrel for a short time in 1986 to $120 per barrel for a brief period in 2007, then back to $25 for a short time in 2010, recently rising to $70 (nominal dollars). So a major challenge for investors in new and uncertain technologies in this market is the decision of when to make long-term commitments—especially commitments to new technology. One solution to this challenge is public policies that directly influence markets (federal tax credits, state

---

[2] Futures markets only work if participants wanting to reduce price uncertainty are matched with roughly equal participation by speculators willing to take the risk of market price movements.

severance tax reductions, and R&D funding from federal and state governments). Such policies serve to spread the investment risk to multiple parties.

The other key public policy was free trade that opened up both markets for export and markets for low-cost production inputs and competitive consumer products from abroad. In 2015 the United States finally removed the prohibition on exporting crude oil that was put in place as part of the government response to the Oil Embargo in 1973. The development of free trade agreements during the last 40 years has created a more efficient world oil market. And, importantly, US public policy that invested in hydraulic fracturing technology ("fracking") and private sector investment by high risk-takers in the industry in this technology has turned the oil industry and other related markets upside down, recapturing the American leadership in world oil markets. The key oil industry development of the century was fracking technology. The development came primarily from Texas, and individually from George Mitchell of Houston who deserves much of the credit for developing the technology.

Another focus for policy that addresses CO_2 emissions is incentives to develop renewable resources for electric power production that replace coal, oil, and natural gas based generation. A companion policy group is so-called "conservation" consisting of behavior changes of consumers and development of improved energy efficiency technology. Conservation includes pricing that discourages electric consumption, especially at peak hours when fossil fuel generators typically come online to meet the demand during hours of the day when renewable generation by wind and solar units are not available.

The current policy debate for addressing global warming is more directly focused on slowing and stopping the CO_2 emissions to the atmosphere via government regulation and less on technology developments that capture and sequester CO_2, or technologies that off-set the CO_2 effects. But since new technology is a major strength of the US economy this approach promises to become the primary means of addressing the CO_2 concentration challenge in the long term. The two approaches are discussed below.

Direct regulation of CO_2 emissions from power plants

The primary focus of a direct regulatory approach to controlling CO_2 emissions is to limit emissions from electric and industrial sector power plants that use coal, oil, or natural gas as fuel. There are two approaches to this policy class. One is so-called cap-and-trade and the other a carbon tax on CO_2 emissions. Cap-and-trade sets an allowable quantitative limit on CO_2 emissions (set for all actors), but allows actors to trade the right to emit. A carbon tax simply taxes measured CO_2 emissions or the carbon content of fuels. It provides a powerful incentive to reduce emissions through a redirection of investment. The tax encourages producers to find a different process for production, and results in other market changes depending how the tax revenues are used. The tax may be used as a credit for investment in alternatives, or passed on to consumers. The main difference in the two approaches—carbon tax or cap-and-trade—is that the carbon tax sets the price of emissions and allows the market to determine the quantity reduction. Cap-and-trade sets the quantity of emission reductions and lets the market set the price (Frank, 2014).

A cap-and-trade form of market-based influences is currently being implemented in nine states in the United States in a collaborative effort known as the Regional Greenhouse Gas Initiative. The implementation is achieved through an auction that is operated for the electric power and industrial sectors for power generation. Active markets allow market participants to trade the right to emit.

Examples of a carbon tax policy approach include tax systems that tax the carbon content of fuels and are currently implemented in Denmark, Finland, Germany, Ireland, Italy, the Netherlands, Norway, Slovenia, Sweden, Switzerland, and the United Kingdom. These systems create a price for emitting CO_2. Carbon tax proposals discussed in the United States usually assume a carbon price of \$30—\$50 per ton and suggest a gradual increase schedule from zero today to such a dollar level by a time certain like 2030. A price on carbon is intended to represent the social costs of the last increment of CO_2 emitted. Such a price, in the natural resource economics professional literature is an addition to market equilibrium that would approximate a fully competitive market solution that would maximize the market net benefits to consumers in a market economy, including the avoidance of CO_2 influence on the consequences of global warming. While the idea is clear, determining a tax level that would result in efficient market solutions is no easy task. For one thing to be effective such a tax would need to implemented around the world. If not there would develop "free loaders" for receiving the benefits while avoiding the costs.

So far the public debate in the United States has favored a cap-and-trade approach over the carbon tax alternative. While there are many players with different objectives represented in the politics of climate change that like this approach, the primary rationale seems to be that the cap-and-trade approach provides more certainty over the quantity of CO_2 that is being taken out of the emissions stream. It is very clear, however, that the tax approach has much lower cost of implementation. In short, the "jury is still out" concerning the choice of tax or cap-and-trade.

Economists address the matter of CO_2-based climate change by examining approaches that force markets to incorporate the "social costs" of CO_2 effects. Economic studies have put this CO_2 social cost at about \$42 per ton of CO_2 in a future period like 2020. Methodologies for estimating such social costs have been developed for use in environmental regulatory policy implemented by the Environmental Protection Agency for quite some time. The social costs are those costs that are not included in competitive markets—so-called "externalities." The basic idea is to force social costs to be recognized in these markets, but use the efficiencies of regular competitive markets to achieve the desired outcomes.

Resources for the Future (RFF), a Washington D.C. based nonprofit organization that analyzes public policy of natural resources development and environmental effects on the US economy is a valuable resource for debates about climate change effects and public policies to address them. The RFF webpage contains the following perspective on the issue of climate change policy:

> *To inform action on climate change, experts at RFF believe that it's essential to identify the actual costs that carbon emissions impose on society, and the social cost of carbon is a way to do just that. The social cost of carbon is an estimate of the dollar value of climate damages caused by every additional ton of carbon dioxide released into the atmosphere— and it affects billions of dollars of regulatory and investment decisions each year. The US federal government, for example, has used the social cost of carbon as a key metric in over 150 proposed and final regulatory measures, including land use decisions and standards for vehicle fuel efficiency, power plant emissions, and appliances.[3]*

http://www.rff.org

Technology

Perhaps the most promising solution to the global warming challenge long term is new technology. The ideas currently in the pipeline of research, development, demonstration, and commercialization span the spectrum from means of capturing CO_2 from the emissions stream of fossil fuel power plants and making usable products with it, to off-setting the effects after CO_2 accumulates in the upper atmosphere and in the oceans or substituting renewable energy for fossil fuels.

Several innovative concepts for beneficial use of captured CO_2 are being tested in Texas. There are only two operational carbon capture and sequestration power plants in the world. One is the Petra Nova coal-fired power plant near Houston, a joint venture of NRG and JX Nippon Oil and Gas Exploration. The 240-megawatt (MW) carbon capture system was added to a 654 MW capacity Unit 8 of the existing W.A. Parish pulverized coal-fired generating plant. The system is designed to remove 90% of the CO_2 emissions from the flue gas slipstream. The captured CO_2 is used in a CO_2 EOR project nearby. The project has been much more economical than its Canadian counterpart and only cost $1 billion, or $4200/kW, aided by $167 million Department of Energy (DOE) grant. The Boundary Dam Integrated Carbon Capture and Storage (CCS) Demonstration near Estevan, Saskatchewan (the first CCS project in the world) cost $1.5 billion with $240 million federal government grants. This

---

[3] Recently, however, the Trump administration issued an executive order to withdraw the previously determined federal estimate of the social cost of carbon ($42 per ton in 2020) and disband the federal interagency working group charged with the development and maintenance of the estimate.

Canadian project has encountered a number of problems and is certainly not economic at today's energy prices (Energy Information Administration, 2017c).

DOE has recently funded seven projects aimed at finding ways of converting captured CO_2 emissions from utility and industrial sources and combining the gas with other chemicals to make useful products. One of the projects is in Texas. These projects were funded with $106 million from the American Recovery and Reinvestment Act of 2009 (ARRA)—matched with $156 million in private cost-share. The demonstration projects were intended to investigate the potential to use CO_2 as an inexpensive raw material to reduce carbon dioxide emissions while producing useful by-products. Such technologies convert captured CO_2 into products such as chemicals, carbonates, plastics, fuels, building materials, and other commodities. The idea is to convert CO_2 into other useful forms that reduce carbon emissions in situations where long-term storage of CO_2 is not practical. Such technologies would be helpful as large volumes of CO_2 become available from fossil fuel—based power plants and other CO_2-emitting industries that are equipped with CO_2 emissions-control technologies required to comply with regulatory requirements. The DOE ARRA funded projects include:

Ramgen Power Systems (Bellevue, Washington)—To support a industrial-sized scale-up and testing of an existing advanced CO_2 compression project with the objective of reducing the time to commercialization, technology risk, and cost (DOE Share: $20 million).

Novomer Inc. (Ithaca, New York)—Novomer and its partners developed a process for converting waste CO_2 into a number of polycarbonate products (plastics) for use in the packaging industry. The project technology enables CO_2 to react with petrochemical epoxies to create a family of thermoplastic polymers that are up to 50% CO_2 (by weight). The project potentially will convert CO_2 from an industrial waste stream turning it into a lasting material that can be used in the manufacture of bottles, films, laminates, coatings on food and beverage cans, and in other wood and metal surface applications. Novomer is located in Rochester, NY, Baton Rouge, Louisiana, and Orangeburg, SC (DOE Share: $18 million).

Touchstone Research Laboratory Ltd. (Triadelphia, West Virginia)—This West Viriginia project tested an open-pond algae production technology that promised to capture at least 60% of flue gas CO_2 from an industrial coal-fired source to produce biofuel and other high value coproducts. Lipids extracted from harvested algae are converted to a biofuel, and an anaerobic digestion process was developed and tested for converting residual biomass into methane. The host site for the pilot project is Cedar Lane Farms in Wooster, Ohio (DOE Share: $6.2 million).

Phycal, LLC (Highland Heights, Ohio)—Phycal completed development of an integrated system designed to produce liquid biocrude fuel from microalgae cultivated with captured CO_2. The algal biocrude can be blended with other fuels for power generation or processed into a variety of renewable drop-in replacement fuels such as jet fuel and biodiesel. Phycal designed, built, and operated a CO_2-to-algae-to-biofuels facility at a thirty acre site in Central Oahu (near Wahiawa and Kapolei), Hawaii. Hawaii Electric Company qualified the biocrude for boiler use, and Tesoro supplied CO_2 and evaluated fuel products (DOE Share: $24.2 million).

Skyonic Corporation (Austin, Texas)—Skyonic Corporation continued the development of SkyMine mineralization technology—a potential replacement for existing scrubber technology. The SkyMine process transforms CO_2 into solid carbonate and/or bicarbonate materials while also removing sulfur oxides, nitrogen dioxide, mercury, and other heavy metals from flue gas streams of industrial processes. Solid carbonates can be stored long-term, in safe aboveground storage without pipelines, subterranean injection, or concern about CO_2 rerelease to the atmosphere. The project team processed CO_2-laden flue gas from a Capital Aggregates, Ltd. cement manufacturing plant in San Antonio, Texas (DOE Share: $25 million).

Calera Corporation (Los Gatos, California)—Calera Corporation developed a process that directly mineralizes CO_2 in flue gas to carbonates that can be converted into useful construction materials. The project used an existing CO_2 absorption facility for the project at Moss Landing, California, for capture and mineralization. The project team completed the design, construction, and operation of a building material production system that at smaller scales has produced carbonate-containing aggregates suitable as construction fill or partial feedstock for use at cement production facilities. The project allows a building material production system to ultimately be integrated with the absorption facility (DOE Share: $20 million; Energy Information Administration, 2017c).

On September 7, 2010, DOE selected an additional 22 projects intended to accelerate CCS research and development from industrial sources. Recovery Act funding of $575 million was dedicated to these R&D projects to complement the industrial demonstration projects already being funded through the Recovery Act.

There are Innovative Concepts for Off-setting the CO_2 Effects in the Atmosphere. CO_2 is widely used for EOR and is piped in to EOR sites from natural geologic formations in other locations. The EOR projects are concentrated in the Permian Basin of West Texas and eastern New Mexico where CO_2 supplies from natural deposits in southern Colorado and northeastern New Mexico are delivered to the projects by pipeline.

A very different technology has been tested in Houston by a company named Net Power. The $150 million project is a power plant that burns fossil fuels with no greenhouse gas emissions, and produces electricity that doesn't cost more than the next best alternative. This technology, called the "Allam cycle" completely changes the way natural gas is used to generate electricity. The "combined-cycle" turbine in a current natural gas generator converts heat energy into mechanical energy which is then converted into electricity. These turbines either use steam or a mix of hot gases to transfer energy from one form to another. Oxygen-containing air enters the chamber where natural gas is burned. Pressure is created as the chamber heats up and the pressure turners a shaft which converts mechanical energy into electricity via a generator. Heat converts water into steam that also boosts the generator. The efficiency of the turbine would be significantly reduced if CO_2 from the combustion was captured at this point as is the usual approach of carbon capture projects. The Allam system instead uses supercritical carbon dioxide to achieve the heat transfer, a process that is much more efficient than that of current turbines.

The Allam turbine is much smaller than the current combined-cycle gas turbine, and it burns natural gas and pure oxygen producing only carbon dioxide and water in a chamber already filled with supercritical carbon dioxide. The combustion produces additional carbon dioxide, water, and lots of heat. The high-pressure mixture then passes through a gas turbine turning a shaft to produce electricity. The cooled mixture is then separated into two parts—some carbon dioxide which is then compressed to become supercritical again where it then enters the initial chamber—and the rest, which is a pure stream of CO_2 is sent to underground storage, or piped to another user. The process also produces water which is discarded.

The Allam system has the potential to revolutionize the electric power generation process that emits zero CO_2 to the atmosphere. The economics is very promising because the overall energy efficiency compares favorably with current combined-cycle gas turbines, and because the fracking technology has impacted the natural gas business so profoundly that the market price is likely going to stay near current levels around $3.00 per mcf making it the fuel of choice in the power industry to be a companion to renewable technologies. The result promises to create an electric market dominated by natural gas, nuclear, wind, and solar, all with near zero CO_2 emissions to the atmosphere (Marshall, 2018, Available at: https://www.eenews.net/stories/1060071081).

NET Power has been cited as an example of industry making innovations on its own, considering it did not receive any US government funding during the pilot project development and operation. However, government support could be pivotal in creating a larger commercial plant. This La Porte, Texas project is also backed by about $140 million of industry money, including $90 million from Exelon Corp. and $50 million from an engineering firm, Chicago Bridge & Iron Company. DOE officials did, however, provide assistance by helping create attention to the project which helped raise investment capital (Marshall, 2018).

While a technology like Net Power would substantially reduce CO_2 emissions into the atmosphere there are other technology approaches to off-set the CO_2 effects in the atmosphere. CO_2 emitted into the atmosphere is for all practical purposes irreversible. Even if we halt carbon dioxide emissions entirely, the elevated concentrations of the gas in the atmosphere will persist for decades. If we find out later, say by 2030 that climate change has become intolerable, cutting emissions alone will not solve the problem. Therefore, "solar shade" ideas are attractive.

For example, there is the prospect of putting various substances into the upper stratosphere that counter the CO_2 effects by scattering and reflecting sunlight back into space and thus reducing global warming. This is a so-called solar shade. The current example that has been most studied requires putting stratospheric sulfate aerosols into the atmosphere from selected locations around the globe. The technology idea is called solar radiation management. Such technology ideas are known as geoengineering. Solar geoengineering is simple and well documented, though often overlooked as means of counteracting the warming caused by atmospheric carbon dioxide buildup.

There are many versions of these basic geoengineering approaches. The science studies of the last two decades propose several possible approaches to off-setting the currently predicted levels of accumulated CO_2. It is recognized that the CO_2 effects could be canceled by preventing about 1% of incoming solar radiation. Sulfate particle emissions into the atmosphere already occur naturally from volcano eruptions. It has been estimated that the Mount Pinatubo eruption in 1991 put about 10 million tons of sulfur into the stratosphere that lowered the world by 0.5°C for a couple of years. So, the prospects that artificial injections could provide much the same result is promising (Teller et al., 1997).

As one might expect there are a number of disadvantages and much uncertainty about these geoengineering approaches. There are also matters of the costs of the various approaches. Teller et al. estimated that the geoengineering approaches might cost about $1 billion per year, a fraction of the costs of the direct control of fossil fuel use. Critics point out that cooling the planet would not resolve the problem of ecosystem damage in the oceans since most geoengineering ideas do not reduce the accumulated level of CO_2 in the stratosphere that will continue to impact global warming because the oceans have, and will continue to absorb CO_2 and warm the oceans and destroy coral reefs and dependent sea life.

Hope around short-term impacts

Much of the current optimism about the need to adopt severe policy restrictions on CO_2 emissions is the indirect result of fracking and its impact—new natural gas reserve estimates not seen in decades and increased production rates resulting in low natural gas prices. The other factor is that electric power consumption/generation growth has flattened so that the rise in natural gas generation and decline in coal-based generation multiplies the decline in CO_2 emissions to levels not seen since 1993 (see Fig. 12.6).

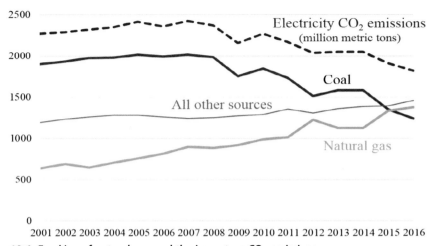

Figure 12.6 *Fracking of natural gas and the impact on CO_2 emissions.*
Raimi, D., 2017. The Shale Revolution: Climate Opportunities and Risks. Available at: http://rff.org/ research/publications/shale-revolution-climate-opportunities-and-risks.

Concerns about long-term impacts

There is concern that the recent CO_2 reductions will be temporary, however, and the reason comes on several levels. First, the low–cost of natural gas also squeezes out wind and solar as well as nuclear. Lower natural gas and electric prices also encourage higher consumption and therefore more CO_2 emissions. Higher natural gas use also results in the release of methane gas into the atmosphere which contributes to global warming. A longer term technology option that promises to reduce CO_2 emissions not explored in this book that may turn out to make a major contribution is small unit, new nuclear technology for power generation called the NuScale reactor. The reactor is about one-tenth the size of traditional nuclear reactors, and importantly is a new technology approach that has a monitor to control the lift of fuel rods. If electric power is lost gravity drops the rods back into the core thus stopping the reactor. Cost and widespread caution will have to improve before this option could materialize on a large scale.

CHAPTER 13

The road ahead: ideas are key

Contents

From then until now

The central theme of this book can be summarized thusly: "If innovation is the engine then ideas are the fuel that makes it go. And if we in Texas and elsewhere in the United States fail to keep this idea generation dynamic engaged and producing new innovation then we will fail to maintain economic growth. The energy sector in Texas is a key part of the equation." To repeat the declaration attributed to Francis Bacon, "He that will not apply new remedies must expect new evils; for time is the greatest innovator."

The term "innovation" seems to dominate the daily news, or at least not a day goes by without many references to it, but it is rarely defined. *The US Advisory Committee on Measuring Innovation in the 21st Century Economy* published in 2008 gave it the old "college try"—defined a working definition of innovation this way:

> *The design, investigation, development and/or implementation of new or altered products, services, processes, systems, organizational structures, or business models for the purpose of creating new value for customers and financial returns for the firm.*

Thought leaders of the past have tried to understand and articulate the innovation process, figure out how to measure and promote it, and then to analyze just how it contributes to economic and cultural well-being. Philosophers have produced interesting thoughts about the good and bad of innovation, sometimes challenging the notion that the human condition is better for having innovation. Economists have tried to use econometric methods (the combination of economic theory, mathematics, and statistics) to explain and model innovation's contribution to economic growth, and to forecast economic growth as influenced by policies that drive it. But it seems clear that technological innovation has become the driver of modern economies in virtually all facets of life. It is in our state and national psyche. The key Texas energy events reviewed in this book clearly shows that innovation has had a major role in past energy sector contributions to our state and national wellbeing. The future economy, national and international politics, and more generally, our way of life is surely to be driven by new ideas, innovation, technologies, and new ways of behaving that we can hardly imagine.

Innovation Dynamics and Policy in the Energy Sector
DOI: https://doi.org/10.1016/B978-0-12-823813-4.00013-0

In one way or another much of our educational systems, government R&D programs, public policy debates and popular media attention today is about technology and innovation much of which is directly or indirectly about energy innovation and technology. In the early 19th and 20th centuries new technology was primarily a machine like the steam engine, the gasoline engine or the cotton gin. In the later part of the 20th century and in all of the 21st century thus far, technology includes machines but also thousands of digital products of the computer and internet era. The new machines in factories, automobiles, trucks, and trains that provide transportation, the electric grid, and farm tractors and harvesters, for example, are integrated with and controlled by computer chips containing operating systems and algorithms for direction by people, and a new player—robots and self-driving cars, trucks, and tractors that perform human-like tasks. In short, technology in the modern world is associated with products of hardware and/or software.

While it is easy to see the contributions of energy sector technologies to our economy and way of life in the past, the future is not so obvious. Past examples abound. There is the substitution of automobiles for horses that eliminated four million pounds of manure from the streets of New York City in the late 1800s; air conditioning of buildings in the 1900s became the technology that made warm climates livable and redistributed population from the northern to the southern states of the United States; and the internet in the last half of the 20th century that has made global communication a daily activity and mobilized global energy markets. Promises for the future include small solar units providing electricity to remote areas of the globe, widespread use of high performance electric vehicles, and a greatly improved battery technology often referred to as the "holy grail" of the future electric grid, and a necessary companion of renewable, clean energy. Artificial Intelligence (AI) promises to dominate much of the evolving digital world. The agricultural industry that both consumes energy and produces other forms of energy is destined to change. The agricultural landscape will likely change dramatically. Change will follow from innovation in genetic engineering, cloning and the artificial leaf. There will also be innovation in the downstream sectors of petroleum refining, chemicals and petrochemicals not addressed in this book.[1]

---

[1] As mentioned in Chapter 1, The Dynamics of Innovation and Technology in a Market Economy, the innovation experience summarized in selected events in this book ignores the petroleum refining and petrochemical industries which have developed in Texas in conjunction with oil and gas production and modern electric generation. These value-added industries (refining, chemicals and petrochemicals) that evolved from the base of oil and gas production have experienced their own set of innovations since Spindletop. These industries transform raw oil and gas products into thousands of final products that are ubiquitous in the modern world. Most of this complex did not exist in 1901 but today Texas is home of 30% of the US petroleum refining sector and 40% of the base petrochemical industry capacity. All of this industry-complex exists in Texas because of the innovations summarized in the prior chapters of this book, complemented by other focus areas not reviewed in the book. The combined economic value added of petroleum refining, base chemicals and petrochemical industries today rival that value added by oil and gas production and electricity sectors amounting to about one third of the combined value added of $422 billion in 2015. Another book would be needed to adequately cover these industries and their innovations.

In contrast with the positive influences of past technologies the current challenge of public policy and/or private sector initiatives is that of speeding and enhancing the innovation process for the future, adequate to the challenges we now face. There is evidence that the past innovation process accelerated the rate of technological change at an increasing rate until about the turn of the century. But an apparent slow-down of innovation coupled with a set of unintended consequences now mean maintaining net positive outcomes will be a challenge in the future. The new challenges come from both the success and failures of our chosen technology path. That is, while advances in technology have greatly improved the human condition, technology often has a dark side. Clean nuclear power plants produce waste products that challenge our ability to isolate deadly toxic waste from the environment. Computer systems can be hacked by intruders, via the internet, who wreak havoc on individuals, companies and nations threatening individual lives, national security and global economic stability. Burning of fossil fuels emits CO_2 into the atmosphere exacerbating global warming. Further, the internet and related digital technologies have awakened the dynamic expectations of underdeveloped countries, though greatly a positive influence long term, create current challenges to stability. Their citizens are more aware of the advantages of the developed world giving rise to unrest and political uprising in their own country, thus creating new challenges to world order.

Texas is the clear national leader of practically all things energy. Our experience has made a major contribution in the past, and because of that experience and the industry and institutional structures that have developed here, the future is both our opportunity, and arguably our responsibility, to lead. Texas is well positioned to lead the national effort.

A central theme of this book is that innovation is driven fundamentally by ideas. Ideas come from many sources—individual intuition, think-tanks, university scholarly activity and R&D efforts, a response to major, unexpected challenges, group thinking from collaborative organizations, war, and pure accidents. Texas has a rich history and diversity—its population, geography, natural resources, climate, and politics—and diversity is an essential element of innovation. But to follow or improve an idea there is often a requirement to integrate across experiences, technologies, industries and cultures. Such integration is perhaps more essential now than in the past. And such integration is occurring here in Texas and elsewhere.

The Texas experience with innovation in the energy sector over the last 117 years has grown increasingly complex. Such complexity is the common experience everywhere across all innovations in virtually all industries and modern economies today, but not in 1901. Spindletop resulted from the primary effort of a single individual, Patillo Higgins, with a simple observation of gas seeping from the ground that was not a scientific matter to say the least. At best he had a hunch and a vision matched with great tenacity, supported with some knowledge base of how to find oil from the previous decades' experience by other pioneers, and the support of a knowledgeable business partner. Captain Anthony Lucas with experience in a related industry (salt mines) joined Higgins and shared the vision. Without a doubt the venture also contained a modicum of good luck. There was also a supporting technological advance that came after failure seemed almost certain—"going down for the count" Higgins lost the first two drilling efforts by well bore cave-ins that trapped the drill stem. Higgins managed to try a third time as he spent every last dime of his available capital, plus all the cash he could raise from friends, family and prospectors. For the third effort the team created a simple form of drilling mud matched with well casing to hold back the sand and gravel down hole that allowed them to complete the well. Modern drilling mud and well casing are essential components of every oil well drilled today. But compared to today the whole process was very simple. The Spindletop effort did not require a drilling permit or environmental impact statement and venture capital to finance the deal was a matter of Higgins' personal salesmanship. The business model was about as simple as one could imagine—bet the farm and hope for the best, making up new approaches on the fly. The technological innovation success also spawned a new business model innovation of independent investors and operators, wildcatters, royalty owners, rough-necks, roustabouts, land-men, and drillers that still exist today.

East Texas in 1930 brought a new challenge for institutional innovation. The discovered oil deposit near Tyler was so prolific that it brought an army of drillers and all of the support structure that turned an oil field into a near battleground of chaos reminiscent of the California gold rush. But the decade that followed developed the rules of the game that guided the industry for the next 40 years. The division of jurisdiction between the federal and state governments was worked out, a permitting process for development and drilling was put in place by the Texas Railroad Commission (TRC) and an additional layer of complexity was put in place that created the modern regulatory system that remains today.

The Permian experience has been unique in a different way. It began in the 1920s with much the same dynamic as Spindletop and East Texas, but the Permian has never been a "flash in the pan." Instead the efforts discovered an enormous resource base, that when matched with ever changing technology just "keeps on giving." When the region began to decline in the 1960s, water flooding innovation extended the life of the fields, then CO_2 flooding did the same thing in the 1980s and 1990s—the new technology for EOR extended the productive life of the fields. Then, once again, a new technology, hydraulic fracturing, has extended the productive life of the resource to levels that 20 years ago, was unimaginable. In this case the complexity of the innovation process reached a new level. The policies of the 1970s—responses to the Embargo of 1973—included encouragement for development of high-cost reserves that (unwittingly) gave EOR a needed boost as new technology was added to the mix. The industry innovators, armed with university R&D support like that of the

Bureau of Economic Geology at UT Austin, perfected CO_2 recovery. While oil and gas prices sometimes gave the technology development encouragement, prices fell dramatically in the 1980s and 1990s, casting doubt on the economics of EOR. But in the meantime the practice of horizontal drilling that lowered the effective cost of wells by expanding the reach to productive geologic zones, built new economic returns. But the most successful innovation came later in the early 2000s—a technology driven by the ideas of George Mitchell and others beginning two decades earlier, supported by federal and state funding and tax incentives, and built on knowledge from other regions, created the "fracking revolution"—a truly unique and powerful innovation that came heavily out of the Texas culture and ecosystem.

The current oil and gas fracking experience is a world away from the Spindletop experience of 1901. Today's market participant operates under long-developed rules of the TRC for approval of, and sighting of a new well, develops contractual arrangements for pipeline extension and access to either intrastate or interstate pipelines to move product to refineries or, the case of natural gas, to a processing plant that separates several products from the raw gas stream for use in the refining and chemical industries, and then sends the remaining "dry" natural gas to a distribution company with access to final consumers. The drilling of new wells is supported by a complex organization of specialized laborers and engineers who operate digitally controlled drilling rigs that directs the drill bit from a vertical to horizontal orientation at specified depths. And, importantly the drilling is informed by 3D seismic technology that locates the right strata and reservoir target. The target is rock, oil bearing formations that release oil when the formation is cracked with a mix of high-pressure water, chemicals and sand. And, of course, the entire operation takes place under a legal framework of leasing to obtain access, and the sharing of risks and payments among land-owners, royalty owners, drillers, and operators. The financing of the entire operation makes use of modern venture capital and banking markets that decide on the mix of equity and loaned funds to finance deals. And, importantly, participants have access to futures markets to hedge their positions and reduce risks of price movements for oil and gas. Speculators join the game by betting on uncertain outcomes; they take the risks that product owners want to avoid—the required other half of futures markets.

The current efforts to create a "smart grid" in the electric sector of Texas is an innovation challenge at the other end of the spectrum from that of Spindletop and East Texas. "Smart" means real-time controls at several levels of the grid, a change from the old one-way flow of electrons from generators to transmission and distribution lines to the consumer to a two-way flow, enabled by and supported by a communication system that provides instant monitoring in near real-time. The new system accommodates many players in the market place. The market structure operates with a competitive generation sector, a regulated "wires" sector and a competitive retail sector. It is also a system surrounded by regulatory requirements and a system-wide grid manager that operates the grid, but does not own any of it, organizes and manages a system that mimics a competitive market and operates a system of advisory/operations committees representing all market participants through volunteer representatives. And the grid manager's operations are monitored by the Public Utility Commission (PUC) directed by commissioners appointed by the Governor with Senate confirmation. Then, of course, the entire arrangement is responsible for reporting to the Texas Legislature which writes the laws and provides oversight. The other special condition is that the electric grid in Texas is mostly operated independently from the federal system of control over interstate commerce. Texas is for the most part a separate market, only connected to other states for emergency purposes via direct current ties in three places at the state border. The regulated wires companies and the deregulated, competitive generation and retail segments participate in capital markets for ownership and for financing investments and operations. Participating companies have to show financial capability that meets minimum standards set by the PUC so that consumers of electricity do not have significant risk of losing power because of a provider suddenly goes out of business. The result of a 10-year-long effort has produced an electric system innovation made up of dozens of technological innovations, and an institutional and business model innovation in the form of a grid manager. The resulting Texas electric market is the envy of other regions in the United States and abroad.

The Texas smart electric grid accommodates many new technologies to monitor and control the electric systems, includes the operation of an internet-based competitive market for bidding by generators, dispatches generators per direction of the grid operator and supports advertising to final consumers by competitive retail participants. Consumers have online access to near real-time information—support for consumers who have to shop for their retail electric provider. The system depends on many new technologies allowing private sector small generators owned and operated by individual consumers to put power into the grid for compensation—either credit against their bills or direct payments—for power provided from privately owned small wind and solar generators on the distribution end of the grid. Then there is a host of new technology-based emergency systems to automatically disconnect and reconnect parts of the grid brought down by storms or other natural or human-caused disasters. And, finally, there is a developing system to protect the grid from cyber-attacks that protects the grid from wholesale outages at the grid operator level, and protection at the individual participant level.

The events reviewed in the prior chapters present a smorgasbord of factors that are important to understanding how it is that innovation happens, and importantly what conditions and public policies encourage it. It seems critical at this stage in world economic growth that we gain better insight into the process of innovation, a significant part of which needs to drive the future of the energy sector. The United States has been a driving force in world innovation during the last 2 centuries, especially since

World War II, and Texas has been in the forefront of it. Our economic and cultural dominance depends on us staying in front of the innovation pack.

There is a continual integration process underway that brings private sector non-profits and government agencies together with the banking industry and stock market that is an entire study of its own. Texas has such an integration network. The most notable institutions past and present that encourage integration are summarized below. The most detail of such entities summarized below is for the Houston innovation entities, but there are also many other institutions operating to support and promote innovation in Austin, San Antonio, Dallas-Ft. Worth and Lubbock. The summary below is only exemplary of the total effort to support innovation across the state, active in the recent past and some that are actively present today.

Rice Alliance: The Rice Alliance for Technology and Entrepreneurship (Rice Alliance) is Rice University's nationally recognized initiative devoted to the support of technology commercialization, entrepreneurship education, and the launch of technology companies (Rice Alliance, Houston, Texas, online at www.alliance.rice.edu, Rice Alliance special program called "The Energy and Clean Technology Venture Forum." September 12–13, 2018 in Houston).

Houston Angel Network (HAN): The HAN is the oldest angel network in Texas and the most active angel network in the United States. Its members have invested more than $73 M in more than 235 deals since its inception in 2001. In 2015, HAN members invested $12 M in 43 deals. The typical individual HAN member is an accredited investor seriously interested in providing capital and coaching to early-stage companies. HAN also has institutional members such as seed funds, accelerators, universities and other networks within the innovation ecosystem. HAN is a nonprofit association that does not charge fees to entrepreneurs; its revenue consists of membership fees and sponsorships (Houston Angel Network at http://houstonangelnetwork.org/han/).

Technology Collaboration Center of the Greater Houston Partnership (TCC): The initial focus on planning for the collaboration center was on building more partnerships within the Space, Medicine and Energy communities in the Houston area, the mission of the TCC was expanded to provide support for any technology and industry, involving organizations from any location. Houston is the Permanent Secretariat of the World Energy Cities Partnership, a collaboration among 19 energy cities worldwide providing a platform for information exchange, networking and public relations (Technology Collaboration Center of the Greater Houston Partnership at https://techcollaboration.center/about/).

The Cannon: a 25 acre campus to serve as the hub for Houston's entrepreneurial community. The Canon states that not only will our campus bring together a diverse business community, but it will provide opportunities for companies to raise capital, participate in an incubator program, and attract top talent (The Cannon, Houston's Campus for Entrepreneurs at https://thecannonhouston.com/).

Houston Exponential: Houston Exponential was created through a combination of Mayor Sylvester Turner's Innovation and Technology Task Force, the Houston Technology Center and the Greater Houston Partnership's Innovation Roundtable, to bolster Houston's innovation ecosystem and drive the region to become a top 10 startup ecosystem by 2022.

Houston Exponential's mission is to accelerate the development of Houston's innovation economy by fostering a robust ecosystem that supports high-growth, high-impact startups. The formation of Houston Exponential comes after a year-long study linking the Roundtable's and Task Force's efforts by the professional services company, Accenture.

Houston Exponential will work toward its mission by convening the ecosystem, helping to build an innovation district in Houston and attracting talent to the region, among other key endeavors. It will also promote Houston's image, both locally and nationally, as a vibrant, innovative economy where startups thrive.

Houston Exponential will also launch and support a Fund of Funds with the primary goal of achieving a compelling return for investors while also attracting leading venture capital firms to bring their expertise and risk capital to the region. The Fund, known as the HX Venture Fund, will also create a key pathway for innovation and information to flow between corporations, startups and Innovators (marketing@houstonexponential.org).

UT Incubator: The Austin Technology Incubator (ATI) is the startup incubator of the University of Texas at Austin. A program of the University's IC^2 Institute, ATI has a 28-year track record of helping founding teams achieve success. TI focuses on helping startups compete successfully in the capital markets. ATI points out that they don't write checks. But they have strong, long term, trust-based relationships with investors—the local angel investors community, local and national venture capital firms, and sources of public funding (ATI Technology Incubator at the IC2 Institute at https://ati.utexas.edu/about-us/).

The UT Bureau of Economic Geology: Established in 1909, the Bureau of Economic Geology in the Jackson School of Geosciences is the oldest and second-largest organized research unit at The University of Texas at Austin. In addition to functioning as the State Geological Survey of Texas, the Bureau conducts research focusing on the intersection of energy, the environment, and the economy, where significant advances are being made tackling tough problems globally. The Bureau partners with federal, state, and local agencies, academic institutions, industry, nonprofit organizations, and foundations to conduct high-quality research and disseminate the results to the scientific and engineering communities as well as to the broad public.

The A&M Innovation Center: The AMIC is a nonprofit 501c(3) corporation formed for the advancement of science, education, entrepreneurship and innovation. The AMIC works in collaboration with Texas A&M University System components headquartered in Brazos County, Texas, the Bryan-College Station community, and private industry ... The AMIC was originally formed in 2007 as a collaborative program between the Texas A&M System's Office of Technology Commercialization, Texas A&M University, the Texas A&M University Health Science Center and the Research Valley Partnership. In 2014 the collaboration partners formed a standalone organization as a 501c(3) entity with initial support from the Texas Emerging Technology Fund, Texas A&M University, Texas A&M Engineering Experiment Station (TEES), the Texas State Energy Conservation Office, the United States Department of Energy and the private sector (Texas A&M University, The AM Innovation Center (AMIC) http://aminnovationcenter.com/about/).

Dallas Innovation Alliance (DIA): The DIA is a coalition of stakeholders from the City of Dallas, corporations, civic and NGO organizations, academia and private individuals who are invested in Dallas' continued evolution as a forward-thinking, innovative, "smart" global city. DIA's working definition of a Smart City is a city where social and technological infrastructures and solutions facilitate and accelerate sustainable economic growth, resource efficiency, and importantly, improves the quality of life in the city for its citizens. Operating from a foundational vision that smart cities are about people and not just technology, DIA is focused on the end user— building a critical mass of the most highly engaged citizens in the country (http://www.dallasinnovationalliance.com/what-we-do/).

SMU Center for Laser-Aided Manufacturing: The Center for Laser-Aided Manufacturing at SMU was created to develop a fundamental understanding of laser-aided intelligent manufacturing. In August 2005, the National Science Foundation (NSF) awarded SMU a grant to establish the Industry/University Cooperative Research Center (I/UCRC) for Lasers and Plasmas for Advanced Manufacturing. SMU is the fourth university site in the multiinstitutional I/UCRC for Lasers and Plasmas for Advanced Manufacturing together with the University of Michigan, the University of Virginia, and the University of Illinois at Urbana-Champaign. In August 2010, NSF granted SMU an extension of the I/UCRC for the next 5 years (https://www.smu.edu/Lyle/Centers/CLAM).

TTU National Wind Institute: Wind Energy is the premiere multidisciplinary program at Texas Tech University developing transformational experts who apply knowledge, skills, and conviction to lead in the advancement of sustainable renewable power solutions with positive regional, national, and global impact (https://www.depts.ttu.edu/nwi/education/BSWE/index.php).

Innovation Hub at Research Park is Texas Tech's center for entrepreneurialism and innovation. The Hub connects entrepreneurs, with Texas Tech and Texas Tech University Health Sciences Center in Lubbock faculty and students to enable collaboration in launching new ventures that develop our intellectual property and to foster public-private partnerships between Texas Tech and industry that builds our technology base and the economic strength of our region as a public good. The Hub is home to a number of programs and facilities to assist entrepreneurs bring their ideas to market.

Texas Emerging Technology Fund (ETF): The Texas Legislature created the State funded program of $200 million (expanded later to $500 million) to promote technology development and commercialization in 2005. It was the state's startup investment fund for 10 years. It was credited with raising close to $1.0 billion in follow-on investment. It focused heavily on medical technologies but also included energy technologies. The ETF was once recognized nationally as the most active early-stage technology venture fund in the country. Under Governor Abbott's leadership this startup investment fund was eliminated and remaining appropriations directed to a new governor-controlled researcher recruitment effort. The transition created the Governor's University Research Initiative, which has $40 million earmarked in the state budget. Abbott has said he plans to use the fund to attract major researchers, including Nobel laureates, to the state's public universities (Texas HB 1765 of the 79th Legislature).

GEAC/TEAC/TENRAC: The Governor's Energy Advisory Council (a commission) was created in 1973 in response to the Arab Oil Embargo. A state agency by the same name followed with a 2-year life with a policy planning and conservation mission, followed by the Texas Energy Advisory Council with a 2-year life with a policy planning and R&D mission, followed by the Texas Energy and Natural Resources Advisory Council with a 4-year life with a policy planning and R&D mission. These entities no longer exist but helped lead the innovation process in energy for Texas.

Center for the Commercialization of Electric Technologies (CCET): A nonprofit collaborative organization was created by Texas electric and technology companies in 2005 for the purpose of jointly funding and commercializing new electric technologies. After managing $30 million of jointly funded demonstration and commercialization projects over a 10-year period this entity was dissolved in 2015.

State Energy Conservation Office (SECO): The first incarnation of this entity was created in the Governor's Office in 1976 as disagreements in GEAC developed over the taking of Federal conservation program funding following the beginning of GEAC as a state agency. SECO has been located at three agency locations since 1976 and now resides in the Texas Comptroller's Office and continues to be funded by Federal conservation program funding.

Texas Public Policy Foundation: The Texas Public Policy Foundation is a conservative think tank based in Austin, Texas. The organization was founded in 1989 by James R. Leininger, who sought intellectual support for his education reform ideas, including public school vouchers.

Center for Public Policy Priorities: CPPP is a nonpartisan, nonprofit policy institute committed to improving public policies to better the economic and social conditions of low- and moderate-income Texans. The center pursues this mission through independent research, policy analysis and development, public education, advocacy, coalition-building and technical assistance.

What comes next

The future of innovation in the energy sector has great promise but, as usual there are many challenges. A few of them are discussed here, and then there is a recommendation for an institutional innovation to help drive future innovation at the end of the chapter.

One innovation that is in the early stage of development and implementation is the internet of things (IoT). The rapid spread of connected devices in concentrated geographical areas like large cities raises the prospect of connecting much of daily life that thrives on instant communication and remote control of devices in the home, the office, in commercial and industrial settings, in automobiles, and as a framework for so-called "smart grids" for the electric systems. IoT promises to create an "always-on" environment. The cell phones, iPads, PCs, and digital watches provide access. The connection of thousands of devices provides the prospect of improving the efficiencies of electric systems through automation of routine functions, for example, as well as providing valuable consumer services. Examples of IoT applications include precision agriculture, remote patient monitoring in the medical industry, and driverless automobiles. While many benefits are envisioned from such a digital world there are significant risks, especially the cyber security risk that spans the spectrum from simple nuisance to interruption of electric service, destruction of computer hard drives and national security risks. These devices and interconnecting systems all run on and/or control energy.

A *second innovation* that promises to disrupt the status quo is AI. This emerging technology refers generally to a computer system and supporting data sets that enable the automation of human-like behavior. The English Oxford Dictionary defines AI as: "The theory and development of computer systems able to perform tasks normally requiring human intelligence, such as visual perception, speech recognition, decision-making, and translation between languages." The idea usually includes the notion of machine learning. Machines can certainly change their process going forward based on outcomes of prior execution of tasks. Many of these functions are operational to some extent today. So-called "cloud" computing and "big data" systems championed by companies like IBM are busy working on such systems that promise to enhance human capabilities, not replace humans in the process, although it certainly will change job definitions and training related thereto. As supporters of these technologies point out, machines will never replace the human brain that allows complex problem solving that makes use of a combination of all of the seven senses and instant access to memory. But AI does make use of disciplinary fields including computer science, mathematics, psychology, linguistics, and philosophy, and applying tools of search and mathematical optimization and methods based on statistics, probability, and economics. The ideas are that AI can improve human conditions but humans become a manager of the systems that can easily add speed and accuracy to problem solving and implementation. Energy production, processing, and delivery systems are obvious sectors for AI development and application, especially in the electric systems where the flow, monitoring and control challenges are of electrons that move around systems with lightning speed. AI needs to routinely employ what is known as blockchain. Blockchain is a methodology that connects potentially endless sets of data but does it securely by a form of encryption. This technology allows protection against, for example, accidental, or malicious changing of data used in algorithms used for computation and machine decision-making.

And then there is AIQ (the combination of portmanteau of AI and human IQ). While it is not exactly clear how this new idea that marries equal partners—AI and human IQ—will be applied to energy challenges, it seems certain that it will. The dual problem of economic markets and environmental degradation comes to mind. Also, the current enormous data sets generated by synchrophasor systems monitoring grid systems 30 times per second (Texas applications covered in the CCET work reviewed in Chapter 8: Upheaval in the Energy Markets: The Arab Oil Embargo and the Iranian Crisis) may use AI to analyze millions of data points and determine that isolation of a part of the grid needs to happen instantly to prevent the entire grid from going down. But matching such a result with human review a few minutes later might make for an improved decision that immediately brings the isolated grid segment back before the interruption makes much of hit on users.

A *third innovation* example that promises great benefits is the world of robots. Robots are already in widespread use in automobile manufacturing, in retrieval of products stored in warehouses and numerous other activities that outperform humans. Robots are currently used in electric utility operations to search for and repair equipment. Robots are being used by the military to search for explosive devices that save lives. The number of uses for robots is just beginning and will no doubt become an increasing resource in the foreseeable future. Many parts of the energy systems will include robots that make the systems more reliable and efficient, but also drive changes in markets and business models.

A *fourth innovation* that is in the pipeline is automated demand response in electric grid operations. Innovation that has been underway for at least two decades in the electric utility industry has been to implement changes that integrate demand responses into the management of the electric grid to make it more efficient. The grid of yesterday is a one-way system that delivers electricity from generator to the user without any participation by consumers. That system of yesterday is being transformed into a two-way system where generation can occur anywhere on the grid, and where consumers participate in demand response. The most important part of this transition is that unaltered electric demand on the hottest day in August requires the highest cost generators to be brought online, and the entire system capacity has to accommodate this instantaneous demand. Demand response systems counter this inefficiency by allowing consumers to recognize and respond to the high cost by changing their demand and by participating in the generation of power for the grid. Automated demand response promises to simplify the whole process and improve the customer experience and at the same time create a more efficient grid.

One of the *more interesting technological innovations* of the promising technology world is the artificial leaf. While a bit removed from the energy sector center, it is an energy technology innovation. If such a technology is developed and commercialized the result would also create major changes in the business model and greatly disrupt the agricultural markets. The energy of growing plants for animal and human consumption is in japery from the idea of an artificial leaf. The idea is to use carbon dioxide, water and sunlight to produce liquid fuels more efficiently than photosynthesis, the process that plants use to create carbohydrates and store energy. The technology would allow directly storing solar energy and producing a carbon-neutral fuel for transportation systems. Scientists are currently working on two necessary processes— one to develop a catalyst that uses solar energy to split water into oxygen and hydrogen, and then creating others to convert hydrogen and carbon dioxide into an energy-dense fuel. The jury is still out on whether these ideas will work—turning sunlight into usable fuels that do not put CO_2 into the atmosphere, but if they succeed they would have a major impact on the global warming challenge.

The infrastructure that will continue to drive the innovation dynamics surrounding the above innovations (and numerous others) is very important. Texas has a network of such institutions that while requiring continual update is currently very robust. The process of innovation is much more integrated today than when Spindletop happened in 1901. The key is networks of integration. What would happen if the list of innovation hubs described above (and others not listed) automatically integrated their processes and memberships? Such integration could leverage the strength of diverse groups in just the same ways the diverse Texas culture and populations have generated ideas and innovation in the past and could openly share their efforts creating a new dynamic—call it the "innovation of innovators."

The future dynamics of policy, technology, and markets

The Texas experience with energy sector innovation has taken the state through stages of innovation development, from the market-driven events led by strong individuals that began with the oil sector in locations like Spindletop near Beaumont in 1901, East Texas near Tyler in 1930 and the Permian Basin at Big Lake in 1923, to a much more complex market, policy and technology combination a 100 years later. The last two decades have produced statewide electric sector deregulation during 1995−2005, wind farm development in West Texas (beginning seriously in 1995−2000) that now leads the national wind power industry, and the phenomenal hydraulic fracturing innovation in the Barnett shale north of Ft. Worth, Texas in 2005. If market disruption is a major source of the next round of (hopefully) beneficial innovation of markets, institutions and technology, as many innovation experts believe, then the fracking revolution among all of the innovation experiences summarized in this book, leads the way to a better future. But this does not negate the ongoing dynamic process that produced Spindletop, East Texas, and the Permian Basin early developments. But the struggle to deal with the aftermath of the Arab Embargo of 1973 and the following Iranian Crisis of 1979, was the disruption that opened the way to serious secondary and tertiary oil and gas recovery, the end to natural gas well head price controls, wind resource development, electric industry restructuring, and fracking of oil and gas reservoirs. These are the big events—there are others.

Prior to the Oil Embargo the experiences in Texas provided early insight to the dynamics of innovation flowing out of rugged individualism (market responses), institutional innovation, new technology, and policy that in many ways came about from the disagreements between federal and Texas government domains like that of the "hot oil" experience of East Texas, and the TRC control of well spacing and ratable production.

Spindletop and East Texas put Texas on the world map as the dominant oil producer from events created by the raw, single-minded tenacity of a few rugged individuals in the oil industry who finally "hit pay-dirt." But chaos came out of these experiences and this rather simple dynamic eventually gave way to an institutional innovation (TRC well spacing and prorationing) which produced a much needed market stabilizing effect. But this institutional innovation did not come easy—there came a federal court challenge based on arguments of TRC price fixing. And, as one might expect in retrospect, this TRC era (1930−72) resulted in hastening the shift of the world oil market power to Organization of the Petroleum Exporting Countries— hence the Arab Oil Embargo.

The last five decades have changed the innovation dynamics from that of oil development in the first quarter of the 20th century to a complex process that today joins public and private sector leaders in a variety of institutions, funding support from both venture capital and public sector R&D sources, and individual enterprisers' contributions from a very diverse, and independent-minded culture. And, while the initiatives of key organizations and individuals have come from within the state, part of which came out of state/federal jurisdiction conflicts, Texas innovation leaders have also taken advantage of the innovations from other states and counties—wind machines from Europe and learning from electric sector deregulation failures in California.

So, how do we make the "mare go" going forward? The short answer is diversity and collaboration to generate ideas, and a securitization process to finance a large number of projects—get Bill Gates and Andrew Lo to join us, bringing with them the Breakthrough Energy Coalition and the Breakthrough Energy Venture fund—they are willing to fund outliers and high-risk ideas. Starting with Romer's understanding and current theory of innovation-driven economic growth, *job one* is to keep the ideas flowing. Such an idea flow process, as understood by Romer, is that ideas flow best in an atmosphere of individual freedom and from large groups of diverse-backgrounds in the population. But in the modern world we now have the contribution of software and computers to test ideas much faster, and often much cheaper than was possible before the computer era. And, because of the digital world and the internet, a much expanded capacity to generate ideas from large, diverse groups.

Texas is blessed by having such diversity, and a culture of freedom memorialized by the individualism of the cowboy that still runs deep in her psyche. Governmental systems led by a "strong man" (dictator types) squelch idea generation—they do not promote innovation well. Going further, we have to think differently about the role of institutions. Institutions of science are fundamentally different from institutions of the market. Science, and scientific contribution to human progress, has to be an open, transparent system. Such transparent systems, although driven by competition among both the individuals within a discipline, and among the disciplines, is essential to perfecting the recognition of facts and the progress of theories (essentials of scientific inquiry). Its individual efforts and products must be shared. Markets, on the other hand, can only function and deliver positive support for both individuals and the common good if fundamental property rights are protected by the legal system. Further, competitive markets only function well to provide the products that sustain us if the rules of the market are protected by the fundamentals of competitive markets enforced by antitrust laws and enforcing agencies. But the Romer era contribution to the theory of economic growth, essential to delivering the economic goods to the world's populations, recognizes that the innovation process must keep turning out ideas that drive policy, markets, and technology.

A recommendation

The innovation process in the modern world is very different from that of 1901, although the process has always begun with ideas. Today's innovation process that drives both new technology, and even new institutions, is a market place made up of a healthy mix of software and hardware, both enhanced by collaborative organizations that broaden access to different, relevant experiences and interests. Software is powerful because it is all about taking an idea, either a simple one or a complex one, and writing a computer program to solve logic problems that have heretofore not allowed alternative futures to be adequately compared. Earlier times, even if clear thinking about alternative futures entered in, were very time-consuming processes requiring physical constructs to be built or geological horizons to be drilled and tested. These limited experiences represented crude representations of the future—all made up of only hardware and physical constructs. The time for testing in a hardware-only world pales in comparison to finding a solution with computer speed. We can now take an idea, program it, and anticipate a conclusion in a market-driven economy without always having to build a physical model or drill a formation of some kind to test it. So, an entity supporting idea creation and funding should include the best evaluation of alternative futures possible to inform decision-making.

So, what shall we do to keep the ideas flowing in Texas? The answer is to identify high priority innovation prospects and fund them with a mix of private and public funds. First, we ought to figure out how to better support existing innovation institutions—but add a joint public/private organization that organizes, funds and oversees a crowd-sourcing process to generate a large number of innovative ideas by leveraging these existing entities. Such an organization would be expected to engage with existing and new entities like those listed above. Such a crowd-sourcing process should invite participants to add new ideas applicable to each of several future time periods, say the next five decades, and covering several functional areas and disciplines. Ideas might be something entirely new or existing ideas that have failed to find a path to the market place—it is unlikely there will be very many ideas that are entirely new.

This recommendation is, *first of all,* intended to envision how to overcome obstacles, and *second*, fund the winning projects by setting up a securitization process based on a mix of venture capital funds and state appropriations. The key is to set up a funding mechanism that does not depend on political support every time the Congress and Legislature meets. The idea is to create stability like that of the Federal Reserve whose members are appointed by the President with Senate confirmation for terms longer than the President's term of office. Attempts to fund ongoing R&D agencies and non-profits in the past (GEAC, TEAC, TENRAC, CCET, and TETF) all existed for a period but died when the crisis or significant policy change that supported their creation ended. The related energy centers at the major Texas universities have been more stable, but also have had difficulty maintaining R&D program funding needed for stability through the ebb and flow of energy market prices and unexpected world crisis events.

The recommended structure could follow that of Breakthrough Energy Collision and its investment arm, Breakthrough Energy Ventures. A Texas version of this recently formed worldwide entity could be the structure while engaging the above Texas innovation entities (see Breakthrough Energy Collision, Accessed September 3, 2018 at http://www.b-t.energy/coalition/). And, we could invite the Breakthrough Energy Ventures to participate in the Texas effort. A beginning list of Texas philanthropic entities that could help create a Texas version of the Breakthrough Energy Collision and the Breakthrough Energy Ventures entities is:

Ed Bass	Red McCombs
Lee Bass	Trevor Rees-Jones
Jeb Bush	Clayton Williams
Caroline Rose Hunt	Boone Pickens
Rodney Lewis	H. Ross Perot
Robert E. Marling Jr.	AEP Foundation
Cynthia and George Mitchell Foundation	

We might also invite collaboration with the NSF or facilitate Texas entities participation in the NSF competition. NSF has an innovation funding program called *Big Ideas Stewardship Funding Model* that focuses on six research-focused Big Ideas with a total 2019 budget request of $180 million. The six areas include (1) Harnessing the Data Revolution, (2) The Future of Work at the Human Technology Frontier, (3) Windows on the Universe, (4) The Era of Multimessenger Astrophysics, The Quantum Leap, (5) Leading the Next Quantum Revolution, Understanding the Rules of Life, and (6) Predicting Phenotype, and Navigating the New Arctic.

We should learn from the experiences of the Texas Emerging Tech Fund, Texas Energy Development Fund, Texas Enterprise Fund and their administration challenges—where did they succeed and where did they fail. Each of them, while driven by well-intentioned missions did not have institutional stability reaching beyond the next Governor/Legislative term. In evaluating the effectiveness of the above institutions we have to look for "measures that mattered" and recognize that "not everything that matters can be measured." Innovation that depends on public/private partnerships cannot be maintained without a measure of institutional stability on the government side. In addition, governments and private entities that have to report in the near term to constituents and stockholders do not fund high-risk ventures. But philanthropic organizations like Bill Gates and Andrew Lo do, so we should create Texas versions of such entities and potentially partner with Bill Gates and Andrew Lo to find high-risk, potentially high pay-off ventures to include in funded ideas leaving less risky ideas for Texas based government/private entity support.

Finally, having generated and funded many innovative ideas, the results can be anticipated by representing the innovations in econometric models of the economy in order to evaluate their effects on long term future economic conditions. By grouping the results into technology advancement expectations, the analyses will inform us about effects on alternative futures, both the good and the bad.

In short, we must institutionalize a stable process of the generation of ideas, including the ideas of economists, engineers, and public policy experts about what drives our economic engine in a capitalistic democracy that has become so heavily dependent on innovation. The outcome of such a process will generate new economic theories about the relationship between innovation and the economy—edifications to the Romer era contribution to the theory of economic growth. Such theories will then become part of the ongoing debate and influence our path forward. A more thorough understanding of the influence of innovation on our future economic and cultural conditions, including both the good and the bad must be a part of the drive for energy innovation. The benefits of a Texas private/public collaborative entity would certainly extend to investors and new and existing companies who have a vested interest in the outcomes. But the mission and structure of the entity must discipline the vested interest without discouraging private company participation. To repeat John Maynard Keynes in **The General Theory of Employment, Interest and Money:**

> *But apart from this contemporary mood, the ideas of economists and political philosophers, both when they are right and when they are wrong, are more powerful than is commonly understood. I am sure that the power of vested interests is vastly exaggerated compared with the gradual encroachment of ideas.*
>
> **Keynes (1965, p. 383)**

We must continue the work of economists and innovation experts to improve the theory and practice of innovation and economic growth, and above all keep the ideas coming.

Appendix A: The principle components of the energy system

There are several viewing perspectives that are commonly used to characterize the energy system. One way to think of energy is from the consumer perspective. The usual practice in data system developments is to distinguish energy use by economic class such as transportation, residential, commercial, and industrial. A second view is consumption from the source including petroleum, natural gas, coal, renewables, and nuclear. The chart below identifies these two consumption perspectives from 1950 to 2018. The most noteworthy change over this 68 year period is the rise of renewables beginning in 1970 and the switch from coal and petroleum to natural gas and renewables beginning in the late 1990s (Fig. A.1).

A second perspective is from the producer view. The primary initial products derived from a barrel of crude oil produced from oil wells and piped to refineries are produced by distillation at temperatures beginning at around 85°F rising in several steps all the way to 1050°F. During distillation the liquids and vapors separate into petroleum components called fractions The process progresses from separation to conversion (usually by a process called cracking) and then treatment before yielding finished products like gasoline. The chart below identifies the seven major products distilled from crude oil. In the cracking distillation unit the liquids and vapors separate into petroleum components (fractions) according to their boiling points. The end

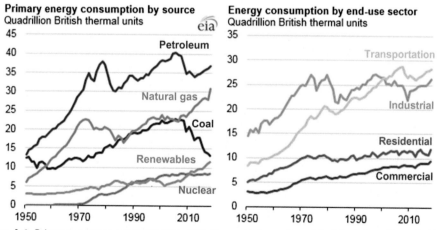

Figure A.1 Primary energy consumption. *U.S. Energy Information Administration, Monthly Energy Review.*

Crude oil distillation unit and products

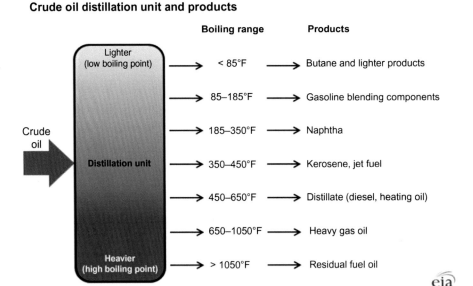

Figure A.2 Crude oil distillation. *U.S. Energy Information Administration.*

products are fuels and feed stocks that are used throughout the economy. Finished products are stored at near at "tank farms" then transported to market by pipelines, trucks, and trains (Figs. A.2 and A.3).

The natural gas system is itself complex because the raw natural gas from wells is processed and used as input to a number of downstream processes. While natural gas is usually produced jointly with crude oil it is often produced from natural gas reservoirs that do not jointly produce crude oil. After separating gas from oil and water that comes to the surface from a well, the natural gas is transported via pipeline to a gas processing plant. From there several products are separated from the raw natural gas stream and the remaining natural gas is transported to distribution companies who sell to residential, commercial, industrial and electric utility consumers. Storage systems provide the capability to allow unconstrained production at natural gas wells but delivered to consumers as needed. Use levels vary for a number of reasons, especially due to seasonal weather conditions. Another useful product is Liquefied Natural Gas that is stored in pressurized, low temperature containers and shipped by tankers at sea to the international market (Fig. A.4).

The coal based energy system is quite different from that of crude oil and natural gas. Coal is mined either from deep mining of coal seams accessed via shafts and tunnels or by surface mining by means of excavation and dragline excavators. In either case the coal is transported to processing plants via conveyors, trams and trucks where it is cleaned and then the cleaned coal from the processing plant is shipped to markets

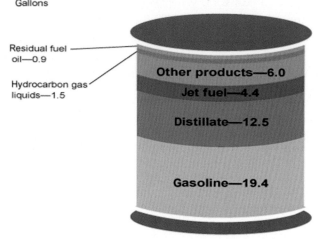

Petroleum products made from a barrel of crude oil, 2019
Gallons

Residual fuel oil—0.9

Hydrocarbon gas liquids—1.5

Other products—6.0

Jet fuel—4.4

Distillate—12.5

Gasoline—19.4

Figure A.3 Petroleum products from a barrel of crude oil. Note: A 42-gallon (US) barrel of crude oil yields about 45 gallons of petroleum products because of refinery processing gain. The sum of the product amounts in the image may not equal 45 because of independent rounding. *U.S. Energy Information Administration, Petroleum Supply Monthly, April 2020, preliminary date. https://www.eia. gov/energyexplained/oil-and-petroleum-products/refining-crude-oil-the-refining-process.php.*

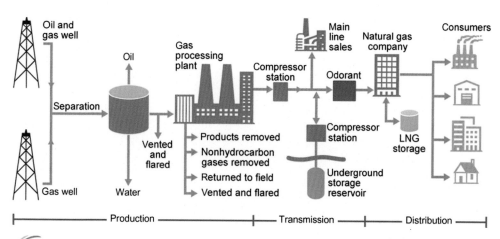

Source: U.S. Energy Information Administration

Figure A.4 Natural gas production and delivery. *U.S. Energy Information Administration available at https://www.eia.gov/energyexplained/natural-gas/.*

around the country by barges, trains and slurry pipelines. The two images in Fig. A.5 depict the two mining types.

The predominant use of coal is for generating steam to drive power generators in the electric power industry and industrial plants. The most dramatic change in the mix of basic fuels for electric power generation is the quickly evolving switch at electric utilities substituting natural gas and renewables for coal (Fig. A.6). The change is being driven by relative prices and environmental impacts of resulting CO_2 emissions.

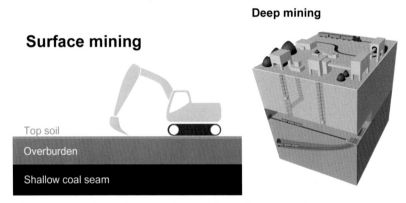

Figure A.5 Energy mining illustration. *U.S. Energy Information Administration.*

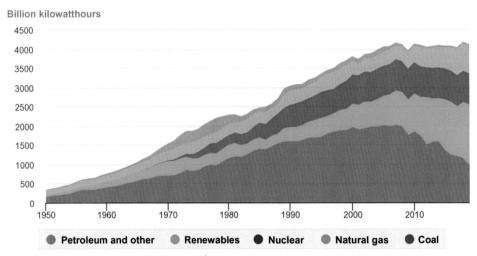

Figure A.6 US electricity generation by major energy source. Note: Electricity generation from utility-scale facilities. *U.S. Energy Information Administration, Monthly Energy Review, Table 7.2a, March 2020 and Electric Power Monthly, February 2020, preliminary date for 2019.*

The fourth, and growing share of basic energy supply is from renewables. As can be seen in the chart below wind and solar have dominated the growth in renewables since year 2000. Texas leads the nation in wind capacity and generation. The contributions from wood, hydro and waste products have declined as a share of the total. Both wind and solar are "intermittent" sources so other resources, especially natural gas generation has to fill in when the wind does not blow and the sun does not shine, sometimes creating a grid management challenge. So far grid managers, including ERCOT in Texas have been able to manage the grid reliably with a growing share of capacity from renewables. Energy storage promises to make the future management easier (Fig. A.7).

The fifth major source of energy for the economy is nuclear. Nuclear power generation has been quite stable for years with little additional new plants, and the slow pace of decommissioning of older units. The gradual decline of nuclear power in the US is primarily due to a public safety concern which arose from the fallout of several nuclear reactor melt-downs in the world over the last several decades, including Three-Mile Island in Pennsylvania in 1979 (see Fig. A.6). Despite recent closures, however, US nuclear electricity generation in 2018 surpassed its previous peak but Energy Information Administration expects the nuclear power generation currently at about 800 megawatt hours to decline by 17% by 2025 (https://www.eia.gov/todayinenergy/images/2019.03.21/chart4.svg).

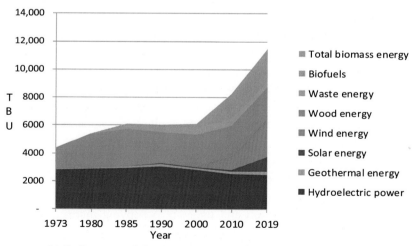

U.S. Renewable Energy Consumption

Figure A.7 US renewable energy consumption. *U.S. Energy Information Administration; EIA Renewable Energy Production and Consumption by Source. https://www.eia.gov/totalenergy/data/monthly/archive/00351901.pdf.*

Another viewing perspective of the energy system is an important time dimension where energy use revolves around seasons of the year and emergencies. Energy use varies greatly as a function of weather conditions, summer travel by consumers and other seasonal influences. A national shutdown of crude oil and product imports could result in great economic adjustment to manage resulting shortages. Such an event occurred during the Arab Embargo of 1973 discussed in Chapter 8, Upheaval in the Energy Markets: The Arab Oil Embargo and the Iranian Crisis. Variable seasonal demand for natural gas and oil products are managed by several types of storage. Natural gas is stored in underground salt domes; oil products are stored in ships, tank farms, and retail level tanks. Coal is stored in above ground mounds at electric power plant locations allowing use as needed.

Electric use in the southern United States usually peaks in the late summer with peak air conditioning. Gasoline use peaks in the summer with vacation travel. Energy storage systems that can store fossil fuels are very important resources for managing the uneven seasonal use of natural gas, coal and petroleum fuels. Someday, perhaps soon, electric storage will provide important capacity to deal with uneven electricity demand patterns, especially the daily variation in use from the day time until night time.

There is still another important energy system perspective—a viewpoint of the flow of energy from production to consumer through the transfer or transportation of energy. There are important technology aspects where energy transport systems overlays public transportation systems—mainly highways and water ways. Pipeline development is surrounded with legal issues of private property rights and waste disposal from, for example, slurry pipelines.

Another energy sector dimension is the view from a market perspective, usually referencing wholesale and retail markets, but also the market functions of banking, insurance, traders, brokers and others. But in the energy world there is also a distinction between intrastate and interstate energy markets and activities; this distinction is peculiar to the United States and is a creation flowing from the US constitution and related statutes and legal treatment. This aspect of the law defines federal and state legal jurisdictions and results in interesting, even perplexing conflicts over state/federal jurisdictions.

Finally, environmental impacts and public policy influences in the energy sector are becoming more important viewing perspectives. CO_2 emissions, nuclear waste disposal, water system pollution, ocean wildlife impacts, and impedance of scenic vistas—all resulting from energy production and use—are occupying increasing components of the public policy debate.

Appendix B: Overview of energy market and regulatory structures and US influence abroad

This appendix is an attempt to shed light on the positive influence that the US and Texas energy sectors have had on the world economy. But clearly an entire book could be written exploring this topic. So the appendix selectively compares the energy/economic systems, focusing on market and regulatory systems, of the United States and Texas (capitalistic democracies) with selected other countries. The comparisons are means of highlighting the essential differences between energy sectors in capitalistic democracies with that of socialistic and authoritarian dictatorships. The comparisons are with Russia, the UK, Germany, Saudi Arabia and Venezuela.

Identifying positive influences among countries by way of activities in the energy sector requires a viewing of economies as driven by a few basic sectors—energy (from mining to refining), transportation, agriculture, manufacturing, finance and services. Agriculture, part of transportation, noninternet based services and energy are mostly confined within country boundaries because they are tied to the land and/or to a labor force that is domestically constrained to the home country. Manufacturing, internet based services, financial services and international travel have become global enterprises in the modern economy, easily organizing across country boundaries.

The sectors that easily operate across country boundaries mostly do so via global value chains (GVC). GVC linkages are especially important for high-tech sectors and these areas involve highly complex value chains encompassing many countries. The energy sector global involvements are different, requiring an invitation of foreign countries for US companies to engage in energy production and export of commodities to consuming countries. In the petroleum industry, for example, the industry is bifurcated into the majors who operate all over the globe and the independents who only operate within the United States. The host country in the case of majors, shares the revenues generated from the production and export of energy. These agreements take the form of concessions as described earlier.

Public policy also plays a major role in explaining the current energy sector dynamic and US influence abroad. In fact it is difficult to fully understand the evolution of US and Texas energy market and regulatory systems and compare them with that of other top energy producing countries of the world without doing so in the context of national security policy and concerns. This is especially true for the crude oil market and governmental systems. Other energy resources are national security concerns indirectly because they are substitutes for oil, especially in the long term. US foreign policy following WW I, and more importantly following WW II, has been driven in significant degree by the need to have secure oil supplies. And, not just for the United States, but for our Allies as well. The US military and diplomatic involvements in the Middle East have been driven importantly by the need for stability among the oil producing countries—totalitarian countries whom control a large share of world oil supplies that are primarily exported to Western democracies.

This appendix reviews the energy market and regulatory structures of the United States and Texas and compares their performance with that of key foreign countries in the context of national security concerns.

Government agency regulatory system in the United States

The US energy systems are structured primarily as private sector activities with government oversight and regulation. The same is true, of course for producing states like Texas. This structure is not universal across energy sectors, however. The oil and gas sectors are structured with private ownership of the resource except for resources located on onshore and offshore public lands. Offshore resources in the Outer Continental Shelf (OCS) are federally owned where access by private companies is carried out by Federal agency leasing extended by way of competitive bidding and auction programs, giving private companies access to the oil and gas deposits in the OCS. The Federal offshore resources are defined as submerged lands lying seaward of state coastal waters which generally extend 3 miles offshore. OCS resources are under US legal and regulatory jurisdiction. The definition of state offshore is uniform in the United States, at 3 miles with the exception of Texas and Florida where the state-owned lands extend to 9 miles offshore. Texas is unique in the 9 mile structure basis dates to agreements made when Texas as an independent country joined the Union (was annexed) in 1845.

There are four federal OCS leasing programs including, Gulf of Mexico, Atlantic, Pacific and Alaska regions. Onshore public lands are also home of important oil and gas resources that are managed by the US Bureau of Land Management (BLM) located mostly inside the Western States. Private companies pay royalties to BLM when energy resources are extracted.

Highly integrated private companies who own energy resources in the United States readily trade produced energy among the several states (cross borders). The exception is the electric sector in Texas where most Texas electricity is marketed and consumed within the state boundaries and within geographical areas defined for the Electric Reliability Council of Texas (ERCOT). The detail discussion of this unusual Texas system is the focus of Chapter 9, Electric Industry Deregulation and Competitive Markets. US companies also establish subsidiaries and operate abroad, ventures that provide benefits across the global.

The Federal agencies that regulate and/or oversee US energy sector activities include the Department of Energy (DOE), Federal Energy Regulatory Commission

(FERC), Nuclear Regulatory Commission (NRC), Department of the Interior (DOI) and the Energy Information Administration (EIA). This current structure evolved over an 8-year period following the Arab Embargo of 1973 (see Chapter 8: Upheaval in the Energy Markets: The Arab Oil Embargo and the Iranian Crisis).

DOE—has activities to develop energy policy, oversee the activities of seventeen National Laboratories, direct a technology and policy development (RD&D) fund and to provide for administration of the EIA. Through initiatives like the Loan Programs Office and the Advanced Research Projects Agency-Energy (ARPA-E), the Department funds cutting-edge research and the deployment of innovative clean energy technologies. The Department also encourages collaboration and cooperation between industry, academia and government to create a vibrant scientific ecosystem (https://www.energy.gov/about-us).

FERC—regulates wholesale market interstate transmission of electricity, natural gas, and oil for resale in interstate commerce; approves siting and abandonment of interstate natural gas pipelines and storage facilities; licenses and inspects private, municipal, and state hydroelectric projects; reviews proposals to build liquefied natural gas (LNG) terminals and interstate natural gas pipelines as well as licensing hydropower projects (https://www.ferc.gov/about/ferc-does.asp).

NRC—the NRC regulates commercial nuclear power plants and other uses of nuclear materials, such as in nuclear medicine, through licensing, inspection and enforcement of its requirements. The regulatory authority includes commercial reactors for generating electric power and research and test reactors used for research, testing, and training, uses of nuclear materials in medical, industrial, and academic settings and facilities that produce nuclear fuel and the transportation, storage, and disposal of nuclear materials and waste, and decommissioning of nuclear facilities from service (https://www.nrc.gov/about-nrc.html).

EIA—the EIA collects, analyzes, and disseminates independent and impartial energy information to promote sound policymaking, efficient markets, and public understanding of energy and its interaction with the economy and the environment. The agency provides a wide range of information and data products covering energy production, stocks, demand, imports, exports, and prices; and prepares analyses and publishes special reports on topics of current interest (https://www.eia.gov/about/).

DOI and BLM—the BLM manages one in every 10 acres of land in the United States, and approximately 30% of the Nation's mineral resources. These lands and minerals are found in every state in the country and encompass forests, mountains, rangelands, arctic tundra, and deserts. The BLM manages the federal mineral estate, which includes oil, natural gas, and coal (https://www.blm.gov/about/what-we-manage/).

Environmental Protection Agency (EPA)—established in 1970, EPA is the chief national environmental agency of the United States. The mission of EPA is to protect human health and the environment by working to ensure that Americans have clean air, land and water through national efforts to reduce environmental risks based on the best available scientific information. The agency declares that environmental stewardship is integral to US policies concerning natural resources, human health, economic growth, energy, transportation, agriculture, industry, and international trade, and these factors are similarly considered in establishing environmental policy. The agency provides national regulatory programs, works with state environmental agencies to implement environmental objectives and carries out major research activities through federal grants to inform policy makers and citizens about environmental technologies and program options (https://www.epa.gov/).

State government agency regulatory systems

Texas

The several states differ somewhat in agency over site and regulation of energy activities. As a general matter the states with large deposits of oil, gas, coal and uranium (the producer states) are structured similar to that of Texas reviewed below. These states include Texas, Oklahoma, Louisiana, New Mexico, California, Wyoming, Colorado, Kentucky, West Virginia and Pennsylvania. Consuming states (all the others) focus primarily on energy efficiency mechanisms and electric utility regulation.

Regulatory jurisdiction over crude oil, natural gas, coal, uranium mining, and electric power, wind and solar is divided among three primary agencies in Texas; the Texas Railroad Commission (TRC), Texas Commission on Environmental Quality (TECQ) and the Public Utilities Commission (PUC). There is also a low-level nuclear waste disposal agency operating under a two-state agreement between Texas and Vermont. Also, there is a state energy conservation agency that administers conservation programs with federal funding. The several regulatory functions are described briefly below.

Texas leads the nation in energy production. The state provides more than one-fifth of United States' domestically produced energy. Texas land area is 800 miles at its widest points from east to west and north to south, and crude oil and natural gas deposits are present across much of that expanse. Texas leads the nation in crude oil production, crude oil refining and natural gas production. Although it is not widely known coal is found in bands across the eastern Texas coastal plain and in other areas in the north-central and southwestern regions of the state. Texas also has abundant renewable energy resources. The state is first in the nation in wind-generated electricity. Although the development is in an early stage, Texas is also among the leading states in solar energy potential. Texas also has geothermal resources in East Texas, and uranium in South Texas. The state has three licensed uranium in situ recovery plants, but with little demand for nuclear fuel there have been no commercial operations for several years.

Because of the state's large population, Texas leads the nation in state residential energy use, but it ranks near the lowest one-fifth of states in per capita residential energy consumption. Texas energy consumption exceeds that of any other State and accounts for almost one-seventh of the US total consumption. The industrial sector, which includes the energy-intensive petroleum refining and chemical manufacturing industries, is the largest energy consuming end-use sector and accounts for half of the state's end-use energy consumption. The second-largest end-using sector is transportation due in large part to the great distances across the state, and the related high number of vehicle miles traveled annually.

The state has more than two-fifths of US crude oil proved reserves and produces two-fifths of the nation's crude oil and also leads the nation in crude oil refining. It has more than one-fifth of the nation's refineries and more than three-tenths of US total refining capacity. The state's 30 petroleum refineries can process a combined total of almost 5.8 million barrels of crude oil per calendar day.

Texas produces more electricity than any other state—almost twice as much as Florida, the second-highest electricity-producing state. Natural gas-fired power plants supplied more than half of the state's electricity net generation in 2019. Coal-fired power plants supplied less than one-fifth of state generation in 2019, down from about one-third in 2014, the result primarily due to lower relative prices of natural gas and the retirement of some coal units.

Wind-powered generation in Texas has rapidly increased during the past two decades. In 2019, wind energy provided more than one-sixth of Texas' generation. The state's two operating nuclear power plants typically supply almost one-tenth of the state's electricity net generation.

Due mostly to the rapid growth of wind generation renewable energy sources contribute nearly one-fifth of the net electricity generated in Texas The state has encouraged renewable energy use by authorizing construction of transmission lines to bring electricity from remote wind farms to urban market centers. The state leads the nation in wind-powered electricity generation, producing almost three-tenths of the US total.

Texas has some of the greatest solar power potential in the nation, and the state was the country's sixth-largest producer of solar power in 2019. The Public Utility Commission of Texas first adopted rules for the state's renewable energy mandate in 1999 and amended them in 2005 to require that 5880 megawatts, or about 5% of the state's electricity generating capacity, come from renewable sources by 2015 and 10,000 megawatts of renewable capacity by 2025, including 500 megawatts from resources other than wind (https://www.eia.gov/state/analysis.php?sid = TX/).

The Texas Regulatory Agencies—The regulatory framework in Texas has evolved from early regulation of oil production through well spacing and production limits first established in the 1930s to a mature regulatory system that covers most all of energy sources, but the energy sector remains greatly driven by competitive market systems. The key regulatory systems are reviewed below.

TRC—The oil and gas industry is regulated in several ways by the TRC. (The name of this important energy agency is left over from the early 1900s when railroads were developed and regulated in the nation, and in Texas. There have been several attempts, over several decades to change the name of the TRC to better depict the current role in oil and gas and other natural resource regulation, but to no avail.)

The historically most important regulatory role of the TRC is regulation of oil and natural gas exploration, production, and transportation. The Oil and Gas Division of the TRC is responsible for regulating oil and natural gas exploration, production, and transportation in Texas. Under state law, the commission is responsible for natural resource management, pollution prevention, and worker safety. The division issues oil and gas permitting and reporting requirements, conducts oil and gas field inspections, monitors industry activities during oil and gas production, and implements programs to convert abandoned wells and sites back to their original conditions by levying fees and taxes on the oil and gas industry. Title 16, Part I of the Texas Administrative Code contains the rules and regulations on energy exploration, production, and transportation (Railroad Commission of Texas, Mining & Exploration, Surface Mining and Reclamation Division, Austin, Texas, 2020).

A most important recent regulatory program of the TRC is that of hydraulic fracking. Fracking regulations in Texas include well construction requirements and steel casing and cement requirements. TRC rules provide for disclosure of chemical ingredients used in hydraulic fracturing fluids. The rule requires Texas oil and gas operators to disclose on the FracFocus website (fracfocus.org), chemical ingredients and water volumes used in hydraulic fracturing treatments. FracFocus is a public Internet chemical registry hosted by the Ground Water Protection Council (GWPC) and the Interstate Oil and Gas Compact Commission (IOGCC). The GWPC is a national association of state ground water and underground injection control agencies. The IOGCC is a national commission of state oil and gas regulators.

Another regulatory function of TRC is uranium exploration and surface mining. TRC is task with several activities in the mining sector designed to prevent the adverse effects to society and the environment resulting from unregulated surface mining operations; to ensure that the rights of surface landowners and other persons with a legal interest in the land or appurtenances are protected from such unregulated exploration and surface mining operations; to ensure that surface mining operations are not conducted where reclamation as required by the TRC is not possible; to ensure that exploration and surface mining operations are so conducted as to prevent unreasonable degradation to land and water resources; to ensure that exploration reclamation and reclamation of all surface mined lands is accomplished as contemporaneously as practicable with the exploration and surface mining operation, recognizing that the extraction of minerals by responsible mining operations is an essential and beneficial economic activity, these sections are promulgated pursuant to the directive and authority of the Texas Uranium Exploration, Surface Mining, and Reclamation Act, Texas Natural Resources Code, Chapter 131, et seq (https://ballotpedia.org/Energy_policy_in_Texas).

Another TRC regulatory duty is ensuring pipeline safety. TRC is responsible for regulating all gas pipeline facilities and facilities used in the intrastate transportation of gas, including LPG distribution systems and master metered systems, onshore pipeline and gathering and production facilities, the intrastate pipeline transportation of hazardous liquids or carbon dioxide and all intrastate pipeline facilities, and all pipeline facilities originating in Texas waters (three marine leagues and all bay areas). These pipeline facilities include those production and flow lines originating at the well. The regulations do not apply to those facilities and transportation services subject to federal jurisdiction (interstate pipelines) (Railroad Commission of Texas, Mining & Exploration, Surface Mining and Reclamation Division, https://www.rrc.state.tx.us/mining-exploration/).

TCEQ—The chief Texas regulatory agency for environmental protection is TCEQ. The agency operates under a Federal program for environmental protection known as a State Implementation Plan (SIP). A SIP is an enforceable plan developed at the state level that explains how the state will comply with air quality standards according to the Federal Clean Air Act. SEP requires facilities that may emit air contaminants to obtain authorization prior to construction (a permit). The type of authorization required will depend on the type of facility and amount of contaminants emitted. Air emissions covered by the SIP includes ground-level ozone, particulate matter, lead, nitrogen dioxide (NO_2), carbon monoxide (CO), and sulfur dioxide (SO_2). Noticeably missing from the list are Green House Gas (GHG) emissions. Texas challenged the EPA ruling about inclusion of GHG under provisions of the Clean Air Act of 1963 per a new rule of EPA in 2010. The EPA implementation of the CAA has provisions for defining technological means for controlling emission for stationary sources called Best Available Control Technology (BACT). The Texas challenged the EPA's rule that included GHG for the first time, without a statutory act making it clear that such is the intent of Congress. After an extended court battle, including the US Supreme Court, GHG are not covered although approval of BACT may restrict CO_2 emissions if other GHGs are in the mix of pollutants along with CO_2. An additional federal program operated by TCEQ is the Texas Emissions Reduction Plan which provides financial incentives to eligible individuals, businesses, or local governments to reduce emissions from polluting vehicles and equipment.

TCEQ also regulates water quality in Texas along with solid waste and low-level nuclear waste. Much of such regulation applies to emissions and waste disposal from electric power plants and industrial operations (https://www.tceq.texas.gov/airquality/sip/sipintro.html). High-level nuclear waste disposal is strictly under the regulation of the Federal government, although the government has tried to solve the sitting of a Federal facility unsuccessfully for over 30 years. Texas was once the target for a high-level waste storage facility in deep salt beds located in the High Plains of Texas. The current preferred site is Yucca Mountain in southern Nevada, but politics has prevented going forward with that site (https://www.tceq.texas.gov/).

Texas PUC—In 1975, the Texas Legislature enacted the Public Utility Regulatory Act (PURA) and created the Public Utility Commission of Texas (PUC) to provide statewide regulation of the rates and services of electric and telecommunications utilities. Although the PUC originally regulated water utilities, jurisdiction was transferred to the Texas Water Commission in 1986. Significant legislation enacted by the Texas Legislature in 1995, along with the Federal Telecommunications Act of 1996 (FTA), dramatically changed the PUC's role by allowing for competition in telecommunications wholesale and retail services, and by creating a competitive electric wholesale market. In 1999, the Texas Legislature provided for the restructuring of the electric utility industry, allowing certain customers electric choice. The PUC's mission and focus have shifted from regulation of rates and services to oversight of competitive markets and compliance enforcement of statutes and rules for the electric and telecommunication industries. Effective oversight of competitive wholesale and retail markets for electric and telecommunication is necessary to ensure that customers receive the benefits of competition. For water and sewer utility service, however, the focus remains on the regulation of rates and services.

The PUC continues to perform its traditional regulatory function for electric transmission and distribution utilities across the state. Additionally, while integrated electric utilities outside of the ERCOT power grid remain fully regulated by the PUC, the PUC is increasingly involved in multi-state efforts to implement wholesale electric competitive market structures and transmission planning in the Southwest Power Pool and Midcontinent Independent System Operator areas (https://www.puc.texas.gov/agency/about/mission.aspx).

State Energy Conservation Office (SECO)—SECO partners with Texas local governments, county governments, public K-12 schools, public institutions of higher education and state agencies to reduce utility costs and maximize efficiency. SECO also adopts energy codes for single-family residential, commercial, and state-funded buildings. The code requires all new single-family homes to comply with energy efficiency and water conservation standards. Also, all commercial, industrial and residential buildings over three stories must meet certain energy efficiency standards (https://comptroller.texas.gov/programs/seco/).

Texas Low-Level Radioactive Waste Disposal Compact Commission—Texas is a member of the Texas Low-Level Radioactive Waste Disposal Compact with the state of Vermont. The compact allows exclusive disposal of the low-level radioactive waste (LLRW) generated within the member states to be disposed of in a LLRW disposal facility constructed within any of those states. Texas is the host state for the compact disposal facility. Nuclear utilities, academic and medical research institutions, hospitals, industry, and the military are the primary producers of non-DOE LLRW in Texas. LLRW typically consists of radioactively contaminated trash such as paper, rags, plastic, glassware, syringes, protective clothing (gloves, coveralls), cardboard, packaging material, organic material, spent pharmaceuticals, used (decayed) sealed radioactive sources, and water-treatment residues. Nuclear power plants contribute the largest portion of LLRW in the form of contaminated ion-exchange resins and filters, tools, clothing, and irradiated metals and other hardware. LLRW does not include waste from nuclear weapons manufacturing or from US Navy nuclear propulsion systems (http://www.tllrwdcc.org/about-the-comission/; Encyclopedia of American Politics, https://ballotpedia.org/Energy_policy_in_Texas).

United Kingdom

The UK is the fifth largest economy among the developed countries of the world as measured by gross domestic product (GDP). Therefore she is also a large energy consumer. The UK has been a large producer of oil and natural gas since the development of the resources in the North Sea. The North Sea is divided into UK, German, Norwegian, Danish, and Dutch sectors. The UK has the largest of these sectors. Among European countries in the Organization for Economic Cooperation and Development (OECD), the UK was the second-largest producer of petroleum and other liquids in 2016 outpaced only by Norway. Three main grades of oil are produced in the UK including Flotta, Forties, and Brent blends. They are generally light and sweet, which makes them attractive to foreign buyers.

The UK was the third-largest producer of natural gas in that year. UK natural gas production peaked in 2000, and consumption peaked in 2004 with both generally declining through 2014. From 2014 through 2016, production has grown at an average rate of 5% per year, while consumption has grown 7% per year.

By the early 1980s Britain had become a net exporter of oil, and by the mid-1990s of gas. Following many years as a net exporter of crude oil and natural gas, the UK became a net importer of both oil and gas in 2004 and 2005, respectively, and in 2013 the UK became a net importer of petroleum products, making it a net importer of all fossil fuels. The net import condition was a dramatic change from a net exporter status that prevailed from the early development of the North Sea oil and gas reserves that began in earnest following the world oil price run-up following the Arab Oil Embargo of 1973.

Renewable energy use has more than doubled in the UK over the past decade (2007—16). But, petroleum and natural gas continue to account for most of the UK's energy consumption. In 2016, petroleum and natural gas each accounted for 38% of total energy consumption. The share of coal in total energy consumption has declined rapidly over the past several years from 19% in 2012 to 6% in 2016 due mostly to increased use of renewables and natural gas (https://www.abdn.ac.uk/oillives/about/nsoghist.shtml/).

UK regulatory authorities

Oil & Gas Authority (OGA)—the OGA has recently succeeded the Department for Energy and Climate Change (DECC) as the regulator of the offshore and onshore oil and gas sector in the UK, including oil and gas licensing, oil and gas exploration and production, oil and gas fields and wells, oil and gas infrastructure and carbon storage licensing. The OGA is a government-owned company charged with both regulating and promoting the oil and natural gas industry in the UK. The OGA was formally established as a fully independent regulator and a government-owned company, with the Secretary of State for Business, Energy and Industrial Strategy (the Secretary of State) as the sole shareholder (www.gov.uk/government/organisations/oil-and-gas-authority).

Department for Business, Energy and Industrial Strategy (BEIS)—BEIS is responsible for setting energy and climate change mitigation policies, and establishing the framework for achieving the policy goals in those areas. (www.gov.uk/government/organisations/department-for-business-energy-and-industrial-strategy).

Office of Gas and Electricity Markets (Ofgem)—is responsible for regulating the downstream gas market, and in particular the monopoly gas transmission and distribution networks. Ofgem also plays a role in enforcing the third-party access regime that applies to downstream gas infrastructure. Offshore Safety Directive Regulator (OSDR) is the Competent Authority responsible for overseeing industry compliance with the EU Directive on the safety of offshore oil and gas operations. OSDR is a partnership between BEIS' Offshore Oil and Gas Environment and Decommissioning Team (OGED) and HSE's Energy Division (ED) (www.ofgem.gov.uk/Pages/OfgemHome.aspx).

Health and Safety Executive (HSE)—the HSE is an independent regulator, responsible for enforcing health and safety legislation in workplaces (www.hse.gov.uk).

Environment Agency (EA)—the EA is the environmental regulator for all onshore oil and gas operations, including shale gas, coal bed methane and underground coal gasification in England. In April 2013, Natural Resources Wales (NRW), a new body formed by the Welsh Government, took over the functions previously carried out by the Environment Agency in Wales (www.environment-agency.gov.uk).

Offshore Safety Directive Regulator (OSDR)—the OSDR is the Competent Authority responsible for overseeing industry compliance with the EU Directive on the safety of offshore oil and gas operations. The OSDR is a partnership between BEIS' OGED and HSE's Energy Division (ED) (www.hse.gov.uk/osdr).

Electric Sector—the UK has a privatized electricity sector, where electric generators and marketers operate in a competitive environment. EDF Energy, a subsidiary of Électricité de France, was the largest supplier of electricity to the national transmission system in 2016, accounting for 24% of generation. The retail electric market is dominated by six large providers, with British Gas accounting for about one-quarter of the total market. The share of independent providers in the retail market is growing, accounting for about 15% of the market at the end of 2016, up from 1% at the beginning of 2012.

The UK electric transmission system is regulated and is managed by independent system operators. The electric transmission systems of England, Wales, and Scotland are fully integrated and are operated as a single market by National Grid Electricity Transmission plc (National Grid), which also owns the electric transmission system in England and Wales. The electricity grid in Northern Ireland is integrated with the grid of the Republic of Ireland, and the combined system is operated by the Single Electricity Market Operator (SEMO). The UK transmission system has two interconnections with the Irish system—one between Scotland and Northern Ireland and one between Wales and the Republic of Ireland. The UK system also has two interconnections with continental Europe—one each with France and the Netherlands (www.legislation.gov.uk).

The UK government's energy policies have long sought to encourage the use of low-carbon sources of energy to generate electricity. The UK's Non Fossil Fuel Obligation (NFFO) was introduced in 1990 to support nuclear and renewables generation.

In 2002, the UK replaced the NFFO with the Renewables Obligation (RO), which continued support for renewables generation but did not include support for nuclear generation. The government is phasing out the RO support system, and beginning in 2017, the main method of renewables support will be via feed-in tariffs (FIT) implemented as contracts for difference. A FIT offers a guaranteed price for electricity generated by qualifying generation projects.

Another government program that supports the use of renewables and other low-carbon sources of electricity generation is the UK's Carbon Price Floor (CPF), established in April 2013 for the 2013—14 tax year. The UK's CPF works in combination with the EU's Emissions Trading System (ETS). If the EU ETS carbon price is lower than the UK CPF, electric generators have to buy credits from the UK Treasury to make up the difference. The CPF applies to both generators that produce electricity for the grid and companies that produce electricity for their own use. Since the beginning of 2012, the EU ETS carbon price has stayed below 10 Euro per metric ton of carbon (below GBP 8 per metric ton or below US$13 per metric ton). The UK CPF has gradually risen to GBP 18 per ton of carbon (about US$25 per ton), where it will remain at least through the 2019—20 fiscal year. The CPF was originally designed to rise to GBP 30 per ton of carbon in 2020 and GBP 70 per ton of carbon in 2030 (about US$40 and US$95 per ton, respectively), but was it capped to limit the impact on businesses.

At the end of 2017, the UK had 15 operating nuclear reactors, with a current capacity of slightly less than 9 gigawatt electric (GWe), according to the World Nuclear Association. All 15 operating reactors are owned and operated by EDF

Energy. Most of the existing nuclear capacity started operations in the 1970s or 1980s and is due to be shut down by 2025. Nuclear power generation accounted for 20% of the country's total gross generation in 2016 (https://www.eia.gov/international/analysis/country/GBR).

Germany

Germany was the largest energy consumer in Europe and the seventh-largest energy consumer in the world in 2015, according to BP Statistical Review of World Energy. It was also the fourth-largest economy in the world by nominal GDP after the United States, China, and Japan in 2015. Its size and location give it considerable influence over the European Union's energy sector. However, Germany must rely on imports to meet the majority of its energy demand.

Germany has begun a long-term initiative to transition to a low-carbon, more efficient energy mix. This initiative, known as the Energiewende, includes ambitious targets for phasing out coal and nuclear and developing renewable energy. Key targets include relying on renewable energy sources for at least 60% of final energy consumption and 80% of electricity consumption by 2050, as well as closing Germany's remaining nuclear power plants by 2022.

Petroleum and other liquids continue to be Germany's main source of energy. In 2015, Germany consumed 2.3 million barrels per day (b/d), making up 34% of the country's total primary energy consumption in 2015. The transportation sector accounts for the largest share of petroleum product demand. The government has a goal of putting 1 million electric vehicles on the road by 2020 and 6 million by 2030. At the end of 2015, there were approximately 25,000 electric vehicles registered in Germany.

With more than 2 million b/d of crude refining capacity at the end of 2015, Germany is one of the largest refiners in the world, and the second largest in Europe and Eurasia after Russia. Germany imports oil through several crude oil pipelines and sea ports. The two largest pipelines are the Druzhba pipeline from Russia via Belarus and Poland, and the Trans Alpine pipeline from Trieste, Italy on the Adriatic Sea. Wilhelmshaven, Germany's sole deepwater port, handles most of Germany's international oil trade.

Germany was the largest consumer of natural gas in Europe in 2015, consuming 7.2 billion cubic feet per day (Bcf/d) of natural gas. Imports account for about 90% of total natural gas supply, and most imports come from three countries: Russia (40% of total imports in 2015), Norway (21%) and the Netherlands (29%), according to the German energy research group, AG Energiebilanzen. Natural gas imports from the Netherlands were down 15% in 2014 versus 2013, as the Netherlands restricted production from its giant Groningen field resulting from concerns about earthquakes in producing areas. Germany plans to completely cease imports from Groningen by 2030.

Germany has no LNG terminals, but is well connected to much of the rest of Europe via natural gas pipelines. Germany imports natural gas from Russia via the Nord Stream pipeline (completed in 2011) under the Baltic Sea and via the Yamal-Europe pipeline running through Belarus and Poland. Germany imports from Norway via Europipe I and II, and via Norpipe. Germany also has pipeline connections with all of its neighbors (Denmark, Netherlands, Belgium, Luxembourg, France, Switzerland, Austria, Czech Republic, and Poland) and either imports or exports natural gas to all of them.

Coal is Germany's most abundant indigenous energy resource, and it accounted for 24% of Germany's total primary energy consumption in 2015. In 2015, Germany was the eighth largest coal producer in the world, producing 203 million short tons of coal. The power and industrial sectors account for most coal consumption in Germany, with lignite-fired generation accounting for about 25% of total electric generation in 2015. According to Germany's Federal Network Agency, there were just over 20 gigawatts (GW) of lignite-fired electric generating capacity operating as of the beginning of 2015. However, in July 2015 the government announced that it would mothball 2.7 GW of the oldest lignite-fired capacity to meet its 2020 climate goals.

Germany was the seventh-largest generator of nuclear energy in the world in 2015 with 86.8 terawatt hours (TWh), accounting for about 16% of total electricity generation, down from 133 TWh generated in 2010 (28% of total generation). Following Japan's Fukushima accident in March 2011, the German government decided to close eight reactors launched before 1980 because of public protests and to close Germany's nine remaining nuclear reactors before 2022.

Germany is a regional and world leader on several categories of renewable energy use. In 2015, 30% of the electricity generated in Germany came from renewables. The German government stated that it will continue to shift its sources of electricity generation from nuclear power to renewable energy sources (https://www.eia.gov/international/analysis/country/DEU).

The governmental laws and systems, including related regulatory agencies, in Germany are somewhat unique among capitalistic democracies. The current focus throughout the government is a very progressive system designed to support new technology in the several parts of the energy sector with a heavy emphasis on moving to a renewable energy world with a low-carbon footprint. The government is organized in the first instance to protect the environment, promote conservation and shield the population from risks. In the second incidence the transportation system by air, water and land are focused on new technologies like driverless vehicles battery powered vehicles and fuel cell powered vehicles fed with hydrogen, and rail operating safety. One of the key principles in German environmental law is the "polluter pays principle," where the polluter, (that is the party causing the environmental damage) is liable to the competent authorities to remedy this damage.

Federal Ministry for the Environment, Nature Conservation, Construction and Nuclear Safety (ENC) (https://www.bmu.de/en/topics/climate-energy/).

Germany supports the Climate Change and the Paris Agreement and much of the Federal Ministry for the Environment, Nature Conservation, Construction and Nuclear Safety is focused on policies and objectives that move the country toward renewable energy and a low-carbon footprint economy. The Federal Cabinet initiated a climate change act contains binding climate targets for every year and every sector. Germany is now the first country with this kind of a binding roadmap toward greenhouse gas neutrality. Should one sector veer off course, a binding adjustment mechanism will take effect and serve as a safety net. Germany's target of becoming greenhouse gas neutral by 2050 will be laid down in law for the first time (Federal Ministry for the Environment, Nature Conservation and Nuclear Safety, https://www.bmu.de/fileadmin/Daten_BMU/Pools/Broschueren/treibhausgasemissionen_en_bf.pdf "Germany's climate targets will be legally binding for the first time. There will be clear provisions for what happens if a sector deviates from the agreed climate action path. Like many others in the Climate Change agreement Germany fell short of its climate target. The German climate change act will fundamentally improve the way the Federal Government cooperates on climate action").

The climate change act, for the first time, sets legally binding limits on how much CO_2 each sector is allowed to emit per year. Precisely quantified and verifiable sectoral targets are defined for every year from 2020 to 2030. Every sector will also be monitored every year to see whether it emits too much CO_2. The German Environment Agency and an independent council of experts will be responsible for monitoring. In the case that a sector deviates from the reduction path, the climate change act obliges the responsible ministries to take immediate action. The climate change act thus guarantees that the cross-sectoral 2030 climate target (-55% CO_2 compared to 1990) will be achieved.

In addition, the act will anchor Germany's target of greenhouse gas neutrality by 2050 in law for the first time. Until now, Germany's 2050 target was to reduce CO_2 emissions by 80%—95%. The new target sends a clear signal to all sectors to prepare in good time for a fossil-free economy. The agreement includes provisions that the Federal Government must define annually falling emission quantities for the post-2030 period, which will plot a course for 2050 greenhouse gas neutrality in detail.

The Federal Cabinet also adopted the Climate Action Programme 2030. This programme is based on the key issues adopted by the Climate Cabinet. The programme includes numerous extensive measures that are to be introduced via cabinet decision. The agreement provides for expert evaluation and public transparency" (https://www.bmu.de/en/pressrelease/minister-schulze-climate-action-becomes-law/).

Federal Ministry of Transport and Digital Infrastructure (https://www.bmvi.de/EN/Home/home.html?https = 1).

"The Federal Ministry of Transport and Digital Infrastructure supports applied research and development in the electric mobility sector, the procurement of electric vehicles and the deployment of charging infrastructure (electric vehicle charging stations and hydrogen refueling points) in Germany on a technology-neutral and cross-modal basis.

The Federal Ministry of Transport provides financial assistance for plug-in hybrids, battery drivetrains and fuel cells:

- in road passenger and freight transport (e.g. for passenger cars, electric buses, commercial and delivery vehicles);
- battery and fuel cell drivetrains in rail transport (diesel hybrid traction);
- battery and fuel cell drivetrains in aviation (for instance fuel cells for on-board energy supply); and
- battery and fuel cell drivetrains in shipping (fuel cells for electricity supply).

Germany supports intelligent transport systems in a European context, meaning a European legal framework for intelligent technologies via aframe work for the deployment of Intelligent Transport Systems in the field of road transport and for interfaces with other modes of transport" (https://www.bmvi.de/EN/Topics/Digital-Matters/Intelligent-Transport-Systems/intelligent-transport-systems.html).

For rail transportation the focus is on safety. Railways must have a safety management system in place and/or employ highly skilled staff, namely operations and safety managers. They receive their license after they have passed a special state examination (Regulations on the Examination of Railway Operations and Safety Managers). Their appointment as an operations and safety manager has to be confirmed by the competent regulatory authority. It can be revoked if the public interest is at risk (https://www.bmvi.de/EN/Topics/Mobility/Rail/Railway-Operating-Safety/railway-operating-safety.html).

"For aviation it is recognized that air transport plays a vital role for the personal mobility of people and the tourism industry in Germany and in other countries- and largely benefits the economy in this field. In order to strengthen Germany's competitiveness as an air transport hub, the coalition parties CDU, CSU and SPD agreed in their coalition agreement to develop an air transport strategy. For this purpose, a stronger role of the Federal Government in the planning of a nation-wide airport network is to be worked out via a dialogue with the federal states and interested members of the public" (https://www.bmvi.de/EN/Topics/Mobility/Aviation/Aviation/aviation.html).

Another focus of the Federal Ministry of Transport and Digital Infrastructure is broadband deployment in Germany. Digitalization is a fundamental technological trend of our time which opens up a wide range of new opportunities for both industry and society. Included is new business models and forms of communication. In the digital world products become more individual, medical diagnoses and treatments better, transport safer, energy consumption lower and our daily life as a whole more comfortable. There is hardly any field of life not affected by digital transformation. The Ministry supports an effort to bring rural and sparsely populated regions up to par with urban areas.

Federal Network Agency

The Bundesnetzagentur (Federal Network Agency) for Electricity, Gas, Telecommunications, Post and Railway (BNetzA) is an independent federal authority based in Bonn. The prime task of the Federal Network Agency is to foster competition in the fields for which it is responsible by regulating these sectors and by ensuring that access to the networks is granted in a way that is both fair and nondiscriminatory.

The central tasks undertaken by the Bundesnetzagentur (Federal Network Agency) with regard to energy regulation notably include the approval of network fees for the transmission of electricity and gas, the removal of obstacles that impede access to the energy supply networks for suppliers and consumers, the standardization of the relevant processes for switching suppliers, and the improvement of conditions under which new power plants are connected to the networks (https://www.bundesnetzagentur.de/EN/Areas/Energy/energy-node.html).

Since 2011, the Bundesnetzagentur (Federal Network Agency) has also been responsible for the faster expansion of the electricity grid through implementation of the Grid Expansion Acceleration Act.

Federal Maritime and Hydrographic Agency

The Federal Maritime and Hydrographic Agency (Bundesamt für Seeschifffahrt und Hydrographie, BSH) is an authority in the division of the Federal Ministry of Transport, Building and Urban Development. As a partner for maritime shipping, protection of the environment and uses of the sea, the BSH, supports maritime shipping and the maritime economy, consolidates safety and the protection of the environment, promotes sustainable uses of the sea, ensures continuity in the measurements, and provides current information about the conditions of the North and Baltic Sea (https://www.bmwi.de/Redaktion/EN/Artikel/Ministry/bundesnetzagentur-bnetza.html).

The Paris climate agreement—Germany support

The Paris Agreement is a milestone in international climate change policy. It sets out a new legal framework for climate change mitigation at global level, with market-based mechanisms playing a key role. Article 6 of the Agreement enables the Parties to use cooperation mechanisms. These in turn allow emission reductions to be transferred between countries. This means that emission reduction activities can be implemented in one country, but a portion of the resulting emission reductions can be counted toward the emission reduction target of another. The Paris Agreement thus lays the foundation for the use of market-based climate change mitigation mechanisms beyond national borders. Many issues of central importance in their implementation remain open, however. These must be addressed and clarified in the course of the coming years.

The Agreement's cooperation mechanisms must be designed in such a way that emission reductions can be accurately recorded and counted toward the national greenhouse gas inventories of the countries involved. This is the only way to prevent double counting of the emission reductions achieved, first by the host country and then again by the country to which the reductions are transferred.

Also, clarification is necessary regarding the relationship between finance provided in relation to the cooperation mechanisms and general climate finance. It is also necessary to prevent double counting—in this case of the funding provided—because the cooperation mechanisms were created explicitly to raise climate change ambition and not to provide an escape hole for countries wanting to duck out of serious climate change mitigation effort.

Finally, the issue of how, in relation to a country's existing emission reduction target, use of the cooperation mechanisms will impact the design and ambition of its future NDCs. The implementation requirements for the cooperation mechanisms and the associated guidelines must ensure that the mechanisms provide no incentive whatsoever for host countries to minimize their own contributions to mitigating climate change and push climate change mitigation ambition off onto others because they prefer to sell their emission reduction potential on the carbon market.

One central challenge in all of this will involve developing the climate change mechanisms at differing levels without approaching each of them in isolation. More and more price-based mechanisms are being planned, developed and introduced—both at national and at sub-national level. The global framework set out by the Paris Agreement must thus be designed in a way that does not detract from these initiatives, but rather supports and harmonizes them while securing the environmental integrity of the system as a whole.

The German government is thus committed to finding a solution to all of the challenges outlined above, the ultimate aim being to ensure that the international cooperation mechanisms contained in the Paris Agreement will secure the environmental integrity of the climate change regime, contribute to greater reduction of emissions and drive sustainable development in countries that implement action to mitigate climate change.

Russia

Russia is the world's largest producer of crude oil (including lease condensate) and the second-largest producer of dry natural gas. Russia also produces significant amounts of coal. Russia's economy is highly dependent on its hydrocarbons, and oil and natural gas revenues account for more than one-third of the federal budget revenues. Most of Russia's oil production originates in West Siberia and the Urals-Volga regions. However, production from East Siberia, Russia's Far East, and the Russian Arctic has been growing.

Sanctions and lower oil prices have reduced foreign investment in Russia's upstream, especially in Arctic offshore and shale projects, and they have made financing projects more difficult.

Russia holds the largest natural gas reserves in the world and is the second-largest producer of dry natural gas. The state-run Gazprom dominates the country's upstream natural gas sector, although production from other companies has been growing.

Russia is one of the top producers and consumers of electric power in the world, with more than 240 gigawatts of installed generation capacity. In 2016, gross electric power generation totaled 1071 billion kWh, and Russia consumed about 900 billion kWh.

In 2016, Russia exported more than 5 million b/d of crude oil and condensate. Most Russian exports (70%) went to European countries, particularly the Netherlands, Germany, Poland, and Belarus. About 36% of Russia's federal budget revenue in 2016 came from oil and natural gas activities. Although Russia is dependent on European consumption, Europe is similarly dependent on Russian oil supply, with more than one-third of crude oil imports into OECD Europe in 2016 coming from Russia.

Russia also exports fairly sizable volumes of oil products. According to Eastern Bloc Research, Russia exported about 1.3 million b/d of fuel oil and an additional 990,000 b/d of diesel in 2016. It exported smaller volumes of gasoline (120,000 b/d) and liquefied petroleum gas (75,000 b/d) during the same year.

Russia has an extensive domestic distribution and export pipeline network. Russia's domestic and export pipeline network is nearly completely owned and run by the state-owned Transneft. One notable exception is the Caspian Pipeline Consortium (CPC) pipeline, which runs from Tengiz field in Kazakhstan to the Russian Black Sea port of Novorossiysk. The CPC pipeline is owned by a consortium of companies, with the largest share (24%) owned by the Russian government (whose interests in the consortium are represented by Transneft). KazMunaiGaz (19%), the state-owned oil and natural gas company of Kazakhstan, and Chevron (15%) are the second- and third-largest shareholders in the consortium. Another exception is the TransSakhalin pipeline, owned by the Sakhalin-2 consortium, in eastern Russia.

The top four Russian ports (Novorossiysk, Primorsk, Ust-Luga, and Kozmino) for crude oil exports together accounted for 84% of Russia's seaborne crude oil exports in 2016. Novorossiysk is Russia's main oil port on the Black Sea coast. It handles petroleum from Central Asian countries as well as from Russia. The Primorsk and Ust-Luga terminals are both located near St. Petersburg, Russia, on the Gulf of Finland. The Primorsk terminal opened in 2006, and the Ust-Luga oil terminal opened in 2009. Both

Primorsk and Ust-Luga receive oil from the Baltic Pipeline System, which brings crude oil from fields in the Timan-Pechero, West Siberia, and Urals-Volga regions. Ust-Luga is also a major port for Russian coal and hydrocarbon gas liquids exports.

In 2016, Russia was the world's second-largest dry natural gas producer (approximately 21 Tcf), surpassed only by the United States (26.5 Tcf). According to Eastern Bloc Energy, most of the country's production comes from the Yamal-Nenets region of West Siberia.

Russia held the world's largest natural gas reserves, with 1688 trillion cubic feet (Tcf), as of January 1, 2017. Russia's reserves account for about one-quarter of the world's total proved natural gas reserves. Most of these reserves are located in large natural gas fields in West Siberia. Five of Gazprom's largest operating fields (Yamburg, Urengoy, Medvezhye, Zapolyarnoye, and Bovanenkovo)—all of which are in the Yamal−Nenets region of West Siberia—together account for about one-third of Russia's total natural gas reserves.

The state-run Gazprom dominates Russia's upstream natural gas sector, producing about two-thirds of Russia's total natural gas output in 2016. While independent and oil company producers have gained importance, upstream opportunities remain fairly limited for independent producers and other companies, including Russian oil majors. Furthermore, Gazprom's dominant upstream position is reinforced by its legal monopoly on pipeline gas exports.

Much like the oil sector, a number of ministries and regulatory agencies are involved in Russia's natural gas sector. The Ministry of Natural Resources and Environment issues field licenses, monitors compliance with license agreements, and levies fines for violations of environmental regulations. The Ministry of Energy develops and implements general energy policy and oversees LNG exports. The Finance Ministry is responsible for hydrocarbon extraction and export taxes, while the Ministry of Economic Development supervises tariffs.

The Federal Antimonopoly Service is the main regulatory agency involved in the natural gas sector. This agency regulates pipeline tariffs and oversees charges of abuse of market dominance, including charges related to third-party access to pipelines.

In Russia, natural gas associated with oil production is often flared. According to the US National Oceanic and Atmospheric Administration (NOAA), Russia flared an estimated 850 billion cubic feet (Bcf) of natural gas in 2016, the most of any country. At this level, Russia accounted for about 16% of the total volume of natural gas flared globally in 2016 from upstream sources. A number of Russian government initiatives and policies have set targets to reduce routine flaring of associated gas. Also, regulatory changes have made it easier and more profitable for third-party producers to transport and market their natural gas. According to the NOAA estimates, from 2012 to 2014, flared natural gas in Russia declined on average 9% per year before growing 8% in 2015 and 14% in 2016.

In 2016, almost 90% of Russia's 7.5 Tcf of natural gas exports were delivered to customers in Europe via pipeline, with Germany, Turkey, Italy, Belarus, and the United Kingdom receiving the bulk of these volumes. Much of the remainder was delivered to Asia as LNG. In 2013, Ukraine was Russia's third-largest importer of natural gas, importing 0.8 Tcf from Russia. In 2016, Ukraine imported a total of 0.4 Tcf of natural gas, none of which was purchased from Russia. Because of a pricing and payments dispute and as part of the wider tensions between the two countries, Ukraine has decreased the volume of natural gas it buys from Russia and increased the natural gas it buys from its western neighbors. However, Ukraine still acts as a transit country for deliveries of pipeline natural gas from Russia to Western Europe, and much of the natural gas Ukraine buys from Western Europe physically originates in Russia.

Russia is one of the top producers and consumers of electric power in the world, with more than 240 gigawatts of installed generation capacity. In 2016, gross electric power generation totaled 1071 billion kWh, and Russia consumed about 900 billion kWh. Fossil fuels (oil, natural gas, and coal) are used to generate about two-thirds of Russia's electricity, with hydropower and nuclear each accounting for about one-sixth of total electric generation. Most of the fossil fuel-fired generation comes from natural gas. Russia's gross electric power generation totaled 1071 billion kWh in 2016, and net electricity consumption was about 900 billion kWh. Russia exported approximately 18 billion kWh of electricity in 2016 and imported about 3 billion kWh of electricity.

Much like the oil and natural gas sectors, a number of ministries and regulatory agencies are involved in the electric sector. The Ministry for Economic Development supervises tariffs and investment in the energy sector. The Ministry of Energy is in charge of general energy policy, including development of the legal framework for the electric sector. The Ministry of Energy also approves investment plans for Russia's electric transmission system.

The main regulatory agency involved in the sector is the Federal Antimonopoly Service, which regulates transmission tariffs and oversees compliance with the unbundling rules and charges of abuse of market dominance in competitive electric markets. The state atomic energy corporation, Rosatom, controls all aspects of the nuclear sector in Russia, including uranium mining, fuel production, nuclear plant engineering and construction, generation of nuclear power, and nuclear plant decommissioning. Russia has seven regional power systems in the electric sector. These systems are: Northwest, Center, South, Middle Volga, Urals, Siberia, and Far East.

The Russian electric sector was restructured in the past decade, and much of it was privatized. The reform required ownership unbundling in the electric sector, separating the industry into largely privately-owned, competitive generation assets and state-controlled, regulated transmission assets. No company is allowed to own both generation and transmission assets. The Federal Grid Company, which is more than 70% owned by the Russian government (directly and through Gazprom), controls most of the transmission and distribution infrastructure in Russia. The grid comprises more than 1.5 million miles of power lines, including slightly less than 100,000 miles of high-voltage cables of 220 kilovolts (Kv) or more. The government has been trying to attract private investment into the wholesale and regional electric generating companies. As part of the market reform, most of Russia's fossil-fueled power generation was also privatized, while nuclear and hydropower remain under state control.

Russia has an installed nuclear capacity of more than 26 million kilowatts distributed across 35 operating nuclear reactors at 10 locations. Nine plants are located west of the Ural Mountains. The exception is the Bilibino plant in the far northeast. Russia's current federal target program envisions a 45%–50% nuclear power share of total generation by 2050 and a 70%–80% share by 2100. To achieve these goals, the rapidly aging nuclear reactor fleet in Russia will need to be replaced with new nuclear power plants. As of July 1, 2017, seven new nuclear reactors were officially under construction across Russia with 5468 MWe net generating capacity (5904 MWe gross). One of the plants under construction is a floating nuclear power plant, which is scheduled to be commissioned by 2019.

With 177 billion short tons of coal at the end of 2016, Russia held the world's third-largest recoverable coal reserves, after the United States and China. Russia produced 425 million short tons in 2016, making it the sixth-largest coal producer in the world behind China, India, the United States, Australia, and Indonesia. Almost 80% of Russia's coal production was steam coal, and slightly more than 20% was coking coal.

In 2016, Russia consumed about 45% of its coal production and exported the rest. Although coal accounts for a relatively modest share of Russia's total energy consumption, coal is a more vital part of consumption in Siberia, where most Russian coal is mined.

More than half of Russia's coal production comes from the Kuzbass basin in central Russia. Kuzbass coal must travel long distances by rail to reach ports in the west or the east of the country for export to European or Asian consumers. This long overland transport generally puts Russian coal at an economic disadvantage to competing sources of coal. Even so, in 2016, Russia was the third-largest coal exporting country in the world, exporting 189 million short tons, seaborne and overland. The top two coal exporters in 2016 were Australia and Indonesia" (Energy Information Administration, International Analysis, https://www.eia.gov/international/analysis/country/RUS) and (Shah, Energy Security https://www.globalissues.org/article/595/energy-security).

Venezuela

Reduced capital expenditures by state-owned oil and natural gas company Petròleos de Venezuela, S.A. (PdVSA) are resulting in foreign partners continuing to cut activities in the oil sector, making crude oil production losses increasingly widespread. With Venezuela's heavy dependency on the oil industry, the country's economy will likely continue to shrink, and that the runaway inflation will remain the mainstay at least in the short term.

Venezuela's revenue from oil exports is severely constricted because only about half of the exports generate cash revenues. US refiners are among the few customers that still remit cash payments. The remaining crude oil exports are sold domestically at a loss or sent as loan repayments to China and Russia (the repayments to Russia are sent to Nayara Energy's (formerly Essar) Vadinar refinery in India to service debt that Venezuela owes to Russian oil company Rosneft, the coowner of the Vadinar refinery).

In January 2018, Venezuela had 302 billion barrels of proved oil reserves, the largest in the world. Venezuela's crude oil production has declined rapidly and has fallen to a 30-year low (excluding the decline in production during the 2002—03 strike). As of May 2018, Venezuela's crude oil production was 1.4 million barrels per day (b/d). Despite its production declines, Venezuela was still the 12th largest producer of petroleum in the world in 2017.

The number of active rigs fell from nearly 70 in the first quarter of 2016 to 25 rigs in September 2018. Reports indicate that missed payments to oil service companies, a lack of working upgraders, a lack of knowledgeable and able managers and workers, and declines in oil industry capital expenditures will continue to affect crude oil production negatively.

Venezuela had 1.3 million b/d of domestic nameplate crude oil refining capacity in 2017, which were all operated by PdVSA. However, actual refining capacity in early 2018 was less than half of its nameplate capacity, estimated at 626,000 b/d.

PdVSA also operates significant refining capacity outside the country. The largest share of Venezuela's foreign downstream operations is in the United States, followed by significant operations in the Caribbean and stakes in Europe.

According to tanker tracking data, Venezuela exported an average of 1.5 million b/d of crude oil in 2017, 10% lower than the 2016 level. In the first quarter of 2018, exports of Venezuela's crude oil fell to 1.1 million b/d, based on tanker loadings data.

Venezuela's crude oil exports to the United States fell from 840,000 b/d in December 2015 to about 506,000 b/d in October 2018—at its peak, US imports of Venezuelan crude oil averaged 1.1 million b/d in 2007. Venezuela was the third-largest supplier of crude oil imports into the United States after Canada and Saudi Arabia.

The United States is the primary destination for Venezuelan crude oil shipments and receives about 41% of Venezuela's total exports. In mid-June 2018, ConocoPhillips' seizure of PdVSA's Caribbean export and storage facilities hampered Venezuela's ability to maintain its export levels until it came to a repayment agreement with ConocoPhillips in August 2018. The country relies on these terminals to send crude oil to Asia.

Venezuela had 203 trillion cubic feet (Tcf) of proved natural gas reserves in 2017. In 2016, Venezuela produced 3.3 billion cubic feet per day (Bcf/d) of natural gas and consumed 3.4 Bcf/d of natural gas. In 2017, more than 38% of Venezuela's total natural gas production was reinjected, according to data published by Rystad Energy.

To attract foreign investment, Venezuela awarded 18 natural gas exploration and production licenses to private companies, but currently only five of those licensees are operating (including three in which PdVSA Gas serves as a minority partner). As of early 2018, these licenses accounted for 860 thousand cubic feet per day (Mcf/d) of natural gas production, according to IPD Latin America. In July 2015, operations began at the offshore Perla field project, where output reached nearly 550 Mcf/d in the first quarter of 2018, according to IPD Latin America.

In 2017, Venezuela generated more than 117 billion kWh of electricity, an increase of more than 5% compared with the previous year, according to data published by BP. The low electric generation in 2016 was primarily the result of extreme drought conditions during the year and lack of sufficient rainfall because hydroelectricity provides most of Venezuela's electricity supply.

More recent declines in generation are the result of technical failures affecting both the hydropower and thermal electric power generation plants. These issues include the government's inability to repair or maintain facilities that are vital to electric power generation.

In February 2018, six states in Venezuela reported power blackouts that lasted as long as 15 hours, affecting large population centers (including the capital city Caracas). Since then, the National Electricity Corporation (CORPOELEC) announced that it is implementing power rationing that affects seven states (Energy Information Administration, International Analysis, https://www.eia.gov/international/analysis/country/VEN; https://www.eia.gov/international/content/analysis/countries_long/Venezuela/venezuela_exe.pdf).

Saudi Arabia

Saudi Arabia is the world's second-largest holder of proved oil reserves, ranking second only to Venezuela, holding roughly 16% of total reserves. In 2016, Saudi Arabia was the largest exporter of total petroleum liquids (crude oil and petroleum products), with exports mostly destined to Asian and European markets. During that same year, Saudi Arabia was the world's second-largest petroleum liquids producer behind the United States and was the world's second-largest crude oil and lease condensate producer behind Russia. During the first half of 2017, Saudi Arabia has maintained this position, despite lower production than in the previous year.

Saudi Arabia's economy remains heavily dependent on petroleum exports, which accounted for nearly 75% of total Saudi export value in 2016. According to the International Monetary Fund (IMF), about 60% of Saudi government's revenues are oil-based, and the real GDP growth rate fell significantly in 2016, as a result of the slowdown in oil-driven growth that year Saudi Arabia's oil revenues have fallen dramatically as crude oil prices have decreased since mid-2014. EIA estimates that Saudi Arabia's net oil export revenues totaled $133 billion in 2016, compared with $159 billion (in 2016 dollars) in 2015 when Saudi Arabia boosted production and exports to record highs. Despite the increase in output, Saudi Arabia's net oil revenues in 2015 saw dramatic decrease compared with 2014, when the country was estimated to have earned $301 billion (in 2016 dollars).

In addition to continued investment in maintaining its crude oil production capacity, Saudi Arabia has also invested considerable resources to develop its natural gas, refining, petrochemicals, and electric power industries. The country's natural gas and electric power industries, in particular, are geared to meet increasing domestic demand. Investments in refining and petrochemical industries aim to improve Saudi Arabia's ability to compete internationally in these sectors.

Saudi Arabia's oil and natural gas operations are dominated by Saudi Aramco, the national oil and gas company, which is the world's second-largest oil company in terms of production (behind Russia's Rosneft). Saudi Arabia's Ministry of Petroleum and Mineral Resources and the Supreme Council for Petroleum and Minerals oversee the oil and natural gas sector and Saudi Aramco.

In 2016, Saudi Arabia announced that it will undertake a national transformation called Vision 2030, encompassing both cultural, governance, and economic aspects of Saudi society. On the economic front, Vision 2030 plans to make the kingdom less dependent on income from oil production by broadening the economic base. The plan outlines far-reaching reforms of the energy sector and includes the partial privatization of state-owned Saudi Aramco, planned for the second half of 2018. Income from Saudi Aramco's initial public offering (IPO) would be used to help finance the economic transition.

Saudi Arabia is located near two of the world's busiest chokepoints, and most of its crude oil and petroleum liquid exports travel through them. The Strait of Hormuz, which connects the Persian Gulf with the Gulf of Oman and the Arabian Sea, is the world's most important chokepoint. The oil flow of 17 million barrels per day (b/d) in 2015 through this strait accounts for about 30% of all seaborne-traded crude oil and other liquids during the year. In 2016, total flows through the Strait of Hormuz increased to a record high of 18.5 million b/d. This strait is an important route for the Persian Gulf countries for oil and liquefied natural gas exports.

Saudi Arabia produces a range of crude oils, from heavy to super light. Of Saudi Arabia's total crude oil production capacity more than 70% is considered light gravity, with the remaining crude oil considered medium or heavy gravity. Lighter grades generally are produced from onshore fields, while medium and heavy grades come mainly from offshore fields. Most Saudi oil production, except for the Arab Light and Arab Super Light crude oil types, is considered sour (containing relatively high levels of sulfur). Saudi Aramco said that its oil fields do not require the use of enhanced oil recovery (EOR) techniques, although fields in the PNZ could require steam flooding. In 2015, Chevron was developing a full-field steam flood injection EOR project to offset field declines and to boost production of the heavy oil play. However, because of the dispute between Saudi Arabia and Kuwait and Chevron's difficulty in securing work and equipment permits, Chevron's activities in the PNZ have stopped. Saudi Aramco continues to drill at existing fields to help compensate for the natural declines from the mature fields.

Another regional chokepoint, Bab el Mandeb, links the Gulf of Aden and the Red Sea. This waterway is a strategic link between the Mediterranean Sea and the Indian Ocean. An estimated 4.8 million b/d of crude oil and refined petroleum products flowed through this waterway in 2016 toward Europe, the United States, and Asia.

Saudi Aramco operates the world's largest oil processing facility and crude oil stabilization plant in the world at Abqaiq, in eastern Saudi Arabia. The plant has a crude oil processing capacity of more than 7 million b/d. The plant processes the majority of Arab Extra Light and Arab Light crude oils, as well as natural gas liquids (NGL). The facility's infrastructure includes pumping stations, gas-oil separation plants (GOSPs), hydro-desulphurization units, and an extensive network of pipelines that connects the plant to the ports of Jubail, Ras Tanura, and Yanbu (for NGL). The Abqaiq processing plant is a vital part of Saudi oil infrastructure. Most of the oil produced in the country is processed at Abqaiq before export or delivery to refineries.

According to the BP Statistical Review of World Energy 2017, Saudi Arabia was the world's 10th largest consumer of total primary energy in 2016 at 266.5 million tons of oil equivalent, of which about 63% was crude oil and petroleum liquids-based.

Saudi Arabia is the largest consumer of petroleum in the Middle East, particularly in the area of transportation fuels and direct crude oil burn for power generation. Although transportation demand is substantial and growing, an increasing share of crude oil demand is attributable to direct crude oil burn for electric power generation, which can reach as high as 900,000 b/d during summer months. Crude oil and petroleum liquids reserves in the country remain plentiful, but Saudi Arabia is looking to diversify its mix of fuels for electric power generation, focusing on natural gas, nuclear, and renewable energy solutions.

Saudi Arabia exported an estimated 7.1 million b/d of crude oil in 2016, according to the Global Trade Tracker (GTT). Asia received an estimated 69% of Saudi Arabia's crude oil exports and most of its refined petroleum products.

Saudi Arabia exported an average of 1.1 million b/d of total petroleum liquids to the United States in 2016. In the first half of 2017, US total petroleum liquid imports averaged slightly higher at 1.2 million b/d. Since 2012, Saudi Arabia has been the second-largest petroleum exporter annually (after Canada) to the United States, when it surpassed US imports from Mexico. In 2016, after the United States, the next four top importers of Saudi crude and petroleum products were Japan (1.2 million b/d),

China (1.0 million b/d), South Korea (0.9 million b/d), and India (0.8 million b/d), according to GTT.

Saudi Arabia's total crude oil export and loading capacity is about 13 million b/d. Most of this capacity comes from its four primary oil export terminals.

The port of Ras Tanura on the Persian Gulf is Saudi Arabia's primary port. This facility, the world's largest offshore oil exporting port, has a combined handling capacity of about 6.5 million b/d. All of Saudi crude oil grades load at this port, along with condensate and products. The port comprises three terminals—Ras Tanura terminal, Ju'aymah crude terminal, and Ju'aymah LPG export terminal.

Saudi Arabia's total crude oil export and loading capacity is about 13 million b/d. Most of this capacity comes from its four primary oil export terminals.

The Ras Tanura terminal, the largest terminal at the port of Ras Tanura, has an average handling capacity of 3.4 million b/d and 33 million barrel storage capacity. The terminal can accommodate tankers up to 500,000 deadweight tons (dwt). All of Saudi Arabia's crude oil grades are loaded at the Ras Tanura terminal.

The Ras al-Ju'aymah terminal has an average handling crude oil capacity of about 3.12 million b/d and because of the availability of six single-point mooring buoys, the terminal can accommodate some of the largest tankers (700,000 dwt) for crude loadings. All of Saudi Arabia's crude grades are loaded at this terminal, along with bunker fuel (at a maximum loading capacity of 120,000 b/d).

The Yanbu King Fahd terminal on the Red Sea, from which most of the remaining volumes are exported, has a loading capacity of 6.6 million b/d. The terminal includes seven loading berths and can accommodate tankers up to 500,000 dwt. Total crude oil storage capacity at the terminal is 12.5 million barrels. Only Arab Light crude oil grade is loaded at the Yanbu terminal. Saudi Aramco operates more than 90 pipelines and 12,000 miles of crude oil and petroleum product pipelines throughout the country, all of which link production areas to processing facilities, export terminals, and consumption centers.

The 746-mile Petroline, also known as the East-West Pipeline, is significant because of its large capacity and because it connects production and processing facilities in the east of the country to export facilities in the west, allowing the crude oil to bypass the Strait of Hormuz. The Petroline system, which runs across Saudi Arabia from its Abqaiq complex to the Red Sea, consists of two pipelines with a total nameplate (installed) capacity of about 4.8 million b/d.

Saudi Arabia generated 330.5 billion kWh of electricity in 2016, 7% higher than in 2015, according to the BP Statistical Review of World Energy 2017. Like many developing countries in the Middle East and North Africa, Saudi Arabia faces a sharply rising demand for power. Demand is driven by population growth, a rapidly expanding industrial sector led by the development of petrochemical cities, high demand for air conditioning during the summer months, and heavily subsidized electricity rates.

Nearly all of the existing generating capacity is powered by oil or natural gas, but Saudi Arabia plans to diversify fuels used for generation, in part, to free up oil for export. Although the Saudi Electricity Company (SEC) plans to continue reducing direct crude burn for electricity generation by switching to natural gas, plans are also in place to develop renewable sources for electric power generation. The King Abdullah City for Atomic and Renewable Energy (KACARE) calls for an additional 41 GW of solar power, 17.6 GW of nuclear power, and 9 GW of wind power by 2032. These goals are ambitious, particularly given the slow pace of progress on currently planned nuclear and renewable projects.

SEC is the largest provider of electricity in Saudi Arabia, with total available generation capacity of 74.3 GW. The SEC is responsible for generation, and the National Grid S.A. Company, SEC's subsidiary, is responsible for the transmission and distribution of electrical power. The state-owned Saline Water Conversion Corporation (SWCC), which provides most of Saudi Arabia's desalinated water, is the second-largest generator of electricity, at 7.4 megawatts (MW) of installed capacity. SWCC plans to rapidly increase its desalination capacity, with an equivalent increase in generation capacity. Privately owned independent water and power plants also provide electricity to the grid. Saudi Aramco continues to build cogeneration plants to generate power for its own needs at various oil facilities. During 2016, Saudi Aramco completed cogeneration projects that added 984 MW of new power capacity.

In 2007, Saudi Arabia began allowing private participation in the electric power sector, approving the first Independent Power Producer (IPP). The first two projects began operations in 2013; the 1204 MW Rabigh 1 project in Mecca and the 1729 MW Riyadh 11 project in Dharma. The Qurayyah project in the Eastern Province was completed in 2006, which added 3927 MW of additional capacity. Several other IPP projects are expected to begin operations in the near term.

Physical improvements are needed to allow more companies to sell power to the grid. SEC has ongoing and planned projects that will link power plants in the eastern, western, and southern portions of the country. To meet peak demand requirements, Saudi Arabia is participating in the Gulf Cooperation Council's (GCC) efforts to link the power grids of member countries. The GCC is an alliance between six Persian Gulf states: Bahrain, Kuwait, Oman, Qatar, Saudi Arabia, and the United Arab Emirates. The alliance seeks to build closer ties with the member countries, including the construction of electric power interconnections. The GCC Interconnection Authority is owned by the six member countries. The GCC member states' grid interconnection was completed in 2011. In 2016, 1.32 million MWh of electric power was traded between five of the six member states (Energy Information Administration, International Analysis, https://www.eia.gov/international/overview/country/SAU).

Appendix C: Recent developments in energy technology and markets

Two rather dramatic events occurring during the first half of 2020 have significantly changed many measures of economic growth and indicators of economic health. The Covid-19 epidemic caused governments across the developed world to temporarily shut down major sectors of the economy, sending unemployment rates to levels reminiscent of the Great Depression. The unusual contagiousness of the virus suddenly caused the medical facilities and personnel to be overrun, or at least highly stressed; airlines almost completely shut down. Restaurants, hotels, sports events, and tourism-related industries shut down. Employees were furloughed or laid off, and new hires were few, if any. Then late in the second quarter racial unrest following the death of George Floyd at the hands of Minneapolis police exploded across the nation, and to some extent in several European countries, resulting in economic disruptions with burning and looting that shut down economic activity in many of our largest cities and even small towns, disruptions lasting for several weeks.

The combined impacts on the economy are only now emerging in the data. It will take some time to provide credible analyses of the long-term impacts, but it seems clear that there will be long-term impacts. First, the economy, which shut down because of the pandemic, cannot return to a near normal condition for some time because it will be a lengthy period, perhaps 12–18 months, before a vaccine is available, if at all. Therefore, many activities that normally require being in close proximity with other people that risk spreading the virus will not return until there is the protection of an available vaccine and time allows it to be widely administered. Further, people have learned to compensate for being locked-in at home, functioning by use of internet-based communication and shopping capabilities not available in any prior episode of similar disruption. The perfection of the capabilities will continue as substitutes for prior behavior, perhaps indefinitely, result in structural change in work and play patterns involving close proximity with other people. So, it is likely that some structural changes in the economy will continue with impacts—impacts that may be negative, or that may be positive. Further, if history is a guide, new behaviors and new markets will emerge and the world will be different, but may be improved in ways not now evident (U.S. Bureau of Labor Statistics News, 2021).

Definitions

Basins collections of organic sediments in depressions of the Earth's surface.

Biofuels liquid fuels created from blending biomass feedstock.

Biomass a form of energy made from organic, biological, nonfossil materials.

BTEXi a chemical compound made of benzene, toluene, ethylbenzene and xylene. BTEX is composed of naturally occurring chemicals that are typically found in petroleum products such as gasoline, home heating oil and diesel fuel.

Carbon dioxide (CO_2) a naturally occurring gas and a byproduct of human and animal respiration as well as coal, oil, and natural gas use. Carbon dioxide is a greenhouse gas in the atmosphere that keeps Earth habitable for life.

Climate change a substantial change in the statistical measures of climate over an extended period of time. These measures include temperature, precipitation, wind patterns, drought, and more.

Coal earth deposits composed of plant materials that have been shaped by pressure and heat under the Earth's surface to become a black or brown rock. Coal is a fossil fuel that is mostly used today to generate electricity.

Coalbed methane the methane contained in coal seams.

Compressed natural gas (CNG) natural gas that has been compressed and stored in pressurized vessels.

Cracking, also known as **catalytic cracking** the process in which the heavy hydrocarbon molecules in crude oil are broken down into smaller molecules using a catalyst. The process allows crude oil extracted from underground reservoirs to be made into a usable fuels and feed stocks.

Crowd-sourcing the practice of obtaining information or input into a task or project by enlisting the services of a large number of people, paid or unpaid

Crude oil a liquid mixture of hydrocarbons mined from under the Earth's surface. Crude oil can be refined into diesel and jet fuel, propane, butane, ethane and gasoline.

Daisy Bradford No. 3 the well that blew in during 1930 near Henderson in East Texas, a well that opened the East Texas Field which became the largest in the world at the time.

DOE US Department of Energy created by Department of Energy Organization Act during the Carter Administration in 1977 responsible for advancing the energy, environmental, and nuclear security of the United States; promoting scientific and technological innovation in support of that mission; sponsoring basic research in the physical sciences; and ensuring the environmental cleanup of the nation's nuclear weapons complex.

Directional drilling drilling that is purposefully nonvertical. A technology that has come to dominate the drilling of oil and gas wells in the United States.

EIA US Energy Information Administration; created during the Carter Administration in 1977is a principal agency of the US Federal Statistical System responsible for collecting, analyzing, and disseminating energy information to promote sound policymaking, efficient markets, and public understanding of energy and its interaction with the economy and the environment. EIA programs cover data on coal, petroleum, natural gas, electric, renewable and nuclear energy.

Energy siting, also known as **energy facility siting** the process of planning and constructing energy-generating facilities.

Ethanol ethyl alcohol, also called "grain alcohol" (CH3CH2OH). Ethanol is a clear, liquid, biofuel that is produced by the fermentation of plant sugars.

Feedstocks raw materials for industrial use derived from residual oil and natural gas streams; also **biomass feedstocks** include prairie grass, soybeans and corn.

FERC US Federal Energy Regulatory Commission.

Flares part of safety systems installed at wellheads, processing plants and gas refineries. If elevated levels of combustible gases are detected, natural gas is piped to a remote location and burned off.

Flowback the mixture of water and chemicals that flows up to the surface after a well has been fracked.

Frac sand one type of proppant used by fracking companies in frack fluid.

Frack fluid the combination of water, chemicals and sand (or proppant) that is injected into a well to release the gas trapped within. Water makes up much of this fluid.

Fracking, or hydraulic fracturing the process of injecting fluid, water and sand mixed with chemicals, into the ground at a high pressure in order to fracture shale rocks to release the hydrocarbons, including natural gas and crude oil.

GEAC Governor's Energy Advisory Commission; created in 1973 by Executive Order of Texas Governor Dolph Briscoe to study and recommend the Texas response to the Arab Oil Embargo of October 1973.

Geothermal energy a renewable energy resource created by using water or steam generated by geothermal reservoirs under the Earth to generate electricity or water pumps.

Global Competitiveness Index the World Economic Forum has published an annual Global Competitiveness Report every year since 2004; the report summarizes the results of ranking 141 economies amounting to 99% of world GDP.

Global Innovation Index (GII) a global innovation rating organization that produces an annual index which developed and updates the index annually; produced by a group at Cornell SC Johnson College of Business, jointly with INSTEAD, WIPO, Confederation of Indian Industry and others.

Green job a term used by some organizations to refer to a job that involves producing goods or services related to an environmental benefit or the conservation of natural resources.

Greenhouse gases heat-trapping gases that absorb infrared radiation in the Earth's atmosphere. These gases include methane, carbon dioxide, nitrous oxide, ozone, and water vapor.

Horizontal drilling a type of well drilling that allows access to sites that are not directly below a drilling site. A technology that has come to dominate the drilling of oil and gas wells in the United States.

Hydrocarbon gas liquids (HGLs) includes propane, isobutane, normal butane, ethane, isobutylene, butylene, propylene and ethylene.

Hydrocarbons chemical compounds containing carbon and hydrogen. Hydrocarbons are organic compounds and can be solids, liquids or gases.

Hydroelectric energy, or hydropower a renewable energy resource that uses the flow of water to generate electricity.

Innovation the design, investigation, development and/or implementation of new or altered products, services, processes, systems, organizational structures, or business models for the purpose of creating new value for customers and financial returns for the firm.

Liquid Petroleum Gases or LPG the hydrocarbon gases that come from refined crude oil and processed natural gas. The primary gases that make up LPG are isobutane, normal butane and propane.

LCRA Lower Colorado River Authority; a Texas regional agency that oversees water development and conservation in the Colorado river basin of central Texas.

MCF an abbreviation for a million cubic feet and is typically used to the quantity of natural gas. One MCF equals roughly 1 million British Thermal Units (BTU).

Megawatt (MW), or one million watts of electricity a unit for measuring the rate at which electricity is consumed or produced.

Megawatt hour (MWh) one thousand kilowatts applied over an hour. This unit measures both how fast energy is used over a given period of time.

Methane (CH4) the main component of natural gas. Methane is an organic compound and a greenhouse gas.

Mineral rights the right to own and mine minerals under the Earth's surface.

Natural gas a traditional energy resource composed of gaseous hydrocarbons, the primary compound being methane.

Natural gas liquids (NGL) include propane, isobutane, normal butane, ethane and other liquefied refinery natural gases.

Natural gas storage refers to the storage of natural gas until needed by electricity suppliers and other consumers. Natural gas can be stored both underground and in tanks above ground in liquid form.

Net metering a billing system where customers who generate their own electricity transfer their excess electricity into the electricity grid.

Nuclear power a way of generating energy by using heat that escapes during the nuclear fission process.

NuScale reactor a natural circulation light water reactor with the reactor core and helical coil steam generators located in a common reactor vessel in a cylindrical steel containment.

Oil Crisis of 1973 Arab members of OPEC imposed an oil embargo on the United States, Canada, Japan, Netherlands, and the United Kingdom on January 10, 1973.

Oil shale, also known as **shale rock** a sedimentary rock that contains an organic material called kerogen, which can be converted to crude oil.

OPEC Organization of Arab Petroleum Exporting Countries

Ozone a gas found in the Earth's stratosphere and absorbs ultraviolet radiation from the sun.

Petrochemicals the chemicals that make up derivatives of crude oil and natural gas and are composed of hydrocarbon molecules.

Petroleum a name for a group of substances that includes several mixtures of liquid hydrocarbons.

Produced water a type of water that occurs naturally in shale formations and has been trapped in these formations for millions of years.

Proppants small, granular materials added to frack fluid to prop open the fissures in the rock.

R&D research and development

RD&D research, development and demonstration

Renewable energy resources resources that naturally replenish and are limited in a way that a specific amount of energy is available per unit of time. These resources include wind, solar, biomass, geothermal and hydroelectric energy.

Sedimentary rocks rocks created through the accumulation of sediments from water and the earth.

Shale gas natural gas that is found in shale plays.

Shale plays areas where given types of hydrocarbons accumulate.

Slick water frac any water used in shales that does not contain high levels of gelling agents and uses friction reducers.

Slurry any mixture of water with insoluble matter. Slurries are often used in the same way as frack fluids and are a mixture of water, other liquids and small amounts of finely ground particles that are injected into a well at high speeds to create fractures in a rock.

Solar energy a renewable energy resource that is generated by the radiant heat of the sun. Once collected, solar energy is used to generate heat or electricity.

Spindletop oil well that blew in on January 10, 1901 near Beaumont, Texas.

TEAC Texas Energy Advisory Council; statutory agency created to continue the role of GEAC and manage a Texas Energy Development Fund created and funded by the Texas Legislature.

TENRAC Texas Energy and Natural Resources Advisory Council; created to continue the role of TEAC but adding other natural resource area responsibilities for environmental, water and geothermal resources.

Texas Railroad Commission the oldest regulatory body in Texas for oil, gas, mining, pipeline safety and alternative fuels; originally responsible for railroad regulation in the early 1900s, hence the name.

Traditional energy resources or nonrenewable energy resources resources that are not replenished over an extended period of time. These traditional energy resources include oil, coal and natural gas.

Transloading the transportation of goods from one mode of transportation to another. This type of transportation is used in fracking to carry bulk resources to and from dispersed fracking sites and to railway stations to be taken to processing facilities.

Vertical drilling a type of drilling that allows users to access sites directly below a drilling site.

Volatile organic compounds (VOCs) vaporize at room temperature. In the context of energy resources, VOC are part of petroleum products.

Wellbore the hole drilled to help find and recover material from a well and becomes an actual well hole.

Wind energy a renewable energy resource that is collected from the kinetic energy of wind and is converted into mechanical and electric energy.

Sources https://ballotpedia.org/Glossary_of_energy_terms and Texas Energy and Natural Resources agency publications personal library.

References

A North Texans for Natural Gas Special Report, 2016. An Energy Revolution: 35 Years of Fracking in the Barnett Shale.

Acts, 44th Leg., regular session, 1935. Chap. 120.

Advanced Resources International, 2006. Basin Oriented Strategies for CO_2 Enhanced Oil Recovery: Permian Basin, prepared for the U.S. Department of Energy.

Advisory Committee on Energy, U.S. Energy, 1971. A General Review, Report to the Secretary of the Interior. U.S. Department of Interior.

Advisory Committee on Measuring Innovation in the 21st Century Economy, 2008. Innovation Measurement: Tracking the State of Innovation in the American Economy. U.S. Department of Commerce, Washington, DC. Available from: <http://www.esa.doc.gov/sites/default/files/innovation_measurement_01-08.pdf/> (accessed 20.12.16.).

Alessi, C., 2018. OPEC Revises Crude Supply Forecasts on Higher U.S. Production. Wall Street Journal, Dow Jones & Company, February 12.

Aramco History, 2018. [online]. Available at: <http://www.saudiaramco.com/en/home/about/history.html> (accessed 30.01.18.).

Arrow, K.J. and Debreu, G., (1954). Existence of an equilibrium for a competitive economy, Econometrica, 22, 3, pp. 265-290. ATI Technology Incubator at the IC2 Institute at https://ati.utexas.edu/about-us/.

Atkinson, R.D., Ezell, S.J., 2012. Innovation Economics: The Race for Global Advantage. Yale University Press, New Haven and London. Available from: <https://www.amazon.com/> (accessed 24.07.16.).

Austin Energy News Release, 2017. [online]. Available at: <https://austinenergy.com/ae/about/news/press-releases/2017/austin-energy-grows-solar-portfolio>.

Ayres, R.U., Warr, B., 2004. Accounting for Growth: The Role of Physical Work. Center for the Management of Environmental Resources, Insead, Fontainebleau, France. <www.elsevier.com/locate/econbase>.

Bailey, R., 2001. New Growth Theory—Interview of Paul Romer by Reason Magazine From Reason Magazine.

Banet Jr., A.C., 1991. Oil and Gas Development on Alaska's North Slope: Past Results and Future Prospects. U.S Department of the Interior, Bureau of Land Management Alaska State Office.

Barta, C., 1996. Bill Clements: Texian to His Toenails. Eakin Press, Austin, TX, pp. 257, 223.

Becker, G.S., 2008. The Concise Encyclopedia of Economics [online]. Available from: <https://www.econlib.org/library/Enc1/HumanCapital.html/> (accessed 10.05.17.).

Berkun, S., 2007. The Myths of Innovation. O'Reilly Media, Inc., Sebastopol, CA.

Binder, A., 2017. Will Trump Deliver a Growth Miracle? Don't Count on It. The Wall Street Journal, January 22, p. 22.

Bower, J.L., Christensen, C.M., 1995. Disruptive technologies: catching the wave. Harv. Bus. Rev. 73 (1), 43—53.

Breakthrough Energy Collision, 2018. Available at: <http://www.b-t.energy/coalition/> (accessed 03.09.18.). Breakthrough Institute, (April 26, 2015). Lessons from the Shale Revolution, Oakland, CA.

Brooks, A.C., 2018. How to Measure the Impact of Ideas. American Enterprise Institute, Washington DC. Available at: <http://www.aei.org/publication/how-a-think-tank-measures-the-impact-of-ideas/>.

Bureau of Economic Geology, 2014. Oil and gas map of Texas. Original text by Kim, E.M., Ruppel, S.C., updated by Smith, D., Ambrose, W.A., Potter, E.C., The University of Texas at Austin, Bureau of Economic Geology, thematic map SM 10, map scale 1 inch = 100 miles.

Bureau of Economic Geology, 2017. Available from: <http://www.beg.utexas.edu/about/who-we-are> (accessed 25.06.17.).

Carbon Capture & Sequestration Technologies, 2016. MIT Energy Initiative. Massachusetts Institute of Technology webpage. Available from: <http://sequestration.mit.edu/tools/projects/tcep.html>.

Carter, J., 1977. President Carter's Address to the Nation on Proposed National Energy Policy: Delivered on 18 April 1977, 8:00 p.m. from the Oval Office at the White House while being broadcast live on radio and television.

Center for the Commercialization of Electric Technologies, 2015. Technology Solutions for Wind Integration in ERCOT: Final Report submitted for Cooperative Agreement DE-OE0000194, Prepared for the U.S. Department of Energy National Energy Technology Laboratory.

CERAWEEK by IHS Markit, 2018. Reported in the Wall Street Journal, Dow Jones & Company, March 7, p. A8. Available at: <www.ceraweek.com>.

Cheek, M., 1938. Legal history of conservation of gas in texas. Legal History. p. 279.

Chevron Corporation. Available at: <https://www.chevron.com/worldwide/united-states>.

Clark, J.A., Halbouty, M.T., 1972. The Last Boom. Random House, New York, p. 109.

Clements, W.P., 1999. Personal Interview in Dallas, Texas.

Clymore Production Co. vs. Thompson, 1935. U. F. Supp. 791, W. D. Tex.; Clymore Production Co. vs. Thompson, 1936. 13 F Supp. 469, W. D. Tex.

CMS Energy. Available at: <https://www.cmsenergy.com/home/default.aspx>.

CNN's New Year's Day Interview with Bill Gates, 2013. [online]. Available from: <https://youtu.be/0K6LNegCaD> (accessed 04.05.17.).

Coastal Corporation, 2017. Texas State Historical Association. Accessed September 4, 2017.

Cochrane, J.L., 1981. Carter energy policy and the ninety-fifth congress. In: Goodwin, C.D. (Ed.), Energy Policy in Perspective. The Brookings Institution, Washington, DC, p. 552.

Collins, R., 2018. Gas Glut in Permian Sparks Dilemma Over How Much to Burn. <https://www.bloomberg.com/news/articles/2018-06-04/a-burning-question-for-texas-what-to-do-with-all-that-gas>.

Cornell SC Johnson College of Business, INSTEAD, WIPO, CII, BSDASSAULT SYSTEMS, SEBRE, and CNI, The Global Innovation Index (GII) 2018.

Cornell SC Johnson College of Business, INSTEAD, WIPO, CII, BSDASSAULT SYSTEMS, SEBRE, and CNI, The Global Innovation Index (GII) 2019.

Covia Holding Company. Available at: <https://www.coviacorp.com/energy/permian/>. Crude Oil Windfall Profit Tax Act of 1980, 96th Congress (1979-1980), H.R.3919.

DiLallo, M., 2016. The 5 Companies Dominating the Niobrara Shale Play. Available at: <https://www.fool.com/investing/2016/08/25/the-5-companies-dominating-the-niobrara-shale-play.aspx>.

Dutta, S., Lanvin, B., Wunsch-Vincent, S. (Eds.), 2015. The Global Innovation Index 2015: Effective Innovation Policies for Development. Cornell University, INSEAD, and WIPO, Ithaca and Geneva.

Eagle Ford Shale News, Market Place and Jobs, 2009−2017. KED Interests, LLC. Available at: <https://eaglefordshale.com/geology/>.

Emergency Petroleum Allocation Act of 1973, (PL 93-159).

Encyclopedia of the New American Nation, 2017. Oil—Oil and World Power. Available from: <http://www.americanforeignrelations.com/> (accessed 20.06.17.).

Energy Information Administration (a), About EIA. Available at: <https://www.eia.gov/about/>.

Energy Information Administration (b), International Analysis. Available at: <https://www.eia.gov/international/analysis/country/VEN>.

Energy Information Administration (c), International Analysis. Available at: <https://www.eia.gov/international/analysis/country/SAU>.

Energy Information Administration (d), International Analysis. Available at: <https://www.eia.gov/international/analysis/country/RUS>.

Energy Information Administration (e), International Analysis. Available at: <https://www.eia.gov/international/analysis/country/DEU>.

Energy Information Administration (f), International Analysis. Available at: <https://www.eia.gov/international/analysis/country/GBR>.

Energy Information Administration (g), SEDS Database. Available from: <www.eia.gov/state/seds/>.

Energy Information Administration (h), State Energy Profile. Available from: <https://www.eia.gov/state/print.php?sid=TX/>.

Energy Information Administration (i), U.S. Natural Gas Wellhead Prices. Available at: <https://www.eia.gov/dnav/ng/hist/n9190us3A.htm>.

Energy Information Administration, 1981. An Analysis of the Natural Gas Policy Act and Several Alternatives; Part I: The Current State of the Natural Gas Market. Natural Gas Division, Office of Oil and Gas, pp. 38–40.

Energy Information Administration, 1986. Annual Energy Outlook, DOE/EIA-0384(86), Washington, DC, May 1987, p. 237.

Energy Information Administration, 2017a. Today in Energy, Principal Contributor, Manussawee Sukunta. Available at: <https://www.eia.gov/todayinenergy/detail.php?id=31072>.

Energy Information Administration, 2017b. Texas Natural Gas Price Sold to Electric Power Consumers Dollars per Thousand Cubic Feet. Available from: <https://www.eia.gov/dnav/ng/hist/n3045tx3m.htm/> and <https://www.eia.gov/electricity/data/>.

Energy Information Administration, 2017c. Today in Energy, Principal Contributor, Kenneth Dubin. Available at: <https://www.eia.gov/todayinenergy/detail.php?id=33552>.

Energy Information Administration, 2018, 2000. Monthly Energy Review, DOE/EIA-0384(86), Washington, DC. Available at: <https://www.eia.gov/totalenergy/data/monthly/>.

Energy Policy, 1981. second ed. Congressional Quarterly Inc., Washington, DC, pp. 21, 22, 150, 178, 181, 227–228.

Energy Policy Papers, 1974. Committee Print, Senate Committee on Interior and Insular Affairs. Serial No. 93–43 (92–72), 93 Cong. 2 sess. Government Printing Office, Washington DC, pp. 350–351.

ERCOT, Inc., 2017a. ERCOT Quick Facts. Available at: <http://www.ercot.com/> and <https://www.smartbrief.com/tags/electric-reliability-council>.

ERCOT, Inc., 2017b. ERCOT Your Power, Our Promise. Available at: <http://www.ercot.com/>.

ERCOT, Inc., 2018, 2017 State of the Grid. Available at: <http://www.ercot.com/content/wcm/lists/144926/ERCOT_2017_State_of_the_Grid_Report.pdf>.

ERCOT, Inc., 2019. IntGenbyFuel. <http://www.ercot.com/gridinfo>.

ERCOT 2019a. State of the Grid Report. Available at: http://www.ercot.com/.

ERCOT Letter Report to the U.S. House Committee on Energy & Commerce (March 18, 2021). Available at http://www.ercot.com/content/wcm/lists/227813/ERCOT_Response_to_House_Energy__Commerce_Committee_3.18.21.pdf.

ExxonMobil. Available at: <https://energyfactor.exxonmobil.com/about-us/>.

Ezell, S., 2014. Comments to White House on Strategy for American Innovation RFI. [online]. Available from: <https://itif.org/publications/2014/09/23/itif-responds-white-house-strategy-american-innovation-rfi> (accessed 23.09.14.).

Federal Energy Administration, 1974. Project Independence Report. Government Printing Office, Washington, DC, pp. 195–204.

Federal Energy Regulatory Commission, 2003. Final Report on Price Manipulation in Western Markets: Fact-Finding Investigation of Potential Manipulation of Electric and Natural Gas Prices (Part I & Part II). Docket No. PA02-2-000.

First Solar. Available at: <http://www.firstsolar.com/PV-Plants/Community-Solar>.

Folger, P., 2014. The FutureGen Carbon Capture and Sequestration Project: A Brief History and Issues for Congress (PDF) (Report). Congressional Research Service. Retrieved 21 July 2014.

Ford, G., 1975a. Public Papers, blc. 1, p. 42.

Ford, G., 1975b. Address to a Joint Session of the Congress on the Economy, October 8, 1974. Public Papers: Gerald Ford, 1974, Government Printing Office, 1975, p. 231.

Frank Jr., C.R., 2014. The Net Benefits of Low and No-Carbon Electricity Technologies. Global Economy & Development, Working Paper 73, Brookings, Washington, DC.

Fri, R.W., 2003. The role of knowledge: technological innovation in the energy system. Energy J. 24, 51–74.

Friedman, T.F., 2016. Thank You for Being Late: An Optimist's Guide to Thriving in the Age of Accelerations. Farrar, Straus & Giroux, New York.

Frontier Associates, LLC, 2008. Texas Renewable Energy Resource Assessment; a Report to the Texas State Energy Conservation Office. Austin, TX, pp. 3–1 through 3–28.

Galbraith, K., Price, A., 2013. The Great Texas Wind Rush: How George Bush, Ann Richards, and a Bunch of Tinkers Helped the Oil and Gas State Win the Race to Wind Power. University of Texas Press, Austin, TX.

Gates, B., 2010. Innovating to Zero. February 10, 2010. Available from: TED. (accessed 15.03.17.).

Gaul, D., Young, L., 2003. U.S. LNG Markets and Uses. Energy Information Administration, Washington, DC.

General Land Office of Texas, 1976. Bob Armstrong, Commissioner, Texas Coastal Management Program: Report to the Governor and the 65th Legislature. General Land Office, Austin, TX, p. 67.

Global Wind Energy Council (GWEC) Organizational Documents. Available at: <https://gwec.net/>.

Goodwyn, L., 1996. Texas American Dreams: A Study of the Texas Independent Producers and Royalty Association. Center for American History by the Texas State Historical Association, Austin, TX.

Governor's Energy Advisory Council, 1975. Executive Summaries of Project Reports of the Council, Governor's Energy Advisory Council. Austin, TX.

Governor's Energy Advisory Council, 1977a. Texas Energy Outlook: The Next Quarter of a Century. Governor's Energy Advisory Council, Austin, TX.

Governor's Energy Advisory Council, 1977b. Texas Energy Policy. Governor's Energy Advisory Council, Austin, TX.

Governor's Energy Advisory Council, 1977c. Texas Energy: A Twenty-Five Year History. Governor's Energy Advisory Council, Austin, TX.

Greenlee, T.A., 2018. Handbook of Texas Online, Public Utility Commission of Texas. Available from: <http://www.tshaonline.org/handbook/online/articles/mdp09> (accessed 14.01.18.).

Griliches, Z., 1994. Productivity, R&D, and the data constraint. Am. Econ. Rev. 84, 1−23.

Hall, R., Jones, C., 1999. Why do some countries produce so much more output per worker than others? Q. J. Econ. 114 (1), 83−116. Oxford University Press.

Halliburton Company. Available at: <https://www.halliburton.com/en-US/about-us/history-of-halliburton-of-halliburton.html?node-id=hgeyxt5y>.

Hammes, U., Hamlin, H.S., Ewing, T.E., 2011. Geologic analysis of the Upper Jurassic Haynesville shale in east Texas and west Louisiana. Am. Assoc. Pet. Geologists Bull. 95 (10), 1643−1666.

Handbook of Texas Online, 2017. Petrochemical Industry. <http://www.tshaonline.org/handbook/online/articles/dop11> (accessed 18.08.17.).

Hess Corporation LP. Available at: <https://www.hess.com/company/hess-history>.

History of the Railroad Commission 1866−1939: Chronological Listing of Key Events in the History of the Railroad Commission of Texas, 1866−1939. Available from: <http://www.rrc.state.tx.us/about-us/history/history-1866-1939/> (accessed 21.11.18.).

Holloway, M.L., 1977. Texas Energy Outlook: The Next Quarter Century. Governor's Energy Advisory Council, Austin, TX.

Holloway, M.L. (Ed.), 1980. Texas National Energy Modeling Project: An Experience in Large-Scale Modeling Transfer and Evaluation. Academic Press, New York.

Holloway, M.L., 1982. Testimony Before Subcommittee on Energy Conservation and Power of the Committee on Energy and Commerce, U.S. House of Representatives Concerning Legislation on High Level Waste Disposal.

Holloway, M.L., 1986. The Texas Natural Gas Market: The Perspective from 1981 and the Current Market Outlook, Prepared for Intratex Gas Company, p. 9−10.

Holloway, M.L., 1995. Community College Return on Investment. Texas Association of Community Colleges, Austin, TX.

Holloway, M.L., 1999. Personal interview with Bill Clements in Dallas, Texas.

Holloway, M.L., 2003. A Study of Alternative Wharfage Rates For a Prospective LNG Project on La Quinta Channel for Port of Corpus Christi Authority, Austin, TX. Also see terminals at: <https://en.wikipedia.org/wiki/List_of_LNG_terminals#United_States_2>.

Holloway, M.L., 2005−2015. Holloway served as President & COO of a Texas non-profit organization of electric utilities and high-tech companies called the Center for the Commercialization of Electric Technologies. <https://www.resource-economics-austin.com/ccet.html>.

Holloway, M.L., King, R.J., Stevens, M., 1979. A Study of the Texas Energy Advisory Council: Its Structure and Functions Relative to State Science, Engineering and Technology Transfer, Texas Energy Advisory Council, Austin, TX. Prepared for the National Science Foundation, Washington, DC, p. 47, A-5.

Houston Angel Network. Available at: <http://houstonangelnetwork.org/han/>.

Hunter, T., 2012. History of Electric Deregulation in ERCOT, presentation to Electric Gas Reliability Workshop, Austin, TX.

Information Technology & Innovation Foundation, 2016. [online]. Available from: <https://itif.org/tech-policy-to-do-list/> (accessed 20.12.16.).

International Association for Energy Economics (IAEE) Organizational Documents. Available at: <https://www.iaee.org>.

International Atomic Energy Agency (IAEA) Organizational Documents. Available at: <https://www.iaea.org/>.

International Renewable Energy Agency (IRENA) Organizational Documents. Available at: <https://web.archive.org/web/20190406045612/> or <https://irena.org/aboutirena>.

International Solar Energy Society (ISES) Organizational Documents. Available at: <https://www.ises.org>.

J.D. Power and Associates Press Release, 2016. Residential Customers Resistant to Changing Retail Electric Providers as Price Gap Closes. Costa Mesa, CA.

Joel Myers of AccuWeather (February 17, 2021). Available at: https://www.accuweather.com/en/winter-weather/accuweather-estimates-economic-impact-of-winter-storms/902563.

Joinerville Texas & Daisy Bradford Discovery Well, 2017. [online]. Available from: <http://www.east-texas.com/index.htm/> (accessed 02.07.17.).

Jones, C.I., 1996. Human Capital, Ideas, and Economic Growth. <https://web.stanford.edu/~chadj/Rome100.pdf>.

Jorgenson, D.W., Ho, M.S., Samuels, J.D., 2014. Long-Term Estimates of U.S. Productivity and Growth. Bureau of Economic Analysis, Washington, DC.

Keynes, J.M., 1965. First Harbinger Edition, Printed in the United States of America, p. 383.

Kiesling, L.L., 2009. Retail restructuring and market design in Texas. In: Kiesling, L., Kleit, A. (Eds.), Electricity Restructuring: The Texas Story. AEI Press, Washington, DC, pp. 154–173.

Kiesling, L.L., Kleit, A.N. (Eds.), 2009. Electric Restructuring: The Texas Story. AEI Press, Publisher for the American Enterprise Institute, Washington, DC.

King, L., Nordhaus, T., Schellenberger, M., 2015. Lessons From the Shale Revolution: A Report on the Conference Proceedings. Breakthrough Institute, Oakland, CA.

Kleit, A.N., 2009. Market monitoring, ERCOT style. In: Kiesling, L., Kleit, A. (Eds.), Electricity Restructuring: The Texas Story. AEI Press, Washington, DC, pp. 174–189.

Kohl, W.L., 1991. After the Oil Price Collapse: OPEC the Exporting States and the World Oil Market. The Johns Hopkins University Press, Baltimore, MD.

Kurzweil, R., 1993. The Singularity Is Near: When Humans Transcend Biology, Mass Use of Inventions, Years Until Use by 1/4 U.S. Population. [online]. Available from: <http://www.kurzweilai.net/the-technological-singularity/> (accessed 22.12.16.).

Kurzweil, R., 2001. The Law of Accelerating Returns. [online]. Available from: <http://www.kurzweilai.net/the-law-of-accelerating-returns> (accessed 22.12.16.).

Leibman, P.R., 1989. Horizonal Drilling Revitalizes the Austin Chalk, Vol. I, EPN01, Petrie Parkman & Co., E&P Notes.

Lester, R.K., Hart, D.M., 2012. Unlocking Energy Innovation: How America Can Build a Low-Cost, Low-Carbon Energy System. The MIT Press, Cambridge.

Lindsey, R. NOAA Climate, Mar 5, 2021a. Available at: https://www.climate.gov/news-features/understanding-climate/understanding-arctic-polar-vortex.

Lindsey, R. NOAA Climate, Mar 11, 2021b. Available at: https://www.weathernationtv.com/news/the-polar-vortex-and-februarys-extreme-cold/.

Lusk, H.F., 1963. Business Law: Principles and Cases. Richard D. Irwin, Homewood, IL, pp. 649–650.

MacAvoy, P.W., 2000. The Natural Gas Market: Sixty Years of Regulation and Deregulation. Yale University Press, New Haven, CT, p. 135.

Magness, B., (February 24, 2021). Review of February 2021 Extreme Cold Weather Event – ERCOT Presentation. Available at http://www.ercot.com/content/wcm/key_documents_lists/225373/2.2_REVISED_ERCOT_Presentation.pdf.

Marathon Oil Corporation. Available at: <https://www.marathonoil.com/about/>.

Marathon Petroleum Corporation. Available at: <https://www.marathonpetroleum.com/>.

Marchi, Nd, 1981a. Energy policy under Nixon: mainly putting out fires. In: Goodwin, C.D. (Ed.), Energy Policy in Perspective. The Brookings Institution, Washington, DC, pp. 409, 527, 530−533.

Marchi, Nd, 1981b. The Ford administration: energy as a political good. In: Goodwin, C.D. (Ed.), Energy Policy in Perspective. The Brookings Institution, Washington, DC, pp. 530−533.

Marshall, C., 2018. E&E News Reporter E&E News, January 16, 2018. Available at: <https://www.eenews.net/stories/1060071081>.

MIT Lab for Innovation Science and Policy, 2018. Available at: <https://innovation.mit.edu/assets/MITii-Lab_Open-Calls.pdf>; <https://innovation.mit.edu/research-policy/lab-innovation-science-policy/> (accessed 22.04.18.).

Mitchell, J., 2012. The Anadarko Basin: Oil and Gas Exploration—Past, Present And Future, SM Energy Co, Tulsa, OK.

Moretti, E., Wilson, D.J., 2014. State incentives for innovation, star scientists and jobs: evidence from biotech. J. Urban. Econ. 79, 20−38.

Murry, W.J., 1983. Report of the Advisory Committee on Natural Gas Proration and Ratable Take. Austin, TX.

National Energy Technology Laboratory, U.S. Department of Energy, 2010. Untapped Domestic Energy Supply and Long Term Carbon Storage Solution. Available at: <www.netl.doe.gov>.

National Enhanced Oil Recovery Initiative, 2012. Carbon Dioxide Enhanced Oil Recovery: A Critical Domestic Energy, Economic, and Environmental Opportunity, pp. 21−22.

National Petroleum Council's Committee U.S. Energy Outlook, 1972. A Report of the on U.S. Energy Outlook.

Nefiodow, L.A., 2006. The Sixth Kondratieff, sixth ed. Allianz Global Investors Capital Market Analysis, Table.

Newell, S., Spees, K., Pfeifenberger, J., Mudge, R., DeLucia, M., Carlton, R., 2012. ERCOT Investment Incentives and Resource Adequacy, Prepared for Electric Reliability Council of Texas.

Nixon, R.M., 1973a. Remarks Announcing Establishment of the Federal Energy Office. Public Papers, December 4, 1973, pp. 990−991.

Nixon, R.M., 1973b. Special Message to the Congress on Energy Policy. Public Papers (Government Printing Office, 1974), April 18, 1973, pp. 302−319.

Nixon, R.M., 1973c. Televised Address to the Nation. The White House, Washington, DC.

North American Reliability Corporation, 2017. 2017 Long-Term Reliability Assessment. Princeton, NJ, pp. 116−390.

Occidental Petroleum Corporation. Available at: <https://www.oxy.com/Pages/default.aspx>.

Olien, R.M., 2016. Oil and Gas Industry, Uploaded on June 15, 2010. Modified on August 19, 2016. Published by the Texas State Historical Association From Texas State Historical Association. Available at: <https://tshaonline.org/handbook/online/articles/doogz> (accessed 15.11.18).

Organization of the Petroleum Exporting Countries (OPEC) Organizational Documents. Available at: <https://www.opec.org/opec_web/en/about_us/24.htm>.

OXY Petroleum Company, 2015. Technical Review of Subpart RR MRV Plan for the Denver Unit. Houston, TX.

Pace Company Consultants & Engineers, Inc. 1974. Petrochemicals in Texas. Prepared for the Subcommittee of the Petrochemical Energy Group, Project No. Special Project C., State of Texas Governor's Energy Advisory Council, pp. 6−8.

Painter, D.S., 2002. Oil. In: second ed. Burns, R.D., et al., (Eds.), Encyclopedia of American Foreign Policy, vol. 3. Charles Scribner's Sons, pp. 1−20.

Patch, B.W., 1931. Development of the Natural Gas Industry. Editorial Research Reports 1931, vol. II. CQ Press, Washington, DC. Retrieved from: <http://library.cqpress.com/cqresearcher/cqresrre1931100500>.

Payward, Inc., 2019. Cryptocurrency Mining: A Primer. Kraken.

Permanent University Fund Financial Statements, Independent Auditors' Report, 2015. p. 8. in 347 U.S. 672, 74 S. Ct. 794, 98 L. Ed. 1035.

Petroleum Technology, 1933. Technology and the "Conroe Crater" American Oil & Gas Historical Society. Washington, DC.

Phillips 66. Available at: <https://www.phillips66.com/>. Phillips Petroleum Co. v. Wisconsin, 347 U.S. 672 (1954).

Power to Choose. <http://www.powertochoose.org/>.

Powerplant and Industrial Fuel Use Act, 1978. (H.R. 5146), 95th Congress.

Prindle, D.F., 1981. Petroleum Politics and the Texas Railroad Commission. University of Texas Press, Austin, TX, pp. 6–9, 23, 56.

Public Citizen's Energy Program, <www.citizen.org/cmep>.

Public Utility Commission of Texas §25.505, 2012. Resource Adequacy in the Electric Reliability Council of Texas Power Region.

Public Utility Commission of Texas Report to the 85th Texas Legislature, 2017. Scope of Competition in Electric Markets in Texas.

Puller, S., 2009. Competitive performance of the ERCOT wholesale market. In: Kiesling, L., Kleit, A. (Eds.), Electricity Restructuring: The Texas Story. AEI Press, Washington, DC, pp. 138–153.

Railroad Commission of Texas, 2019. Abandoned Mine Land Section, Surface Mining and Reclamation Division. Austin, TX. <smrdmaps@rrc.texas.gov>.

Railroad Commission of Texas, 2020. Mining & Exploration, Surface Mining and Reclamation Division. Austin, TX. <https://www.rrc.state.tx.us/mining-exploration/>.

Railroad Commission vs. Flour Bluff Oil Co., 1949. 219 S. W. 2d 506 (Tex. Civ. App.), error ref'd., p. 508.

Rice Alliance, Houston, TX. [online]. Available at: <www.alliance.rice.edu>.

Rose, C., 2016. [online]. Available from: <https://charlierose.com/videos/28411> (accessed 30.06.16.).

Rostow, W.W., 1978a. A Policy for the Fifth Kondratieff Upswing. University of Texas Press, Austin, TX and London.

Rostow, W.W., 1978b. The World Economy: History & Prospect. University of Texas Press, Austin, TX and London.

Saucier, H., 2017. Texas' Austin Chalk Booms While Shale Plays Remain Mostly Dormant. Available from: <https://www.rigzone.com/news/texas_austin_chalk_booms_while_shale_plays_remain_mostly_dormant-07-dec-2017-152720-article/>.

Schlumberger. Available at: <https://www.slb.com/who-we-are/technology-development>.

Schlumberger, 2020. <https://www.glossary.oilfield.slb.com/en/Terms/c/concession.aspx/>.

Schumpeter, J.A., 1942. Capitalism, Socialism, and Democracy. Harper, New York.

Schwab, K. (Ed.), 2019. The Global Competitiveness Report 2019. World Economic Forum. Geneva, Switzerland.

Sempra Energy. Available at: <https://www.sempra.com/history#time-line-slide-out-1374>.

Sherlock, M.F., 2011. Energy Tax Policy: Historical Perspectives on Current Status of Energy Tax Expenditures. Congressional Research Service.

Silverstein, A., 2017. Resiliency, Reliability and the DOE, Husch Blackwell and Texas Renewable Energy Industries Alliance Webinar Series. Available from: <https://event.on24.com/> (accessed 03.12.17.).

Smith, K.R., 1977. The Intersection of Time and Technology: Propositions Suggested by an Examination of Coal and Nuclear Power, Hazard Indices, the Temporal Judgments of Law and Economics and the Place of Time in Mind and Myth (Ph.D. thesis). University of California, Berkeley, CA.

Smith, J.C., 2018. Handbook of Texas Online, Panhandle Field. <http://www.tshaonline.org/handbook/online/articles/dop01> (accessed 20.11.18.).

Solow, R.M., 1956. A contribution to the theory of economic growth. Q. J. Econ. 70 (1), 65–94.

Solow, R.M., 1957. Technical change and the aggregate production function. Rev. Econ. Stat. 39 (3), 312–320.

Spaulding, W.C., 2020. Economics. Available at: <http://thismatter.com>.

Spence, D., Buch, D., 2009. Why does ERCOT have only one regulator? In: Kiesling, L., Kleit, A. (Eds.), Electricity Restructuring: The Texas Story. AEI Press, Washington, DC, pp. 9–21.

Stevens, M., Cummings, G., 1977. Texas Energy: A Twenty-Five Year History. Governor's Energy Advisory Council, Austin, TX.

Stockton, Henshaw, Graves, 1952. Gas in Texas, pp. 231–236.

Summers, J., 2018. U.S. Crude Output on Track to Rival Giants Russia, Saudi Arabia. Available at <https://www.bloomberg.com/news/articles/2018-01-09/> (accessed 03.08.18.).

Sweeney, J.L., 2008. The California Electricity Crisis: Lessons for the Future. National Academy of Engineering, Washington, DC.

Taylor, M.Z., 2016. The Politics of Innovation: Why Some Countries are Better Than Others at Science and Technology. Oxford University Press, New York. TEAC, 1977. Energy Policy Planning Act of 1977, S.B. 1172, 65th Texas Legislature.

Taylor, M.Z. 2016. The Politics of Innovation; Oxford University Press, New York, NY.

Technology Collaboration Center of the Greater Houston Partnership, 2018. Available at: <https://tech-collaboration.center/about/> (accessed 01.05.18.).

Teller, E., Wood, L., Hyde, R., 1997. Global Warming and Ice Ages. 22nd International Seminar on Planetary Emergencies, Erice, Italy.

Tenaska Trailblazer Partners, LLC, 2011. Bridging the Commercial Gap for Carbon Capture and Storage. Global CCS Institute, p. 5.

Texas A&M University, 2017. [online]. Available from: <http://geoweb.tamu.edu/> (accessed 25.06.17.).

Texas A&M University, 2018. The AM Innovation Center (AMIC). Available at <http://aminnovation-center.com/about/> (accessed 22.04.18.).

Texas Energy Advisory Council, 1979. Texas Energy Policy. 1978 Update, Texas Energy Advisory Council, Austin, TX.

Texas Energy Advisory Council, 1983. Texas Energy Development Fund Biennial Report. Texas Energy Advisory Council, Austin, TX.

Texas Energy and Natural Resources Advisory Council, 1982. Impact of the NGPA on Current and Projected Natural Gas Markets: Docket No. RM82-26-000. Comment Pursuant to the Notice of Inquiry, Submitted by Governor William P. Clements, Lt. Governor William P. Hobby and Milton L. Holloway, Ph.D. Executive Director, Austin, TX.

Texas Energy and Natural Resources Advisory Council, 1983a. Texas Energy Development Fund Biennial Report. Texas Energy and Natural Resources Advisory Council, Austin, TX.

Texas Energy and Natural Resources Advisory Council, 1983b. Texas Five Year Energy Research Plan: 1984–1988. Austin, TX.

Texas Energy and Natural Resources Advisory Council, 1983c. Texas Five Year Energy Research Plan: 1984–1988. Technology Development Division.

Texas Legislative Council, Research Division, 2015. Amendments to the Texas Constitution Since 1876, Austin, Texas. Available at: <http://www.tlc.texas.gov/const_amends.htm>.

Texas Railroad Commission, 2013. Natural Gas Trends.

Texas Railroad Commission, 2016.

Texas Reliability Entity response to NERC Request for Follow up Actions To 9/8/2011 SW Blackout – August 1, 2012. Available at: http://www.ercot.com/calendar/2012/7/30/33465-ROS-TRE.

Texas Tech University in Lubbock, 2017. [online]. Available from:<http://www.geosciences.ttu.edu/> and <http://www.depts.ttu.edu/coe/> (accessed 25.06.17.).

The Cannon, Houston's Campus for Entrepreneurs. Available at: <https://thecannonhouston.com/>.

U.S. Bureau of Labor Statistics News, County Employment and Wages News Release, Wednesday, February 24, 2021 available at County Employment and Wages News Release (bls.gov).

U.S. Department of Energy, National Energy Technology Laboratory, 2010. Carbon Dioxide Enhanced Oil Recovery Untapped Domestic Energy Supply and Long Term Carbon Storage Solution. Available from: <https://energy.gov/fe/science-innovation/oil-gas-research/enhanced-oil-recovery> (accessed 31.07.17.).

U.S. Department of Energy, Office of Fossil Energy, 2017. Enhanced Oil Recovery. Available from: <https://energy.gov/fe/science-innovation/oil-gas-research/enhanced-oil-recovery> (accessed 17.07.17.).

U.S. Department of Energy, Solar Energy Technologies Office, 2018. Funding Opportunity Announcement: FY 2018 Solar Energy Technologies Office. Available from: <https://www.energy.gov/eere/solar/funding-opportunity-announcement-fy-2018-solar-energy-technologies-office> (accessed 29.04.18.).

U.S. Department of Interior, Bureau of Mines, 1974. Minerals Yearbook. U.S. Government Printing Office, Washington, DC.

U.S. Energy Information Administration, 2017. Natural Gas Monthly. Available at: <https://www.eia.gov/todayinenergy/detail.php?id=34032>.

U.S. Energy Information Administration, 2018a. Annual Energy Outlook 2018. Petroleum and Other Liquids Supply and Disposition. Available at: <https://www.eia.gov/outlooks/aeo/data/>.

U.S. Energy Information Administration, 2018b. Drilling Productivity Report. Available at: <https://www.eia.gov/petroleum/drilling/pdf/dpr-full.pdf>.

Vertrees, C.D., 2017. Handbook of Texas Online, Permian Basin. Texas State Historical Association. <http://www.tshaonline.org/handbook/online/articles/ryp02> (accessed 18.08.17.).

Warr, B., Ayres, R., 2006. Economic growth, technological progress and energy use in the US over the last century: Identifying common trends and structural change in macroeconomic time series. Paper for session 49 in Helsinki 2006: INSEAD, Fontainebleau, France and IIASA, Laxenburg, Austria.

Whitaker, J.C., 1976. Striking a Balance: Environment and Natural Resources Policy in the Nixon-For Years. American Enterprise Institute for Public Policy Research, pp. 61–68.

Wooster, R., Sanders, C.M., 2009. Handbook of Texas Online. Texas State Historical Association, Austin, TX, <https://tshaonline.org/handbook/online/articles/dos03>.

World Coal Association (WCA) Organizational Documents. Available at: <https://www.worldcoal.org/>.

World Fact Book (a), Central Intelligence Agency Library. Available at: <https://www.cia.gov/library/publications/resources/the-world-factbook/fields/261rank.html>.

World Fact Book (b), Central Intelligence Agency Library. Available at: <https://www.cia.gov/library/publications/resources/the-world-factbook/fields/269rank.html>.

World Fact Book (c), Central Intelligence Agency Library. Available at: <https://www.cia.gov/library/publications/resources/the-world-factbook/fields/252rank.html>.

World Fact Book (d), Central Intelligence Agency Library. Available at: <https://www.cia.gov/library/publications/resources/the-world-factbook/fields/265rank.html>.

World Fact Book (e), Central Intelligence Agency Library. Available at: <https://www.cia.gov/library/publications/resources/the-world-factbook/fields/260rank.html>.

World Petroleum Council (WPC) Organizational Documents. Available at: <https://www.world-petroleum.org/>.

Yergin, D., 1991. The Prize: The Epic Quest for Oil, Money, and Power. Simon and Schuster, New York, pp. 593, 594, 595, 613, 614, 620, 625, 701, 702, 704.

Zagorski, W.A., Wrightstone, G.R., Bowman, D.C., 2012. The Appalachian Basin Marcellus gas play: its history of development, geologic controls on production, and future potential as a world-class reservoir. In: Breyer, J.A. (Ed.), Shale Reservoirs—Giant Resources for the 21st Century, vol. 97. AAPG Memoir, pp. 172–200.

Further reading

Acts, 74th Leg., regular session, 1995. Chap. XXX, SB 373, AN ACT relating to the continuation, operations, and functions of the Public Utility Commission of Texas and the Office of Public Utility Counsel; providing penalties. Full text of SB 373 is available at the website of the Texas legislature. <http://www.legis.state.tx.us/Billlookup/Text.aspx?LegSess=74&Bill = SB3743>.

Alcoa Rockdale Texas, 1954. Alcoa Rockdale Works Dedication Program. <Milamcountyhistoricalcommission.org/>.

Allianz Global Investors, 2010. The Sixth Kondratieff-Long Waves of Prosperity.

American Recovery and Reinvestment Act of 2009 (ARRA) (Pub.L. 111–5).

Annual Crude Oil Prices by State, 2015. [online]. Available from: <http://eia.gov/dnav/pet/hist/LeaHandler.ashx> (accessed 16.01.17.).

ARI, 2011. Improving Domestic Energy Security and Lowering CO2 Emissions with "Next Generation" CO2-Enhanced Oil Recovery (CO2-EOR).

Arrow, K.J., 1996. The economics of information: an exposition. Empirica 23 (2), 119–128.

Austin American Statesman, 2018. Austin, TX.

Avangrid Rewables. Available at: <www.avangrid.com>.

Baldick, R., 2018. Debate on "Comparing the US and EU Electricity Markets." Available at: <http://rossbaldick.com/comparing-the-us-and-eu-electricity-markets/> (accessed 20.04.18.).

Becker, G.S., 1975. Human Capital: A Theoretical and Empirical Analysis, With Special Reference to Education, second ed. Columbia University Press for NBER, New York.

Black, J., 2013. European union energy regulation. OECD, International Regulatory Co-operation: Case Studies, Vol. 2: Canada-US Co-operation, EU Energy Regulation, Risk Assessment and Banking Supervision. OECD Publishing.

Clark, J.H., 1954. Three Stars for the Colonel. Random House, New York, p. 9.

Clymore Production Co. vs. Thompson, 1936. 13 F Supp. 469 (W. D. Tex.).

Compliance in ERCOT. Available at: http://www.ercot.com/mktrules/compliance.

Congressional Quarterly, Inc., Energy Policy, 1979. Washington, DC, p. 3-A.

Crenshaw, C.E., The Regulation of Natural Gas.

Dallas Innovation Alliance. Available at: <http://www.dallasinnovationalliance.com/>.

Dutton, W.H., Kahin, B., O'Callaghan, R., Wyckoff, A.W. (Eds.), 2004. Transforming Enterprise: The Economic and Social Implications of Information Technology. MIT Press, Cambridge, MA and London, England.

Breakthrough Institute, 2015. Lessons From the Shale Revolution. Oakland, CA.

East Texas Oil, 1934. p. 71. (In Prindle, David, F., 1981. pp. 25−26).

Encyclopedia of American Politics, <https://ballotpedia.org/Energy_policy_in_Texas>.

Energy Factor. Available at: <https://energyfactor.exxonmobil.com/category/science-technology/>.

ERCOT 2019b. State of the Grid Report. Available at: http://www.ercot.com/content/wcm/lists/197391/2019_ERCOT_State_of_the_Grid_Report.pdf.

ERCOT Market Components descriptions. Available at: https://en.wikipedia.org/wiki/Electric_Reliability_Council_of_Texas.

Federal Energy Regulatory Commission, What FERC Does. Available at: <https://www.ferc.gov/about/ferc-does.asp>.

Federal Ministry for the Environment, Nature Conservation and Nuclear Safety. Available at: <https://www.bmu.de/en/pressrelease/minister-schulze-climate-action-becomes-law/> and <https://www.bmu.de/fileadmin/Daten_BMU/Pools/Broschueren/treibhausgasemissionen_en_bf.pdf/>.

Go Big in Texas. Available at: <https://businessintexas.com/industries/petroleum-refining-chemical-products>.

Handley, M., 2013. An Energy Lifeline: Fracking a Game-Changer for U.S. Economy. U.S. News & World Report.

Hardwicke, R.E., 1955. The rule of capture and its implications as applied to oil and gas. Tex. Law Rev. 13.

Holloway, M.L., 1987. A National Energy Policy for the United States, prepared for Commissioner Kent Hance. Railroad Commission of Texas, Austin, TX.

Holloway, M.L., Vetter, E.O., Fisher, W.L., Rostow, W.W., Rollins, M.M., 1982. A National Energy Policy for the United States. Governor of Texas, Austin, TX, Proposed by William P. Clements, Jr.

Howell, J., 2016. World Patent Marketing, Oil Drilling Inventions That Changed the World: Oil Drilling Inventions, The Power Behind; Cars, Planes, Plastics, Fertilizers, and Chemistry. [online]. Available from: <https://worldpatentmarketing.net/oil-drilling-inventions/#respond/> (accessed 26.01.16.).

Hutton, W.H., Kahin, B., O'Callaghan, R., Wyckoff, A.W., 2005. Transforming Enterprise: The Economic and Social Implications of Information Technology. MIT Press, Cambridge, MA and London, England.

Jamrisko, M., Lu, W., 2018. The U.S. Drops Out of the Top 10 in Innovation Ranking. <https://www.bloomberg.com/news/articles/2018-01-22/>.

Kennedy, J.F., 2020. ScienceFocus. Available at: <https://www.sciencefocus.com/space/we-choose-to-go-to-the-moon-read-jfks-moon-speech-in-full/> (accessed 03.07.20.).

Kinder Morgan CO_2, Website at: <https://www.kindermorgan.com/pages/business/co2/>.

Logan, J., James, T.L., 2009. A Comparative Review of a Dozen National Energy Plans: Focus on Renewable and Efficient Energy: Technical Report NREL/TP-6A2−45046.

Lucas Gusher, 2006. [online]. Available from: <https://commons.wikimedia.org/w/index.php?curid=3425193/> (accessed 20.12.16.).

Making Sense, 2018. Available at: <https://www.pbs.org/newshour/show/can-finance-cure-cancer> (accessed 10.02.18.).

Moore's Law, 1965. [online]. Available from: <http://www.intel.com/content/www/us/en/silicon-innovations/moores-law-technology.html> (accessed 22.03.17.).

Murry, W.J., 2004. Austin American Statesman, Obituary.

Natural Gas Policy Act, 1978. Public Law 95−621, 95th Congress (H.R. 5289).

Nelson, V., Gilmore, E., 1975. Potential for Wind Generated Power in Texas. The State of Texas Governor's Energy Advisory Council, Austin, TX.

Nuclear Regulatory Commission, Backgrounder on the Three Mile Island Accident. Retrieved March 19, 2018. Available at: <https://www.nrc.gov/reading-rm/doc-collections/fact-sheets/3mile-isle.html>.

Oil Concessions in Five Arab States 1911−1953: Kuwait, Bahrain, Qatar, Trucial States and Oman, 1989. Cambridge Archive Additions Near and Middle East Titles. Available at: <http://archiveeditions.co.uk/titledetails.asp?tid=71>.

Paleontological Research Institution, 2006. [online]. Available from: <https://commons.wikimedia.org/w/index.php?curid=3425193> (accessed 12.03.06.).

Pierobon, J., 2003. George P. Mitchell: Founder of Shale Gas—Here's How He and His Team Did It. The Energy Fix. Available at: <http://www.theenergyfix.com/2013/08/05/george-p-mitchell-founder-of-shale-gas-heres-how-he-and-his-team-did-it/#sthash.6dvUj61E.dpbs>.

Potomac Economics, Ltd, letter to the Public Utility Commission of Texas, March 11, 2021.

Raimi, D., 2017. The Shale Revolution: Climate Opportunities and Risks in Resources. RFF's Online Magazine, issue 196.

Romer, P.M., 1990. Part 2: The problem of development: a conference of the institute for the study of free enterprise systems. J. Political Econ. 98 (5), S71−S102.

Rotman, D., 2013. A Cheap and Easy Plan to Stop Global Warming. MIT Technology Review. [online]. Available at: <https://www.technologyreview.com/topic/sustainable-energy/> (accessed August 2018).

Shah, A., Energy Security. Available at: <https://www.globalissues.org/article/595/energy-security/>.

SMU University, Dallas, Texas, Center for Laser Aided Manufacturing. Accessed at: <https://www.smu.edu/Lyle/Centers/>.

Solman, P., 2018. Interview with Andrew Lo on The PBS NewsHour.

State Gross Domestic Product, 2016. [online]. Available from: <https://bea.gov/iTable/iTable.cfm/> (accessed 17.01.17).

State Historical Association. Uploaded on June 15, 2010. Published by the Texas State Historical Association.

State of Texas Energy Policy Partnership, 1992, Texas Energy Policy: Texas Energy Coordination Council Final Report.

Texas A&M University (a), Mays Business School Innovation Research Center (MIRC). Available at: <https://today.tamu.edu/2017/09/05/multimillion-dollar-gifts-establish-new-innovation-research-center-at-texas-am-mays-business-school/> (accessed 22.04.18.).

Texas Commission on Environmental Quality (TCEQ). <https://www.tceq.texas.gov/>.

Texas Comptroller of Public Accounts, State Energy Conservation Office. Available at: <https://comptroller.texas.gov/programs/seco/>.

Texas Energy and Natural Resources Advisory Council, Austin, TX.

Texas Low-Level Radioactive Waste Disposal Compact Commission. Available at <https://www.tceq.texas.gov/permitting/radmat/licensing/low-level_rad_waste.html/> and <http://www.tllrwdcc.org/about-the-comission/>.

Texas Reliability Entity. Available at: https://en.wikipedia.org/wiki/Texas_Reliability_Entity.

Texas State Historical Association. <https://tshaonline.org/about-tsha/28991>.

Texas Tech University Innovation HUB at Research Park. Available at: <https://www.depts.ttu.edu/vpr/researchers/researchguide/innovationhub.php> (accessed 22.04.18.).

Texas Tech University National Wind Institute. Available at: <https://www.depts.ttu.edu/nwi/education/BSWE/index.php> (accessed 02.05.18.).

The Concise Encyclopedia of Economics, 2008. [online]. Available from: <http://www.econlib.org/library/Enc/HumanCapital.html> (accessed 16.01.17.).

The University of Texas at Austin: Department of Petroleum and Geosystems Engineering, 2017. [online]. Available from: <http://www.pge.utexas.edu/> (accessed 11.05.17.).

Thornley, D., 2008. Texas Wind Energy: Past, Present, and Future. Texas Public Policy Foundation. Available at: <www.TexasPolicy.com>.

U.S. Bureau of Economic Analysis, 2016. [online]. Available from: <https://bea.gov/iTable/iTable.cfm/> (accessed 20.12.16.).

U.S. Cybersecurity & Infrastructure Security Agency (CISA), Presidential Policy Directive 21 (PPD-21): Critical Infrastructure Security and Resilience. Available at: https://www.cisa.gov/critical-infrastructure-sectors.

U.S. Dept of Commerce, National Oceanic and Atmospheric Administration, National Weather Service, 1325 East West Highway, Silver Spring, MD accessed March 8, 2021.

U.S. Department of Energy, Science & Innovation. Available at: <https://www.energy.gov/science-innovation> (accessed 14.07.2020.).

U.S. Department of Interior, Bureau of Land Management (BLM). Available at: <https://www.blm.gov/about/what-we-manage>.

U.S. Energy Information Administration, Crude Oil Proved Reserves 2016 (CVS download). Retrieved 28 May 2017.

University of Texas at Austin (a), Entrepreneurship & Innovation. Available at: <https://www.utexas.edu/campus-life/entrepreneurship-and-innovation> (accessed 22.04.18.).

University of Texas at Austin (b), IC2 Institute. Available at: <http://ic2.utexas.edu/about/> (accessed 22.04.18.). Uploaded on June 15, 2010. Published by the Texas State Historical Association.

Wall Street Journal, Saudis Struggle Over Where to List Aramco, January 29, 2018.

Wikipedia, 2006a. [online]. Available from: <https://en.wikipedia.org/wiki/Permanent_University_Fund/> (accessed 15.03.17.).

Wikipedia, 2006b. [online]. Available from: <http://www.east-texas.com/joinerville-texas-daisy-bradford.htm/> (accessed 15.03.17.).

Wikipedia, 2016a. [online]. Available from: <https://en.wikipedia.org/wiki/Zeitgeist/> (accessed 20.12.16.).

Wikipedia, 2016b. [online]. Available from: <https://en.wikipedia.org/wiki/Innovation_economics/> (accessed 20.12.16.).

Wikipedia, 2016c. [online]. Available from: <https://en.wikipedia.org/wiki/Innovation_economics#Historical_origins/> (accessed 20.12.16.).

Wikipedia, 2018a. Operating solar farms that are 150 MW or larger. [online]. Available from: <https://en.wikipedia.org/wiki/List_of_photovoltaic_power_stations> (accessed 28.04.18.).

Wikipedia, 2018b. Operational solar thermal power stations (of at least 50 MW capacity). [online]. Available from: <https://en.wikipedia.org/wiki/List_of_solar_thermal_power_stations#Operational> (accessed 28.04.18.).

Wikipedia (a). [online]. <https://en.wikipedia.org/wiki/German_Renewable_Energy_Sources_Act/> and <https://en.wikipedia.org/wiki/Green_paradox/>.

Wikipedia (b). [online]. <https://en.wikipedia.org/wiki/Russia_in_the_European_energy_sector/>.

Wilcox, S., 2007. National Solar Radiation Database 1991−2005 Update: Users Manual, Technical Report, NREL/TP-581−4134.

World Economic Forum. Available at: <https://www.weforum.org/>.

Zimmermann, E.W., 1957. Conservation in the Production of Petroleum. Yale University Press, New Haven, CT, p. 145.

Index

Note: Page numbers followed by "*f*" and "*t*" refer to figures and tables, respectively.

Printed in the United States
by Baker & Taylor Publisher Services